D0374549

Using Oil Spill Dispersants on the Sea

Committee on Effectiveness of
Oil Spill Dispersants

Marine Board
Commission on Engineering and Technical Systems
National Research Council

NATIONAL ACADEMY PRESS
Washington, D.C.   1989

National Academy Press • 2101 Constitution Avenue, N.W. • Washington, D. C. 20418

NOTICE: The project that is the subject of this report was approved by the Governing Board of the National Research Council, whose members are drawn from the councils of the National Academy of Sciences, the National Academy of Engineering, and the Institute of Medicine. The members of the panel responsible for the report were chosen for their special competences and with regard for appropriate balance.

This report has been reviewed by a group other than the authors according to procedures approved by a Report Review Committee consisting of members of the National Academy of Sciences, the National Academy of Engineering, and the Institute of Medicine.

The National Academy of Sciences is a private, nonprofit, self-perpetuating society of distinguished scholars engaged in scientific and engineering research, dedicated to the furtherance of science and technology and to their use for the general welfare. Upon the authority of the charter granted to it by the Congress in 1863, the Academy has a mandate that requires it to advise the federal government on scientific and technical matters. Dr. Frank Press is president of the National Academy of Sciences.

The National Academy of Engineering was established in 1964, under the charter of the National Academy of Sciences, as a parallel organization of outstanding engineers. It is autonomous in its administration and in the selection of its members, sharing with the National Academy of Sciences the responsibility for advising the federal government. The National Academy of Engineering also sponsors engineering programs aimed at meeting national needs, encourages education and research, and recognizes the superior achievements of engineers. Dr. Robert M. White is president of the National Academy of Engineering.

The Institute of Medicine was established in 1970 by the National Academy of Sciences to secure the services of eminent members of appropriate professions in the examination of policy matters pertaining to the health of the public. The Institute acts under the responsibility given to the National Academy of Sciences by its congressional charter to be an adviser to the federal government and, upon its own initiative, to identify issues of medical care, research, and education. Dr. Samuel O. Thier is president of the Institute of Medicine.

The National Research Council was organized by the National Academy of Sciences in 1916 to associate the broad community of science and technology with the Academy's purposes of furthering knowledge and advising the federal government. Functioning in accordance with general policies determined by the Academy, the Council has become the principal operating agency of both the National Academy of Sciences and the National Academy of Engineering in providing services to the government, the public, and the scientific and engineering communities. The Council is administered jointly by both Academies and the Institute of Medicine. Dr. Frank Press and Dr. Robert M. White are chairman and vice-chairman, respectively, of the National Research Council.

The program described in this report is supported by Cooperative Agreement Nos. 14-12-0001-30301 and 14-12-0001-30360 between the Minerals Management Service of the U.S. Department of Interior and the National Academy of Sciences.

Library of Congress Cataloging-in-Publication Data

National Research Council (U.S.). Committee on Effectiveness of Oil Spill Dispersants
    Using oil spill dispersants on the sea / Committee on Effectiveness of Oil Spill Dispersants, Marine Board, Commission on Engineering and Technical Systems, National Research Council.
    p. cm.

    Bibliography: p.
    Includes index.
    ISBN 0-309-03882-0 (paper); ISBN 0-309-03889-8 (cloth)
    1. Oil pollution of the sea.   2. Dispersing agents.   3. Oil spills—Environmental aspects.   4. Dispersing agents—Environmental aspects.   I. Title

    TD427.P4N38   1989
    626.1'6833—dc 19   88-38879
                                       CIP

Cover: The slate-blue color common to the surface of the northern seas and windrows—or parallel mases—of undispersed oil are reflected in the artist's rendition on the cover of this report.

Printed in the United States of America

Acknowledgments

The extensive reference list on which this report is based could not have been compiled without the assistance of many scientists and engineers throughout the world. Members of the committee obtained and provided unpublished reports and reports of work in progress that were essential for our discussions.

In particular, Mervin Fingas of Environment Canada, Ottawa, provided the committee with a full shelf of reports and symposium proceedings. The Environment Canada Library, Dartmouth, and the Fisheries and Oceans Library at the Bedford Institute of Oceanography, also in Dartmouth, Nova Scotia, helped with literature searches.

Two members of the committee served as vice-chairman in charge of a working group: Colin Jones chaired the group concerned with effectiveness and participated in the Beaufort Sea dispersant tests conducted in August 1986 by Environment Canada. Peter Wells, as vice-chairman of the biological working group, acquired and assessed an immense body of published and unpublished literature, and summarized it for the committee. These activities represented considerable extra commitment, and are especially appreciated.

James N. Butler
Chairman

MARINE BOARD

SIDNEY WALLACE, *Chairman,* U.S. Coast Guard (retired), Reston, Virginia

RICHARD J. SEYMOUR, *Vice-Chairman,* Scripps Institution of Oceanography, La Jolla, California

ROGER D. ANDERSON, Cox's Wholesale Seafood, Inc., Tampa, Florida

KENNETH A. BLENKARN, Amoco Oil Company (retired), Tulsa, Oklahoma

DONALD F. BOESCH, Louisiana Universities Marine Consortium, Chauvin

C. RUSSELL BRYAN, U.S. Navy (retired), St. Leonard, Maryland

F. PAT DUNN, Shell Oil Company, Houston, Texas

JOHN HALKYARD, Arctec Offshore Corporation, Escondido, California

EUGENE H. HARLOW, Soros Associates Consulting Engineers, Houston, Texas

DANA R. KESTER, University of Rhode Island, Kingston

JUDITH T. KILDOW, Massachusetts Institute of Technology, Cambridge

WARREN G. LEBACK, Consultant, Princeton, New Jersey

WILLIAM R. MURDEN, Murden Marine Ltd., Alexandria, Virginia

EUGENE K. PENTIMONTI, American President Lines, Ltd., Oakland, California

JOSEPH D. PORRICELLI, ECO, Inc., Annapolis, Maryland

ROBERT N. STEINER, Atlantic Container Line, South Plainfield, New Jersey

BRIAN J. WATT, TECHSAVANT, Inc., Kingston, Texas

EDWARD WENK, JR., University of Washington (retired), Seattle

Staff

CHARLES A. BOOKMAN, Director

DONALD W. PERKINS, Associate Director

DORIS C. HOLMES, Administrative Associate

DELPHINE GLAZE, Administrative Secretary

AURORE BLECK, Senior Secretary

GLORIA B. GREEN, Senior Secretary

CARLA M. MOORE, Senior Secretary

Preface

Exploring the potential of chemical dispersants to combat open-sea oil spills has been attractive to both government and industry. However, an unanswered question is whether testing and technical advances in the laboratory and in the field, and progress in deployment, justify the use of dispersants as a primary response technique.

Much of the dispersant research in the United States has been conducted by private industry. In addition, research was conducted by the Interagency Technical Committee for the Oil and Hazardous Materials Simulated Environment Test Tank (OHMSETT) facility. The committee included the Minerals Management Service of the Department of the Interior, the U.S. Coast Guard, the U.S. Navy, the Environmental Protection Agency, and Environment Canada. Further, a major amount of dispersant research has been done in Canada, sponsored by Environment Canada. Significant research also has been conducted by industry and by others abroad, most notably in France, Norway, and the United Kingdom. In view of the extent of recent research, U.S. and Canadian cooperating agencies requested that the National Research Council (NRC), "review the state of knowledge in toxicity, effectiveness of application techniques, and effectiveness of commercially available dispersants." In response, the Commission on Engineering and Technical Systems of the NRC convened the Committee on Effectiveness of Oil Spill Dispersants.

Members of the committee were selected for their experience in and knowledge about the fates and effects of petroleum in the marine environment, physical chemistry of pollutants, physical and chemical oceanography, marine biology and zoology, oil and hazardous materials cleanup techniques and management, oil spill contingency planning and response, marine engineering and field deployment, and public interest in environmental protection. A balance of technological, scientific, and geographic perspectives was a major consideration in the selection of the committee. Consideration of public concern about environmental safety was another major consideration.

SCOPE OF STUDY

The committee was charged to assess the state of knowledge and practice about the use of dispersants in responding to open-ocean oil spills. This assessment will guide federal and local governments and industry in both the United States and Canada, in defining the role of dispersants in oil spill response and implementing the use of dispersants. Equally important is the charge to identify gaps in knowledge where research is especially needed. The committee was specifically asked to

• determine the effectiveness of dispersants and identify the best techniques for their use;
• identify the possible impacts of dispersants and dispersed oil on marine and coastal environments; and
• provide guidance about when and where dispersants should or should not be used.

The committee interpreted these charges broadly, and addressed two fundamental questions: Are dispersants effective—that is, do the chemicals remove oil from the surface of the sea and disperse it into the water column? Is the biological impact of dispersed oil greater or less than that of untreated oil? In answering these questions, the committee also identified where knowledge is lacking and what research is fundamental to improving effective and responsible use of dispersants.

While the focus of the assessment was on the use of dispersants in the open ocean, the committee examined the possibility of impacts on important characteristics of the marine environment—tropical, temperate, and polar coasts, areas important to marine life cycles, and ocean areas frequented by marine mammals and bird life.

Government policies were addressed only to the extent that they affect optimum use and deployment of dispersants and international cooperation in their assessment and application.

The committee determined that a detailed assessment of comparative costs for using dispersants for treating a spill and for mechanical cleanup and disposal would be beyond the charge given to the committee. However, since this topic could be a significant element in the decision-making process about the choice of response techniques, a summary giving a range of costs and important references has been provided in the text for the interested reader.

METHOD OF STUDY

Data and information about dispersant use and its effects were acquired from Canadian and U.S. sources through the auspices of the sponsoring agencies, with extensive input by Environment Canada. A significant contribution was the evaluation of literature on *Comparative Field Effects of Dispersed and Undispersed Petroleum Oils in Coastal Marine Environments*, prepared at the request of the committee by P. Lane and Associates Limited, Halifax, Nova Scotia. This work was conducted under the guidance of Peter Wells, a member of the committee.

Canadian, European, and U.S. experiences and data were also acquired through background papers written by committee members and reviewed during the Workshop on Oil Spill Dispersant Effectiveness held March 24-26, 1986, at Reston, Virginia. Participants included guests from British, Canadian, French, and U.S. research institutions and industry. They were Gerard Canevari, Exxon Research and Engineering Company, Houston, Texas; Jean Croquette, Centre de Documentations sur le Pollutions Accidentelles de Eaux (CEDRE), Brest, France; Merv Fingas, Conservation and Protection, Environment Canada, Ottawa, Ontario; Richard Golob, World Information Systems, Inc., Cambridge, Massachusetts; David Kennedy, National Oceanic and Atmospheric Administration, Seattle, Washington; Gordon Lindblom, Exxon Chemical Company, Houston, Texas; Donald MacKay, University of Toronto, Ontario, Canada; and Joseph Nichols, The International Tanker Owners Pollution Federation Ltd., London, England. Information was also received from D. Cormack, Warren Spring Laboratory, Stevenage, England.

The assessment process of the report development was undertaken in two subsequent meetings and was based on test data as well as on observations of field experience, including field trials conducted by Environment Canada, August 1986, in the Beaufort Sea, Northwest Territories. These field trials were witnessed by Colin Jones, a committee member, at the invitation of the Canadian government. The committee report and its conclusions and recommendations reflect a broad data base and assessment in the marine toxicological and biological area. A more limited data base supports the field applications and experience area; however, field observation enhances the experience basis for the committee's conclusions and recommendations in that area.

Contents

Executive Summary

Concern for the possible aesthetic, ecological, and economic impacts of oil spills in the ocean, and the adequacy of technologies for controlling them, led to a request from the U.S. and Canadian governments for an assessment of the effectiveness and use of dispersants. The assessment was sought to establish an improved basis for making decisions about when and how to clean up oil spills.

Dispersants, which are solvents and agents for reducing surface tension, are used to remove oil slicks from the water surface. The treated oil enters the water column as fine droplets where it is dispersed by currents and subjected to natural processes, such as biodegradation. If this process is effective, the oil may thus be prevented from moving into sensitive environments or stranding onshore, thereby eliminating or reducing damage to important coastal habitats, marine life, or coastal facilities.

This study addresses two questions about the use of dispersants: Do they do any good? and Do they do any harm?

DO THEY DO ANY GOOD?

In a few carefully planned, monitored, and documented field tests, as well as in laboratory tests, several dispersants have been shown to be effective—that is, they have removed a major part of the oil from the water surface when properly applied to oils that were dispersible.

However, at other field tests and at accidental spills, dispersants have been reported to have low effectiveness. The latter results may have been due to the use of inadequate application techniques, such as poor targeting and distribution of aerial sprays, as well as the possibility that the oils were not dispersible, some dispersants were poorly formulated, or the results were inconclusive (differing reports from several observers who used different monitoring techniques). Resolution of these ambiguities will require the correct choice of dispersants; accurate calibration in field tests; training and competence in application technique; and the development of more objective and reliable monitoring methods.

Chemical dispersants have been used extensively on a few large oil spills—*Ixtoc I, Torrey Canyon,* and Main Pass Block 41 *Platform C*—but no systematic documentation of effectiveness has been published.

Much is known about why and how dispersants work. This knowledge is largely based on empirical tests and observations made in the laboratory and during sea trials. The general mechanism by which surfactant chemicals disperse crude oils and some refined products (diesel, heating oils, and bunkers) into the water column is qualitatively understood, and quantitative studies and models are being developed to describe this process. The necessary conditions for effective dispersal of oil in the laboratory have been established, and dispersant compositions, which could be effective under the proper conditions, have been developed. In addition, larger-scale laboratory tests and field experiments at sea have more closely simulated the action of wind, waves, and other aspects of the marine environment, and have helped to define their influence on effectiveness.

One important aspect inadequately understood is the interaction of various physical and chemical processes involved in oil dispersion. Few published studies exist on the fundamental science concerning how dispersants act on oil in water. Most of the published studies on oil spill dispersants describe laboratory and sea tests of commercially available products. These tests generally were conducted to evaluate how well the products worked, and few studies have investigated the interactive phenomena of surfactants (surface tension reducing chemicals), crude oil, and water.

When dispersants are sprayed from boats or aircraft, how well they work depends on sea conditions and application techniques as well as on the chemical nature of both the dispersants and the oil. Application techniques have improved with experience, but are still far from routinely optimal. A critical factor in the strategy of dispersant application is that the viscosity of oil increases rapidly with weathering. Because more viscous oil is more difficult to disperse, response within a few hours is generally essential to high effectiveness. Some light crude oils and refined products remain dispersible for a longer time. In addition, oil slick areas tend to expand linearly with time, and current shear may later increase slick areas even more rapidly; this expansion rate further emphasizes the need for early treatment response.

DO THEY DO ANY HARM?

Concern that chemical dispersants could be toxic to marine life has led to considerable caution in authorizing their use at spill sites. Laboratory studies of dispersants currently in use have shown that their acute lethal toxicities are usually lower than crude oils and their refined products.

A wide range of sublethal effects of dispersed oil has been observed in the laboratory. These occur in most cases at concentrations comparable to or higher than those expected in the water column during treatment (1 to 10 ppm), but seldom at concentrations less than those found several hours after treatment of an oil slick (less than 1 ppm). The times of exposure in the laboratory (24 to 96 hr) are much longer than predicted exposures during slick dispersal in the open sea (1 to 3 hr), and the effects would be expected to be correspondingly less in the field. Direct application of dispersant to marine life (as when birds and fur-bearing mammals or their habitats accidentally are sprayed) is to be avoided, because dispersants destroy the water-repellency and insulating capacity of fur or feathers, and various components may disrupt the structural integrity of sensitive external membranes and surfaces.

Laboratory bioassays have shown that acute toxicity of dispersed oil generally does not reside in the dispersant, but in the more toxic fractions of the oil. Dispersed and untreated oil show the same acute toxicity, a conclusion obscured in much of the literature by the large number of studies that quote oil concentration as being the total oil per unit volume of the experimental system, rather than the actual measured dissolved and dispersed hydrocarbon concentrations to which organisms are exposed.

The immediate ecological impact of dispersed oil varies. In open waters, organisms on the surface will be less affected by dispersed oil than by an oil slick, but organisms in the water column, particularly in the upper layers, will experience greater exposure to oil components if the oil is dispersed. In shallow habitats with poor water circulation, benthic organisms will be more immediately affected by dispersed oil. Although some immediate biological effects of dispersed oil may be greater than for untreated oil, long-term effects on most habitats, such as mangroves, are less, and the habitat recovers faster if the oil is dispersed before it reaches that area.

RECOMMENDATIONS FOR USING DISPERSANTS

The committee recommends that dispersants be considered as a potential first response option to oil spills, along with other response options. Implementation of this recommendation must consider spill size, logistical requirements, contingency planning, equipment and dispersant performance and availability, appropriate regulations, and personnel training.

Sensitive inshore habitats, such as salt marshes, coral reefs, sea grasses, and mangroves, are best protected by preventing oil from reaching them. Dispersion of oil at sea, before a slick reaches a sensitive habitat, generally will reduce the overall and particularly the chronic impact of oil on many habitats. This has been shown by research studies that compared the biological effects of untreated and dispersed oil released on water over the intertidal or immediate subtidal zones at Baffin Island (arctic); Long Cove, Maine (north temperate); and Panama (tropical). Although these studies generally showed that dispersed oil caused less chronic environmental damage than oil alone, the committee recommends that additional ecological studies be conducted, under controlled or naturally established water circulation regimes in shallow environments, to help define the conditions under which dispersant use will be effective and environmentally safe.

Because the principal biological benefit of dispersant use is prevention of oil stranding on sensitive shorelines, and because dispersability of oil decreases rapidly with weathering, prompt response is essential. Therefore, the committee recommends that regulations and contingency planning make rapid response possible; this includes prior approval to field-test a dispersant immediately after a spill to establish dispersability in cases of doubt.

SUGGESTED RESEARCH

The committee recommends:

• Assessment of ecological effects of dispersed oil on marine life in shallow-water environments, and other habitats having restricted water exchange, to better define conditions under which dispersant use can be environmentally safe.

• Research on the interaction of oil and dispersed oil with suspended particulates, sediments, plankton, and benthic organisms to establish a better quantitative basis for comparing the adhesion properties of untreated oil and dispersant-treated oil.

• Research in the mechanics of dispersed oil resurfacing and spreading to guide improvement in dispersant application strategy and to reduce possible impact of dispersed oil on fish larvae and marine birds and mammals.

• Laboratory studies to determine the effect of dispersed oil, under exposure conditions comparable to those expected in the field, on water-repellency of fur and feathers and on the hatchability of seabird eggs after adult birds are contaminated.

• Long-term studies of the recovery of selected ecosystems from exposure to oil and dispersed oil. Investigations at sites where the effects of oil and dispersed oil have already been studied on a shorter-time scale would be particularly useful.

• Additional investigations on the toxic effects of both untreated oil and dispersed oil on surface-dwelling organisms that could be affected by oil slicks.

• Laboratory and field research to analyze how turbulent diffusion, surface circulation, and wave motion affect dispersed oil distribution as a function of depth, time, and volume of spilled and treated oil.

• A program of additional research to explore the mechanisms by which dispersant droplets contact an oil film, mix, and penetrate into it. The research should examine how surfactants interact with the oil and migrate to the oil-water interface, and the microscopic processes by which emulsions actually form.

The committee's formal conclusions and recommendations are contained in Chapter 7.

1
Use of Oil Dispersants:
History and Issues

Decisions about using dispersants are made in the context of the reasons for treating spilled oil. The image of a black, sticky mass covering a once clean shore or yacht hull is dramatic enough, but there are other less visible incentives to respond to a spill, such as threats to valuable habitat and nursery areas or to vulnerable early stages of marine life. Other concerns are: How great a threat do spills pose in quantity and frequency? What are the usual sources of oil in the sea and what can be done to counter a spill? These concerns establish the setting for using chemical dispersants and are discussed here. This chapter also reviews how dispersants have been used and developed and what their use is expected to accomplish.

The potential use of chemical dispersants raises some primary questions, principally: Are they effective? and Do they reduce the potential damage caused by spilled oil? These issues are discussed generally in this chapter and in more detail in later chapters.

REASONS FOR TREATING OIL SPILLS

Aesthetic and Ecological Damage

Introduction of petroleum into the marine environment is a direct consequence of the production and transportation of crude oil and refined products. Even though natural seeps of crude oil occur in areas of the ocean floor and stable biotic communities are associated with them, the sudden introduction of high concentrations of hydrocarbons can kill or cause sublethal effects in some marine organisms.

In the 1960s and 1970s, more and more oil was transported by sea and—with the introduction of supertankers—in larger vessels. The increased risk of a major oil spill was displayed dramatically when the first major tanker catastrophe—the *Torrey Canyon* spill—occurred.

Public outcry from the *Torrey Canyon* spill and similar incidents stimulated development of a variety of response techniques to contain or remove spilled oil before it could harm property or the environment. Unfortunately, some early attempts to clean up the oil for aesthetic reasons caused more ecological harm than good (National Research Council [NRC], 1985; Smith, 1968; Southward and Southward, 1978; Spooner, 1969).

Perhaps the most dramatic symbol of the consequences of an oil spill is an oiled seabird. Attempts by volunteers to rescue birds have been a response to many oil spills, yet despite enthusiastic care, the survival percentage is low. Diving birds, such as auks, are especially vulnerable. Some marine mammals may also be affected under certain circumstances.

Nearshore marine waters and shallow fishing banks are also rich in a variety of less visible organisms. Some areas, such as salt marshes, are among the most productive ecosystems known. The impact of spilled oil on these areas is obvious and often severe, and public awareness of these ecosystems has expanded concern about the danger from oil spills. The concern that oil spills pose a threat to commercial fisheries has been extended to open-ocean life in general (Boesch and Rabalais, 1987).

Economic Damage

In addition to aesthetic and ecological concerns, coastal regions can suffer economically from damage done by oil spills to recreation areas, harbors and vessels, commercial shellfish grounds, and intake sources for desalination and power plants. During summer months, beaches along the coasts of most maritime countries are crowded with people on weekend outings and vacations. Thus, there is considerable economic incentive in coastal recreation areas to protect beaches from spills or to clean them up quickly. Cleanup of contaminated boats, seawalls, and harbor equipment can be expensive. In a marina where boat owners pride themselves on the appearance of their craft, pressure for cleanup is likely to be even more intense than in an industrial harbor.

Because hydrocarbons from spilled oil can become entrained in sediments inhabited by many commercial shellfish species, shellfish can become contaminated with oil. They may then be unfit for human consumption due to high levels of hydrocarbons. Contamination can close a commercial shellfish bed for years, resulting in a considerable financial loss to shellfish producers.

In some regions of the world—such as Bermuda, several Caribbean islands, and the Middle East—desalination plants provide an important source of drinking water for the surrounding sea. Oil, even when naturally dispersed, can enter the intake of such a plant and threaten to contaminate an entire drinking water supply with hydrocarbons (Kruth et al., 1987). Power-plant cooling water intakes are similarly vulnerable.

There have been several attempts to identify and analyze the costs for mechanical cleanup and disposal as well as for treatment with chemical dispersants. These analyses have produced a wide spectrum of results, reflecting many variables that are difficult to measure. For example, when oil is recovered by mechanical means what is usually measured includes large amounts of water and debris. Another variable is the cost of transporting equipment to the cleanup or staging site. Usually the cost of moving boats and other equipment for mechanical cleanup is omitted, but aircraft and equipment transport expenses are commonly included in evaluations of dispersant operations.

Mechanical costs of oil cleanup range from $65 per bbl to $5,000 per bbl. Dispersant costs range from $15 per bbl to a maximum of $65 per bbl (Lasday, 1985). These ranges are related to spill sizes of 10,000 to 100,000 bbl of oil. These statistics and other analyses, such as provided by White and Nichols (1983), show that the costs of oil spill cleanup are high, and that they are an order of magnitude more for mechanical cleanup than for treatment using chemical dispersants and still another order of magnitude more for cleanup of the immediate and obvious damage done once the oil has come ashore.

Long-term residual effects have been identified in several reports (Teal and Howarth, 1984; White and Nichols, 1983), but attempts to place a monetary value on the loss of or damage to natural resources and to establish replacement or restoration costs have been based on sweeping assumptions about natural processes and usually lack long-term environmental data.

Safety Hazards

Crude oil contains a substantial fraction of volatile hydrocarbons, as do refined petroleum products such as gasoline and diesel oil. Spilled oil evaporates, producing high concentrations of flammable hydrocarbon-air mixtures. If such a mixture ignites, fire can spread to the bulk oil container and consume it. Although a fire may not pose a major hazard to aquatic systems, air pollution from partially burned crude oil can pose severe public health and ecological hazards (Gundlach and Hayes, 1977; Thebeau and Kana, 1981). Therefore, prevention of fires—better yet, prevention of spills—is a desirable goal.

Some volatile components of crude and refined oils, such as benzene, are very toxic to humans. Indeed, a human health concern is that cleanup crews can receive doses of toxic fumes high enough to cause nausea and possibly other health effects.

POTENTIAL SOURCES OF SPILLED OIL

The worldwide input of petroleum to the marine environment was estimated by the National Research Council (1985), which noted the sources, probable ranges, and selected the best single number for the sources as shown in Table 1-1. The estimates, based on data acquired from 1971 through 1980, are provided only as indicators of major sources and their importance. Input of petroleum fluids from municipal and industrial wastes (urban and river runoff, sewage, refineries, and industrial sources), atmospheric-borne particulates, and natural sources (marine seeps, sediment erosion) is mostly continuous and, while the environmental effects are of serious concern, control or treatment technology usually differs greatly from that related to marine oil spill response. However, some accidental spills on inland rivers and waterways may be amenable to common marine oil spill response.

Marine transportation—tankers, barges, and lighters—and offshore operations for the exploration and production of hydrocarbons are the key sources of interest in regard to marine oil spills. On a global basis, marine transportation accounts for the largest single portion of petroleum inputs, more than 40 percent on the basis of

TABLE 1-1 Input of Petroleum Hydrocarbons Into the Marine
Environment (Million Barrels Per Year)[a]

Source	Probable Range	Best Estimate
Transportation		
Operational	4.4-15.0	7.2
Accidents	2.2- 3.0	2.9
Subtotal	6.6-18.0	10.1
Municipal wastes and runoff	4.0-21.5	8.1
Atmosphere	0.3- 3.4	2.1
Natural sources	0.2-17.3	1.7
Offshore production	0.3- 0.4	0.3
Total	11.4-60.6	22.3

[a]One barrel (bbl) equals 42 U.S. gallons or about 35 Imperial
gallons. This report will refer to barrels as the standard
volume measurement in reference to petroleum transport and
spills.

SOURCE: Converted from NRC (1985) estimates based on metric
tons per year.

"best estimates" (NRC, 1985) and 30 to 50 percent depending on
the ranges of other sources. However, the bulk of this input is due
to routine operational discharges. About 12 percent is due to accidents, but this can vary from year to year by factors of 2 or 3. In
contrast, offshore oil and gas operations are estimated to contribute
only a little more than 1 percent of the oil accidentally put into the
sea. A single incident can dramatically alter these data, however; for
example, the largest spill to date, the *Ixtoc I* well blowout, put 3.1
million bbl in the Gulf of Mexico in 1979 and 1980.

Apart from oil discharges caused by military action in the Per-
sian Gulf, the incidence of major tanker spills elsewhere has been
dramatically reduced from 1974 to 1986 (Table 1-2). The majority
of tanker spills over 5,000 bbl were from collisions and groundings.

All discharges reported in waters under the U.S. Coast Guard's
jurisdiction or by the U.S. Environmental Protection Agency (U.S.
EPA) from 1974 through 1983 range from 200,000 bbl to more than
500,000 bbl per year (Figure 1-1). Of the reported discharges, only
about 5 percent (about 11,000 bbl in 1983) occurred in offshore
waters, that is, in the territorial sea (shore to 3 mi), the contiguous
zone (3 to 12 mi), and the high seas (beyond 12 mi). River channels,

ports and harbors, and open sheltered waters receive most spillage, in contrast to offshore areas.

In U.S. waters, waterborne transport accounted for about 28 percent of the total volume reported spilled in 1983. Most of this volume was attributed to tank barges, which accounted for 22 percent (43,000 bbl) of all oil spilled. Much of this discharge occurred in ports and harbors. As more and more oil is refined outside of the United States, ocean transport of refined products to and along the U.S. East Coast by barge is increasing.

About 35 to 50 percent of spillage in U.S. waters is crude oil. Spill volumes and number of incidents in U.S. offshore waters are shown in Figures 1-2 and 1-3. These statistics cover both transportation and offshore oil and gas operations in the three major U.S. coastal areas. Canadian offshore oil spills during the period from 1974 to 1984 averaged about 10,000 bbl per year, but the definition of offshore includes inlets and coastal areas, while the U.S. offshore definition excludes harbors and sheltered waters. Most Canadian discharges were fuel oil and distillates, in contrast to U.S. experience, but statistical comparisons of incidents and volumes between the United States and Canada would require access to the raw data.

Most spills are small and most spilled oil volume comes from only

TABLE 1-2 Number of Oil Spills
From Tankers Worldwide, 1974-1986

Year	Barrels	
	50-5,000	>5,000
1974	92	26
1975	98	23
1976	67	25
1977	66	20
1978	57	24
1979	56	36
1980	50	13
1981	50	5
1982	45	3
1983	53	11
1984	25	7
1985	29	8
1986	23	7

SOURCE: International Tanker Owners
Pollution Federation Ltd.

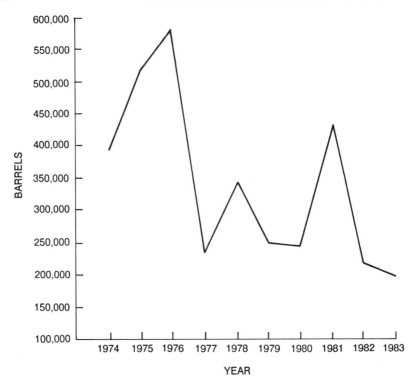

FIGURE 1-1   Volume of oil and the substances discharged, 1974-1983. All volumes have been converted from U.S. gallons to barrels. The 1983 figure includes 55,000 bbl attributed to "other," a category not previously used in Coast Guard reporting. Source: U.S. Coast Guard, 1987.

a few large incidents. For example, in 1983* the U.S. Coast Guard reported the following spill statistics in and around U.S. waters:

• Small spills (less than 12 bbl)—5,923 spills (63 percent) accounted for 6,200 bbl (3 percent of total volume). These spills are usually not treated with dispersants nor cleaned up by mechanical means.

• Moderate spills (12 to 1,200 bbl)—617 spills (6.5 percent) accounted for 60,000 bbl (30 percent of total volume). Efforts to clean up these spills usually use mechanical means, but in some cases such spills could be treated with dispersants.

---

*Although Coast Guard statistics for 1984 are available, they are preliminary. Variations between preliminary and final statistics can be significant, thus the figures for 1983 are the latest final data reported here (U.S. Coast Guard, 1987).

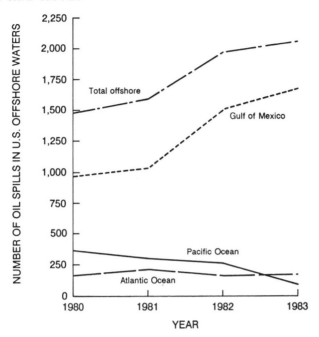

FIGURE 1-2 Number of oil spills in U.S. offshore waters (territorial, contiguous, and high seas). Source: U.S. Coast Guard, 1987.

• Large spills (more than 1,200 bbl)—19 accidents (0.2 percent) released 133,000 bbl (67 percent of the total volume). These spills may require major response effort, including aerial application of dispersants.

TREATMENT METHODS

There are four major options for responding to oil spills: mechanical containment and collection; use of chemical dispersants; shoreline cleanup; and natural removal (no cleanup action). Countermeasures that are less widely used or have major limitations are burning, sinking, gelling, and enhanced biodegradation. In determining the best possible countermeasure for a given situation, availability and applicability must be carefully weighed against potential environmental damage (Table 1-3).*

---

*Comprehensive presentations of oil cleanup methods are provided in U.S. EPA (1981) and Wardley-Smith (1976).

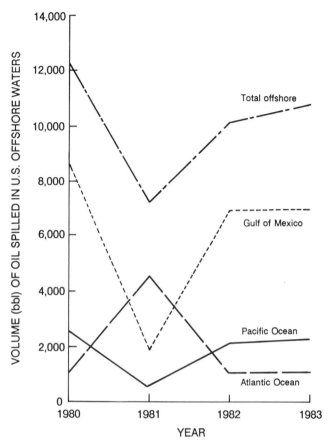

FIGURE 1-3   Volume of oil (bbl) spilled in U.S. offshore waters (territorial, contiguous, and high seas). Source: U.S. Coast Guard, 1987.

Mechanical Containment, Recovery, or Removal

Mechanical means for oil spill mitigation include barriers (booms) to contain or divert oil, and skimmers or sorbents to recover or remove it from the water surface. In addition to mechanical barriers (which include sorbent booms), surface-collecting agents ("oil herder" chemicals), water jets, air jets, and air bubble barriers have been used to contain or divert spilled oil. Performance of any of these methods, however, can be severely limited by oceanic conditions and weather, including currents, waves, and wind, and by the nature of the oil slick.

TABLE 1-3 Cleanup Control Options and Their General Application

Cleanup Operations and Options	Application			
	Water Surface	Under-water	Along-shore	Shore-line
Containment and diversion				
Major options				
Booms	x		x	
Sorbent booms	x		x	
Surface-collecting agents	x		x	
Water jets	x			
Air jets	x			
Other possibilities				
Air barriers	x			
Gelling	x			
Viscoelastic additives	x			
Underwater containment		x		
Removal				
Major options				
Skimmers	x			
Sorption	x		x	
Sorbent booms	x			
Biodegradation	x		x	
Other possibilities				
Burning	x			
Sinking	x			
Dispersants	x		x	
Cleanup of stranded oil				
Major options				
Flushing				x
Beach cleaning				x
Manual	x	x		x
Vegetation cropping	x	x		
Organism rehabilitation	x			
Other possibilities				
Substrate removal		x		x
Burial				x
Sandblasting				x
Steam cleaning				x
Natural Cleansing	x	x		x

Full containment using a boom cannot be assured unless maximum water speed at right angles to the boom is less than 1 kn. In faster currents, the barrier must be placed at an angle to the current to guide the floating oil toward slower current areas. Air and water jets and bubble barriers are sometimes used in restricted areas; in general, the current has to be fairly slow and the waters quiescent. Although surface-collecting agents do not move oil relative to underlying water, they can limit its spread regardless of water speed.

Maximum wave height for complete containment should be less than equal to the freeboard of the barrier, or waves will splash water and oil over it. Mechanical barriers are thus severely limited on open seas. In practice, most barriers lose effectiveness in wave heights greater than 1.3 m (4 ft). In experiments, surface-collecting agents have kept oil from spreading in 1 to 2 m (4 to 6 ft) seas, but duration of effectiveness decreases with increasing sea state.

Limitations to operation of most containment options are related to current and wind-induced wave heights. Air and water jets are seriously affected by winds, especially when they blow toward the jet. Surface-collecting agents cannot move floating oil against the wind, but the ability of a surface-collecting agent to keep oil from spreading is not affected. Wind stress on boom barriers can be appreciable, and the boom must be strong enough to withstand such forces.

The rate of oil recovery by any mechanical device decreases with decreasing oil thickness. Because the recovery rate of most skimmers is negligible at thicknesses of less than about 1 mm, booms are often used in conjunction with skimmers. This approach is limited, however, by problems of maneuvering, anchoring, and coordinating multiple vessels needed to handle such arrays.

Mechanical collection devices have environmental limitations similar to those of containment devices because skimmers are affected by wind, waves, and currents in much the same way as barriers. Oil may pass under or by a skimmer and not be collected if the skimmer is moving through the water faster than 1 kn. Waves higher than 1.3 m may cause oil to splash over skimmers or otherwise cause them to lose effectiveness. More robust skimmers and barriers have been designed for the U.S. Coast Guard to use in the open ocean, and others for use in the North Sea. Some of these devices have successfully recovered oil in seas higher than 4 ft. In large spills, the effectiveness—the percent of surface oil removed by treatment—has been low for mechanical cleanup systems; for example, possibly 10 percent during the *Ixtoc I* blowout (Teal and Howarth, 1984).

Shoreline Cleanup

If oil strands on a shoreline, attempts are usually made to remove it using mechanical means, by flushing, by manual pickup, or by physically removing the substrate. In some cases, oil-soaked vegetation has been cropped, but this method is usually regarded as environmentally undesirable. Hard surfaces, such as rocks or bulkheads, have been sandblasted or steam cleaned. In most cases, shoreline cleanup is expensive and may be environmentally damaging. However, the only method for cleaning and restoring public beaches and easily accessible shorelines near fisheries and industrial areas is removal of oil.

Natural Removal

Oil left alone is eventually removed from water surfaces and shorelines by a variety of natural means, including evaporation, photooxidation, solution, physical dispersion, sedimentation on particulate matter, and biological degradation. Although these natural processes may be slow, possibly as long as several years, they are generally conceded to be environmentally acceptable, and in some cases may be preferable to using active countermeasures.

Other Countermeasures

In addition to the mitigation measures discussed above, a number of other countermeasures are available or have been proposed, including burning, sinking, and gelling. All of them have some limitations, however. A major limitation to burning oil at sea is that oil tends to spread on water and the sea is a very effective heat sink; it is difficult to raise the temperature of a thin layer of floating oil high enough to permit ignition. However, where spilled oil cannot flow well, as in the Arctic and on ice, there has been effective use of burning as a cleanup technique. Even so, combustion is never complete and air pollution is a real concern. The addition of chalk or treated sand has been used or proposed as a means of sinking oil. However, sinking is seldom completely effective initially, and some oil tends to resurface. Moreover, oil that sinks to the bottom contaminates benthic life and degrades more slowly than when floating, dispersed, or dissolved in water.

Several gelling formulations have been proposed but have yet to be demonstrated in practice. Most formulations would require substantial mixing energy to make them effective, which is not practical under actual spill conditions. A further limitation is that gelling formulations are expensive; the cost of cleanup using gelling agents is likely to be far higher than conventional means. A possible exception is a viscoelastic additive, however, it has yet to prove itself in a real spill; thus there is no basis as yet for judging its utility.

Because constituents of oil degrade naturally when attacked by bacteria, algae, protozoa, and marine fungi, enhancement of biological degradation has been proposed using specially chosen or bioengineered microbes. However, microbes that degrade hydrocarbons are readily available everywhere in nature, except in polar waters where the rates of breakdown are very slow and variable (Atlas, 1985). It does not appear necessary in most cases to enhance their action.

ROLE OF DISPERSANTS

Much of the biological and ecotoxicological research on dispersants, oils, and dispersed oils has been conducted in support of chemical dispersant response to oil spills in coastal waters. Real or perceived impacts on biological populations, habitats, water sources, and recreational areas have focused the attention of resource managers, cleanup specialists, policymakers, marine scientists, and the public on the need for better understanding of oil spill effects and development of techniques for dealing with them.

Rationale for Dispersant Use

An initial reason for using dispersants is to respond to public and governmental concerns by preventing potential damage to birds, fish, marine mammals, and other natural resources; fouling of shorelines and boats; and contamination of drinking water sources. Dispersing an oil spill will make it less visible, and may reduce its economic and ecological impact—provided the water volume, which it disperses into, is great enough. If the oil is dispersed into a small volume of water with poor circulation, the ecological impact may in fact be increased. Some specific cases have been studied in adequate detail (see Chapter 4, "Biologically Oriented Mesocosm and Field Studies"). Rapidly dispersing oil into the water column will, in most cases, be less costly environmentally than manual shoreline cleanup.

Sea and Weather Conditions

Dispersants may be especially valuable when other countermeasures fail, for example when an open-ocean spill is moving oil onshore, but waves are too high to permit the use of booms and skimmers. Another example is if tidal currents are so strong that oil would be carried under a boom, resurface, and threaten a sensitive area. Because severe weather enhances natural dispersion of oil, the results of chemical dispersant use may not be apparent as in calm seas. Nevertheless, reduction of interfacial tension by dispersants is likely to hasten dispersion.

Logistics

Application of dispersants can be accomplished more quickly than recovery of spilled oil by mechanical means. Aerial spraying of chemical dispersants is usually the preferred method of application to use at sea, because it is more efficient and provides a wider range of coverage than application from boats. Like vessel spraying methods, aerial application is limited in high winds, fog, or darkness. The speed of aerial response itself is not the key goal, but rather the imperative to ensure protection of sensitive environments and public amenities.

Protection of Ecologically Sensitive Areas and Organisms

Marine ecosystems, such as salt marshes, mangroves, and coral reefs, and bird nesting areas are extremely sensitive to damage by oil, but the use of dispersants raises questions about the relative environmental effects of oil and dispersed oil (see Chapters 3 and 4). Dispersants may be applied when it is judged that the impact of dispersed oil on organisms, habitats, and ecological processes will be less than that of oil alone. Determining relative environmental effects requires objectively assessing alternative treatments in terms of effectiveness, acute toxicity, ecological effects, and cost, including potential loss of natural resources.

Accurate exposure assessment for surface and subsurface oil is critical to estimating the hazards of dispersed oils to seabirds, whose protection is a major reason for dispersant use over open waters (Peakall et al., 1987).

Protection of Fisheries Resources

Fisheries resources are vulnerable to damage by oil, but the nature of this biological damage is thought to be largely indirect and seasonal, via planktonic eggs and larvae and through the contamination of fish habitats. If a major spill occurs where a limited fish population is spawning, an entire year-class could be exposed to damage. Although long-term damage could be ameliorated by strong previous or subsequent year-classes, losses to commercial fisheries might be substantial for a species under pressure from fishing, predation, or stresses from chronic pollution (Boesch and Rabalais, 1987; Howarth, 1987; Longhurst, 1982; NRC, 1985). Using dispersants may mitigate this situation by moving oil into the water column and lowering its concentration and the time of exposure of gametes, developing embryos, and larvae that dwell on or near the surface. Dispersant use may also limit the damage to sensitive nearshore nursery waters, especially if oil is effectively dispersed in areas remote (i.e., where natural processes can treat oil) from the sensitive waters.

Protection of Shoreline Amenities

Beaches, harbors, marinas, and other shoreline amenities can suffer costly damage from oil spills. The same strategies suggested for protection of sensitive ecological areas can be applied to commercial areas by applying dispersants before a slick approaches too close to shore. In addition, booms may be set to protect lagoons and harbors.

Dispersants as an Aid to Natural Cleanup

Microbial degradation of oil appears to be enhanced by dispersal because of the larger surface area available. However, to some degree, the lower concentration of nutrients available in open water may limit the potential for growth of the hydrocarbon-utilizing microbial species.

Laboratory and mesocosm studies show increased oil biodegradation rates when dispersants are used. Temporary inhibition of biodegradation with dispersed oil also has been observed in the laboratory, but appears to occur only at dispersed oil concentrations higher than would be observed in the field. Mesocosm studies in ponds suggest that when effectively used, dispersants should enhance the rate of biodegradation of oil. Unfortunately, supportive field observations or case histories are not available. Finally, available information suggests that some compounds, such as tar balls, would remain undegraded regardless of the addition of dispersants (Lee and Levy, 1986). Prevention of the formation of tar balls and large oil-in-water emulsions accumulations—called mousse—is a potentially important advantage of chemically dispersing oil. These forms of oil, with low surface area, tend to be resistant to biodegration, largely because there are few microbes and nutrients present in the mass.

HISTORY OF DISPERSANT USE

The *Torrey Canyon* Spill

Before about 1970, dispersant formulations were basically degreasing agents that were developed to clean tanker compartments, bilges, and engine rooms. A number of "detergents," as they were called, were used to attempt to disperse the nearly 1 million bbl of crude oil that spilled from the tanker *Torrey Canyon* off the English coast in 1967. The extreme toxicity of these agents to marine life was attributed primarily to alkylphenol surfactants and the aromatic hydrocarbons in the solvent(Portmann, 1970).

Over 14 days following the *Torrey Canyon* spill, about 10,000 bbl of various dispersants were sprayed on the water and along the shore. Investigations showed that the solvents evaporated (90 percent in 100 hr), but the denser surfactants did not evaporate, mix with, or dissolve in seawater. Instead, they formed stable "detergent-oil" emulsions, the most stable of which were produced by the most toxic dispersants—Houghton, BP1002, and Gamlen. Smith (1968:22) noted:

> As to spraying at sea, we have no information about its eventual effectiveness. It was generally agreed by those taking part in the sea operations that dispersal was often achieved in the immediate neighborhood of spraying. However, despite the large quantities of detergents used, large areas of undispersed oil persisted for weeks as extensive and discrete patches.

The biological impacts along rocky shores were highly visible and devastating. Evidence of mortality was repeated all along the affected shoreline: empty limpet seats were conspicuous in pools, dead barnacle shells persisted for some time, and mussel shells gaped. The rotting flesh did not take long to disappear, but even when the shells had broken away, clumps of short straw-colored byssus threads persisted for a few weeks. Although the official position was that "the effects have not been catastrophic" (Smith, 1968:178), adverse publicity during and after the *Torrey Canyon* incident gave dispersants a bad reputation. Indeed, the experience led to a very cautious attitude toward dispersant use among several industrialized nations.

Development of New Formulations

The main concerns supporting the cautious attitude toward dispersant use after the *Torrey Canyon* spill were

- toxicity of the products themselves, and
- concern that effective dispersants would make oil constituents

more available to biota and thus enhance toxicity of oil components.

There was further concern about greater quantities of hydrocarbons entering and persisting in shallow, low-energy marine environments when dispersed with chemicals. Biological concerns focused on species important to fisheries, young life stages of marine animals, and littoral or shallow sublittoral habitats. Despite these concerns, the United Kingdom and several other European countries continued to test dispersants in a series of government-sponsored programs, and Canada and several other countries conducted research and prepared guidelines for acceptability and use.

Second-generation dispersants were produced; generally they were much less acutely toxic than the earlier formulations but of variable effectiveness. In addition, a third generation of dispersants, with concentrates consisting mostly of surfactant with little solvent, was introduced to reduce volume for storage and transportation. The concentrates were intended to be diluted with seawater during spraying from boats, which was supposed to make spraying more uniform and increase operating time at sea (Morris and Martinelli, 1983). Laboratory tests suggested, however, that dilution reduced effectiveness by as much as a factor of 5 (Crowley, 1984a,b; Doe and Wells, 1978; Cormack, private communication). Experience at sea did not agree with the laboratory prediction, however, and the large effect experienced may have been specific to the formulation used in the tests (Lindblom, private communication). Undiluted dispersant is applied from aircraft, but some equipment on boats use water-diluted dispersant.

Concerns about the environmental impact of dispersants stimulated considerable research in Canada, Europe, and the United States in the 1970s and 1980s. The Conference of the American Society for Testing and Materials (ASTM) in 1977 stimulated U.S. work leading to major worldwide studies funded by the American Petroleum Institute (API) and Exxon (Koons and Gould, 1984; McCarthy et al., 1978), and the 1980 Toronto dispersant conference (Mackay et al., 1981) brought most of the major researchers together to consider new work. In part sparked by the Toronto discussions, a Canadian research and development program was continued, which built on earlier work, to gain better understanding of the fate and effects of chemically dispersed oils in cold, temperate, and arctic waters. This program continued into the 1980s, in particular with the Arctic Marine Oilspill Program (AMOP), which supported the annual AMOP Technical Seminars and the Baffin Island Oil Spill Project (BIOS) study. Considerable research was initiated in Norway and Sweden and continued in France, the Netherlands, and the United Kingdom, among others. Despite these efforts, many key questions asked 15 to 20 years ago are still the subject of controversy and research today.

Development of Equipment

Boat spraying systems were developed by the United Kingdom's Warren Spring Laboratory (WSL) shortly after the *Torrey Canyon* incident. They were designed to make dispersant chemical spraying more controllable, and were especially useful for applying undiluted hydrocarbon-based dispersants at very high doses, 327 to 889 l/ha (35 to 95 gal/acre) (Cormack, 1983c). Early dispersant formulations required agitation to promote dispersion. This was done using trailing wooden "breaker boards," which required that the spray boom be mounted toward the rear of the vessel, often aft of where the bow wave breaks from the hull. This configuration caused some dispersant spray to miss the oil that was pushed away. Development of dispersant concentrates eliminated this problem because they reduced the need for externally applied agitation and made breaker boards obsolete. The dispersant concentrates also reduced the amount of dispersant needed to treat a given quantity of oil, and therefore reduced the application rate and greatly extended the spraying time for a vessel or aircraft.

A fixed-wing aircraft system, which uses undiluted concentrate, was developed by the WSL from crop-spraying equipment in the 1970s. Additional extensive experimental and developmental work on aerial spraying was done in Canada and the United States in the 1970s and 1980s (Cormack and Parker, 1979; Lindblom and Barker, 1978; Nichols and Parker, 1985; Ross et al., 1978). Helicopter spraying systems were also developed in the late 1970s and were used to spray dispersants on the slick from the *Hasbah 6* blowout along the coast of Saudi Arabia in 1980 (Martinelli, 1980; van Oudenhoven, 1983).

Case Studies

Dispersants have been used at many spills (more than 50 have been recorded), including the very large operations such as the *Ixtoc I* blowout in the Gulf of Mexico (Nelson-Smith, 1980, 1985). A few of these spills are described briefly in Appendix B. However, lack of controls, ad hoc observations, poor documentation, and lack of objective criteria for effective dispersal have made these situations less informative than might be expected. Planned dispersant field trials are reviewed, assessed, and summarized in Chapter 4.

ISSUES AND QUESTIONS

This report focuses on two main themes that are amplified in the form of more detailed concerns or questions in this section, along with a brief discussion of related problems:

• Dispersant effectiveness: Do dispersants do any good? Are current formulations and techniques effective? Can more effective techniques be developed?

• Harmful effects of dispersants: Do dispersants do any harm? Is the toxicity of dispersant formulations significant to marine species and, if so, under what environmental conditions? Is the biological impact of dispersed oil greater or less than that of untreated oil?

Using Chemical Dispersants to Remove Oil From the Surface of the Water

Developing optimum techniques for applying dispersants under various conditions requires an understanding of the numerous factors affecting oil dispersion. To achieve this, a definition of effectiveness is required, as well as a knowledge of the physical and chemical mechanisms of dispersion, techniques and logistics of dispersant application, and possible methods for modeling effectiveness quantitatively.

Defining Effectiveness

A commonly accepted definition of effectiveness of spill treatment, and the one used in this discussion, is the fraction of oil removed from the surface of the water. Alternative definitions have been devised, based on laboratory measurements, in an attempt to provide an easily measured surrogate (Chapter 2, "Laboratory Studies of Effectiveness"). Field studies have measured the concentration of dispersed oil in the water column under a slick, to find how much oil might be encountered by organisms (Chapter 4, "Physical and Chemical Studies").

Effectiveness depends on many factors. The following are discussed in detail later in this report: the physical and chemical interaction of oil and dispersants to reduce interfacial tension (Chapter 2, "Composition of Dispersants"), slick structure, sea state, turbulent mixing of oil droplets in the subsurface water column (Chapter 2, "Fate of Oil Spilled on Open Water" and "Behavior of Oil-Dispersant Mixtures"), and efficiency of application techniques (Chapter 5, "Design of Dispersant Application Systems").

Identifying Physical and Chemical Factors That Influence Effectiveness

Many factors influence the dispersibility of oil and the effectiveness of treatment. Specific questions include: How is effectiveness influenced by

- oil composition (amounts of aromatic and aliphatic hydrocarbons, asphaltenes, and metalloporphyrins)?
- dispersant composition (hydrophilic-lipophilic balance [HLB]) of surfactant(s), kind of surfactant(s), and type of solvents?
- dispersant-to-oil ratio?
- energy input (breaking waves, subsurface turbulence, and mechanical mixing)?
- water salinity and temperature?
- oil viscosity?
- oil slick thickness and distribution on the water surface?
- weathering of the oil (loss of volatile hydrocarbons, photooxidation, and water-in-oil emulsion formation [mousse])?

Viscosity, slick distribution, and oil weathering are time-dependent variables. These environmental and time-dependent factors are discussed in Chapter 2.

Factors Affecting Dispersant Application Techniques

New dispersant formulations have been developed through the years. In addition, the techniques of dispersant testing and application are continuously revised with experience. New developments range from testing new formulations for effectiveness and toxicity in the laboratory, to the design of spray equipment for boats and aircraft, to the training of skilled response personnel in operating such equipment effectively. New formulations are usually proposed when a manufacturer is convinced of their effectiveness in treating oil spills. Government agencies normally test, or oversee the testing, of such new products for effectiveness and toxicity (see Chapter 3). Questions raised are:

- Can one dispersant be used with a wide variety of oils with expectation of reasonably equal effectiveness?
- Are there significant differences between the expected performance characteristics of different dispersants used with the same oil?
- How important is it to apply dispersants before the oil weathers, and what factors control this?
- At the time of a spill, is it important to test the proposed dispersant on a sample of the oil before proceeding to apply the dispersant? What are the alternatives?
- Is it important to spray dispersant uniformly?
- How should dispersant effectiveness be measured?
- What are the concentrations and persistence of dispersed oil under chemically dispersed and untreated oil slicks? What are the concentrations and persistence of dissolved hydrocarbons under slicks, and how do they differ from the C_{11+} fraction?
- Are current spray systems (circa 1987) adequate to control oil spills?

Factors Affecting Toxicity of Chemical Dispersants and Dispersed Oil

As discussed in later chapters, there is a wide diversity of use of chemical dispersants for the control of oil spills in various countries. A range of opinion exists concerning the biological effects resulting from the use or proposed use of chemical dispersants. The following questions (discussed in Chapters 3 and 4) have been raised, based on some real observations or on perceived concerns:

- How toxic are dispersants alone, and chemically dispersed oil, to marine plants and animals in laboratory studies?
- How do laboratory measured toxicity exposures compare with exposures measured in field studies?
- Does chemically dispersed oil pose more or less hazard than untreated oil to organisms at the sea surface and in the immediate subsurface waters?

- Does chemically dispersed oil pose more or less hazard than untreated oil to benthic organisms?
- How do offshore biological effects compare with effects on shorelines habitats, including the intertidal and immediate subtidal zones?
- Are there chronic effects from dispersants used to control oil spills?
- Is the oil biodegradation rate increased after chemical dispersion?
- Is photooxidation greater or less after chemical dispersion?
- Is there need to further refine and standardize toxicity testing in laboratory studies?
- Does chemical dispersion increase or decrease the amount of oil that attaches to solids in the water column and the amount of oil that enters the sediments?
- Will chemical dispersion of the oil slick protect seabirds and mammals?
- Should a more effective dispersant (less applied) be used even if it is more toxic?

Answers to the above questions are addressed principally in Chapters 3 and 4.

2
Chemistry and Physics of Dispersants and Dispersed Oil

Most oils spilled on water rapidly spread into a slick, with thickness from several millimeters down to one micrometer depending on the oil type and the area available for spreading. Wind-driven waves and other turbulence can break up the slick, producing more or less spherical droplets ranging in size from a few micrometers to a few millimeters. Sometimes, these droplets can be stabilized by natural surface-active agents (surfactants) present in the oil or contributed by the sea-surface microlayer in the region where the oil was spilled. These surfactants stabilize the droplets by orienting in the oil-water interface with the hydrophobic part of the surfactant molecule in the oil phase and the hydrophilic part in the water phase, thereby diminishing the interfacial tension.

Applying chemical dispersants to an oil slick greatly increases the amount of surfactant available and can reduce oil-water interfacial tension to very low values—it therefore takes only a small amount of mixing energy to increase the surface area and break the slick into droplets (Figure 2-1).

Dispersants also tend to prevent coalescence of oil droplets. The interface, stabilized by the surfactant, permits droplets to survive despite frequent collisions with adjacent droplets. The same stabilizing factors reduce adherence to hydrophilic solid particles, such as sediments, as well as other solid surfaces (discussed later in Chapter 4).

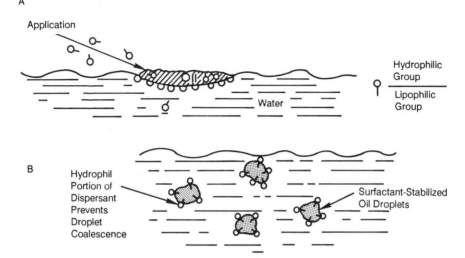

FIGURE 2-1 Mechanism of chemical dispersion. A. Surfactant locates at oil-water interface. B. Oil slick is readily dispersed into micelles or surfactant-stabilized droplets with minimal energy. Source: Derived from Canevari (1969).

During the past 20 years, significant reviews and descriptions of dispersant chemistry include those by Poliakoff (1969), Dodd (1974), Canevari (1971, 1985), Wells (1984), Pastorak et al. (1985), Wells et al. (1985), API Task Force (1986), and Brochu et al. (1987).

COMPOSITION OF DISPERSANTS

The key components of a chemical dispersant are one or more surface-active agents, or surfactants—sometimes loosely called "detergents." They contain molecules with both water-compatible (hydrophilic) and oil-compatible (lipophilic or hydrophobic) portions. Most formulations also contain a solvent to reduce viscosity and facilitate dispersal.

Chemistry of Surfactants

The behavior of a surfactant is strongly affected by the balance betweeen the hydrophilic and lipophilic groups in the molecule. Griffin (1954) defined the hydrophile-lipophile balance. The useful range of this parameter is from 1 (most lipophilic) to 20 (most hydrophilic). Many organic compounds, like hexane, with no hydrophilic groups could have HLB as low as zero, and would not be surface active. In the HLB range of 1 to 4, the surfactant does not mix in water; above 13, a clear solution in water is obtained (Rosen, 1978).

Bancroft's rule states that the dominant group of a surfactant tends to be oriented in the external phase (Bancroft, 1913). Thus, a predominantly lipophilic surfactant (HLB, 3 to 6) would stabilize a water-in-oil emulsion, and a predominantly hydrophilic surfactant (HLB, 8 to 18) would stabilize an oil-in-water emulsion. Surfactants used in oil spill dispersants tend to be of the latter type. Natural surfactants, which promote mousse (water-in-oil emulsion), tend to be predominately lipophilic.

HLB is important in determining the effect of salinity on dispersant performance, since hydrophobic portions of the surfactant molecule tend to be salted out. Laboratory measurements on weathered crude oil with a dispersant sensitive to salinity showed that, at a comparable treatment rate and mixing energy, the amount of oil dispersed is approximately 58 percent in seawater compared to 1 percent in fresh water. This formulation, balanced for effective performance in seawater, is too hydrophilic for freshwater service (Canevari, 1985).

Surfactants are also classed by charge type, as noted below (a list of formulations is given in Appendix A):

• *Anionic.* Examples include sulfosuccinate esters, such as sodium dioctyl sulfosuccinate (e.g., Aerosol OT). Other examples are oxyalkylated C_{12} to C_{15} alcohols and their sulfonates.

• *Cationic.* An example is the quaternary ammonium salt $RN(CH_3)3^+Cl^-$, but such compounds are often toxic to many organisms and are not currently used in commercial dispersant formulations (Lewis and Wee, 1983).

• *Nonionic.* These are the most common surfactants used in commercial dispersant formulations. Examples are sorbitan monooleate (HLB, 4.3), sold as Span 80, and ethoxylated sorbitan monooleate (HLB, 15), sold as Tween 80. In addition, polyethylene glycol esters of unsaturated fatty acids and ethoxylated or propoxylated fatty alcohols are used.

• *Zwitterionic or amphoteric.* These molecules contain both positively and negatively charged groups, which may balance each other to produce a net uncharged species. An example would be a molecule with both a quaternary ammonium group and a sulfonic acid group (refer to Appendix A), but such compounds are not found in current commercial formulations.

As surfactants become more concentrated, the interfacial tension between oil and water decreases until a critical micelle concentration (CMC) is reached. Micelles are ordered aggregates of surfactant molecules, with the hydrophobic portions of the molecules together at the interior of the micelle and the hydrophilic portions facing the aqueous phase. Above the CMC level, there is little change in interfacial tension, and additional surfactant molecules form new micelles. Below the CMC, additional surfactant molecules accumulate at the water-air or oil-water interfaces. The CMC can be estimated from the concentration at which a change in slope of a plot of interfacial tension (as measured by the drop weight technique) versus dispersant concentration occurs (Figure 2-2).

Some formulations can reduce interfacial tension to a few percent of the value without surfactant addition. For example, in a specially adapted Wilhelmy Plate instrument, the initial oil-water interfacial tension of 18 dyn/cm was reduced to the minimum detectable value, approximately 0.05 dyn/cm, within 1 min after dispersant was added (Ross and Kvita, unpublished data).

Current Dispersant Formulations

Early dispersant formulations were derived from engine room degreasers, and some were highly toxic (Chapter 1). To reduce toxicity, nonaromatic hydrocarbons (or water-miscible solvents such as ethylene glycol or glycol ethers), as well as less toxic surfactants, have been used in more recent formulations (Chapter 4 and Appendix A).

Figure 2-3 illustrates how different surfactants become oriented at the oil-water interface. Compound A is sorbitan monooleate (HLB, 4.3), predominantly lipophilic. Compound B is Compound A that has been ethoxylated with 20 mol of ethylene oxide, rendering it more hydrophilic (HLB, 15). A dispersant containing both A and B, with a larger amount of B, can stabilize an oil-in-water emulsion.*

A blend of surfactants with different HLB, giving a resultant HLB of 12, will be more effective than a similar quantity of a single surfactant with HLB of 12. This is shown in Figure 2-3: the hydrophilic groups of B penetrate farther into the water phase, per-

---

*As discussed earlier, the dominant group of a surfactant tends to be oriented in the external phase (Bancroft, 1913).

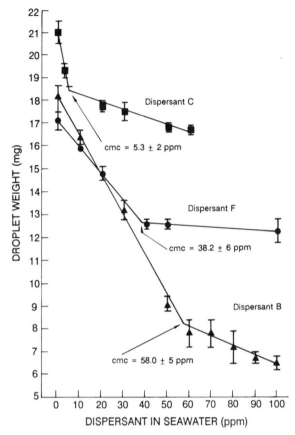

FIGURE 2-2   Interfacial tension as a function of dispersant concentration showing the
discontinuity in slope at the critical micelle concentration (cmc). Light Arabian crude
with three dispersant formulations at 28°C and 38 percent salinity; interfacial tension
is measured by the drop-weight method. Source: Rewick et al., 1984. Reprinted, with
permission, from the American Society for Testing and Materials. © 1984 by ASTM.

mitting closer physical interaction between the lipophiles of both A
and B. The overall result is to provide a stronger interfacial surfactant
film and resistance to coalescence of dispersed oil droplets.

A review of the patent literature (Appendix A), combined with
discussions with several major suppliers of dispersants, indicates that
a limited number of surfactant chemicals are used in the dispersant
formulations most widely available today. The exact details of dispersant formulations are proprietary, but the chemical characteristics of
these formulations are broadly known (Canevari, 1986; Brochu et
al., 1987; Wells et al., 1985; Fraser, private communication). Thus,

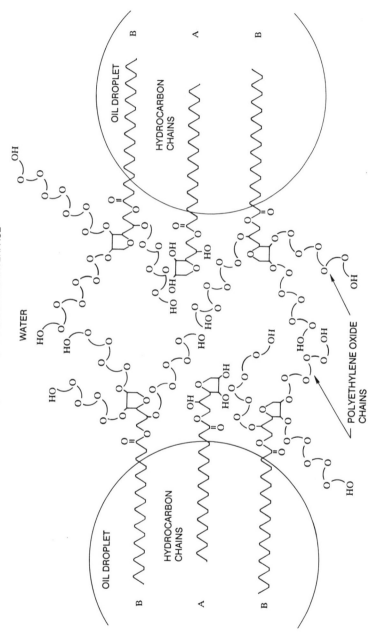

FIGURE 2-3 Hypothetical orientation of different surfactants at oil-water interface. Source: Adapted from Canevari (personal communication) and Brochu et al. (1987).

modern dispersant formulations containing one or more nonionic surfactants (15 to 75 percent of the formulation) may also contain an
anionic surfactant (5 to 25 percent of the formulation) and include
one or more solvents.

The surfactants used include the following:

- nonionic surfactants, such as sorbitan esters of oleic or lauric
acid, ethoxylated sorbitan esters of oleic or lauric acid, polyethylene glycol esters of oleic acid, ethoxylated and propoxylated fatty
alcohols, and ethoxylated octylphenol; and
- anionic surfactants, such as sodium dioctyl sulfosuccinate and
sodium ditridecanoyl sulfosuccinate.

Dispersant formulations also contain a solvent to dissolve solid
surfactant and reduce viscosity so that the dispersant can be sprayed
uniformly. A solvent may be chosen to promote rapid solubility of
the surfactant in the oil and to depress the freezing point of the
dispersant mixture so that it can be used at lower temperatures.

The three main classes of solvents are: (1) water, (2) water-
miscible hydroxy compounds, and (3) hydrocarbons. Aqueous solvents permit surfactants to be applied by eduction into a water
stream. Hydrocarbon solvents enhance mixing and penetration of
surfactant into more viscous oils. Examples of hydroxy-compound
solvents are ethylene glycol monobutyl ether, diethylene glycol mono-
methyl ether, and diethylene glycol monobutyl ether. An example
of a hydrocarbon solvent is a low-aromatic kerosene. High-boiling
solvents containing branched saturated hydrocarbons are also used
since they are less toxic than aromatics.

Appendix A gives a list of chemical formulations for use on
oil discharges listed in 1987 by the U.S. Environmental Protection
Agency in its National Contingency Plan Product Schedule (Flaherty
and Riley, 1987). Composition of some of these formulations have
been disclosed in patents, but all are proprietary. Only U.S. EPA
listed formulations may be used to treat oil discharges in U.S. waters
(Chapter 6).

Matching Dispersant Formulations With
Oil Type for Increased Effectiveness

Because oils vary widely in composition, it is reasonable to hy-
pothesize that a particular dispersant formulation could be more
effective with one oil than another; indeed that a dispersant could

be matched to a particular oil for increased, possibly optimal, effectiveness. This is not possible at present for a number of reasons. Only a limited number of different dispersants have been used or are available. Such a narrow subset of all possible formulations has resulted from a convergent evolution in the industry. Furthermore, limited use has been made of chemical dispersants during accidental spills in the United States, and only a few research spills have been conducted using different dispersants with the same oil.

Because it is not possible to conduct field tests of all dispersants with all oils, a more practical approach has been taken: formulations that perform best in laboratory studies are used on oil spills. If a dispersant works, it continues to be used.

It should be noted, however, that the effect of dispersant application may be delayed, as was observed in Lichtenthaler and Daling's (1985) Norwegian offshore research studies. Effectiveness also may be reduced by oil resurfacing later (Bocard et al., 1987). Temperature can affect dispersant performance in ways other than by changing the viscosity of the oil. The solubility of ethoxylated surfactants in water increases at lower temperature. For example, at similar dispersant-oil ratios, water salinity, and wave energy, dispersant OSD-1 was found to be 100 percent effective (all oil was removed from the water surface) in laboratory tests at 15°C, but only 56 percent effective at 5°C. In contrast, dispersant EXP-A was 100 percent effective at both temperatures. This difference was explained by changes in water solubility and HLB of the dispersant (Becker and Lindblom, 1983).

Fate of Surfactants and Solvents in the Aqueous Environment

Surfactants are used for many purposes other than treating oil spills, and their degradation in the aqueous environment has been a concern since the 1950s, when synthetic alkylbenzene sulfonate (ABS) detergents were found to be resistant to biodegradation and produced persistent foam on waters receiving domestic sewage effluent. This problem was solved by replacing the ABS detergents with more readily biodegradable surfactants. Manufactured surfactants in 1983 totalled about 24 million metric tons worldwide, most of which were employed as household detergents and industrial cleaners (Layman, 1984:49). It is recognized that surfactants used in detergents may contribute to stream pollutant discharges that are continuous and often affect a large area over many years. In contrast, dispersants and dispersed oil inputs to the sea are usually rare or infrequent events that could cause temporary effects in the open sea, but cause large, short-term disruptions in restricted areas. Therefore, a direct comparison of continuous surfactant discharge from industrial and household use and dispersant surfactant loading is tenuous, at best, in regard to environmental effects of the two sources.

Some discharged surfactants are biodegraded in sewage treatment plants, but many are not because much of the sewage is not treated. Some of the linear alkylbenzenes (LAB), used in the manufacture of linear alkyl sulfonates (LAS) remain as an impurity in LAS, and are found in suspended particles and sediments surrounding municipal waste discharges. Eganhouse et al. (1983) used LAB as tracers and stated that they appear to be preserved in sediments for 10 to 20 years.

Some work has been done on the rate of breakdown and environmental concentrations of surfactants in the water column (Kozarac et al., 1983; Lacaze, 1973, 1974; Penrose et al., 1976; Una and Garcia, 1983). Surfactants are also transferred from water to air via sea spray, and increase the production of marine aerosol. They may thus encourage the transfer of oil slick components into the atmosphere (Fontana, 1976). Adsorption of surfactants onto sediments is discussed later in this chapter (see also Inoue et al., 1978).

Surfactants can also become bioconcentrated and metabolized in the tissue of fishes and invertebrates (Comotto et al., 1979; Kimerle et al., 1981; Payne, 1982; Schmidt and Kimerle, 1981). Metabolic breakdown of the surfactants is rapid (85 percent in 4 days).

The fate of solvents used in dispersant formulations might also be of concern. Hydrocarbon solvents are similar to portions of the oil being dispersed and tend to suffer a similar fate. Glycol ether solvents are likely to be more readily biodegradable than the oil being dispersed, but nothing appears in the literature about the toxicity of their degradation products. However, the concentrations are usually small and decrease rapidly owing to dilution and mixing.

FATE OF OIL SPILLED ON OPEN WATER

Slick Thickness

Slick thickness is an important parameter in predicting optimum dispersant dosage (Chapter 5), but the thickness of an oil slick at sea cannot be readily determined. Reliable measurements of thickness over the whole area of a slick have rarely been made (Hollinger and

Menella, 1973; Lehr et al., 1984; O'Neill et al., 1983). Infrared remote sensing provides an image of thick slicks, 10 to 50 μm or greater (O'Neill et al., 1983), and ultraviolet sensing can measure slicks down to the submicron range. Combined, infrared and ultraviolet remote sensors can be used to calculate areas and ratios of thin to thick slicks. The limits to infrared detection are unknown, however, and certainly vary with environmental conditions and oil type. Varying intensity levels in the infrared have been processed to yield additional contours, but assignment of thickness to such contours, although attempted, has been only relative (Ross, 1982).

Actual slicks at sea are nonuniform in thickness and distribution on the surface. The thickest portion of a slick can be as great as several millimeters and the "sheen" (microlayer) only 1 to 10 μm. In one experiment, infrared thermography showed that thin areas of the slick were 10 to 20 μm and thick areas were 150 to 200 μm. The thicker portion contained 28 of the 42 bbl of spilled oil (Bocard et al., 1984), but covered only about one-fifth of the slick's surface area.

Most estimates of slick thickness are averages based on the visual appearance of the oil or calculated by dividing the total volume of oil by its observed area (International Tanker-Owners Pollution Federation [ITOPF], 1982). Some investigators believe that using average thickness, although formally consistent, is misleading and obscures one of the most important aspects of an oil slick from the cleanup team's point of view—its nonuniformity. Also, from a biological point of view, exposures under a nonuniform slick are likely to be patchy.

Nevertheless, many spills of widely varying size tend to reach a similar average thickness of about 0.1 mm rather quickly and this rule of thumb is widely employed by dispersant application specialists. Over several days, as the slick spreads, average thickness may decrease to 0.01 mm (API Task Force, 1986; McAuliffe, 1986). The following is some evidence for such a generalization:

• The Chevron Main Pass Block 41 C blowout, released oil at 1,680 to 6,650 bbl/day. The slick size varied, but was about 1-km wide and 10-km long, with thicker oil near the platform. The average thickness was 0.02 to 0.09 mm (McAuliffe et al., 1975; Murray, 1975).

• The *Hasbah 6* oil well blowout in the Arabian Gulf released a viscous oil slick that extended for many miles. After 2 weeks, 9,930 bbl of oil were skimmed from an area of 11.3 km^2. The slick was thus estimated to be at least 0.13-mm thick (Cuddeback, 1981).

• In the API-EPA research spills, 20 bbl were released over 5 to 10 min, corresponding to rates of 3,000 to 6,000 bbl/day. After 15 to 30 min, before dispersant spraying began, the slicks covered 20,000 to 30,000 m². Average thickness was therefore 0.1 to 0.2 mm (Johnson et al., 1978; McAuliffe et al., 1980, 1981).

• In a Norwegian test spill of 700 bbl over 2 hr (corresponding to 8,400 bbl/day), the thick part of the slick, containing 90 percent of the oil, covered 1 km² after 8 hr and remained at 1.5 to 2 km² for the next 5 days. The average spill thickness decreased from 0.06 to 0.013 mm (Audunson et al., 1984).

• Data from the 1983 Halifax trials showed average thicknesses of untreated slicks was of the order of 47 μm after 1 hr, decreasing to 40 μm after 2 hr. The thin portion of the slicks was estimated to be of the order of 1-μm thick (Canadian Offshore Aerial Applications Task Force [COAATF], 1986).

A large release over a short time, such as would occur from a tanker accident, initially produces much thicker slicks near the release point. For example, release of 200,000 bbl into an area 100 m in diameter could create an average thickness of 20 mm, but if the slick spreads, the average thickness decreases. In cold climates and waters, however, the higher viscosity of oil can cause a stable slick to be thicker than 0.1 mm. In addition, slick thickness could increase as the oil becomes emulsified to form mousse, which occurred during the *Amoco Cadiz* disaster and many other incidents.

Slick Spreading

Oil slicks are usually nonuniform in thickness because of the interaction of interfacial tension, gravity, and viscosity in spreading processes, the accumulation of oil at downwelling convergence zones produced by water movement, and the formation of high-viscosity water-in-oil emulsions (mousse). Furthermore, oils have different spreading tendencies, particularly on cold water (Tramier et al., 1981). The work of Fay (1971) provided a mathematical prediction for spreading under the influence of interfacial tension and gravity. His model predicts an area increasing to a maximum value proportional to the 3/4 power of the volume of oil spilled. Although the model included water viscosity, it did not include the effects of oil viscosity, emulsification, and evaporation, and considered only a calm water surface. Nevertheless, the model was successful in predicting results of laboratory experiments and was subsequently used in more elaborate models (Huang and Monastero, 1982).

However, field tests revealed the limitations of Fay's approach. Jeffery (1973) reported the formation of an elongated slick with the major dimension increasing linearly with time over 4 days. The minor dimension increased rapidly during the first few hours, then remained constant. In the experiments reported by Cormack et al. (1978), the minor dimension increased at a rate in accordance with the Fay equations, but the major dimension increased at 10 times the expected rate. Attempts to deal with factors not addressed by the Fay approach include constraining the slick to be elliptical (Lehr et al., 1984) or imposing artificial nonhomogeneity on the slick by dividing it into "thick" and "thin" portions (Mackay et al., 1980a).

Recent studies have shown that the dominant mechanism of oil spreading is the interaction of the oil droplets with diffusive and current shear processes in near-surface currents (Elliot, 1986). Work by oceanographers, such as Bowden (1965), Fisher et al. (1979), and Okubo (1967), have shown that a patch will elongate in the direction of flow. The length scales of the patch in the along- and cross-flow directions are approximated by equations that suggest that initally the spreading of oil droplets will be Fickian*—that is, it is dependent on the spatial concentration gradients—and that spreading will grow linearly with time. These expressions also show that, with time, velocity shear is more influential.

Subsurface release from a well blowout produces a thinner slick than a release directly onto the water surface because of the entrainment of water (and oil) in rising gas bubbles (Fannelop and Sjoen, 1980). At the surface, water and oil flow away from the center of the plume, and the oil slick spreads faster than oil released on a quiet surface.

A model of the *Ixtoc I* blowout developed on these principles estimated that, at 1 km from a 30,000 bbl/day subsurface discharge, slick thickness would be 0.06 mm (Fannelop and Sjoen, 1980). This may be compared with an average thickness of 0.07 mm calculated by Jernelov and Linden (1981) and McAuliffe (1986). For a surface discharge, the calculation of Fannelop and Sjoen would indicate a thickness of 1.3 mm under the above conditions.

---

*Fick's first law states that the flux, or rate of diffusion, of a material (e.g., oil particles, chemicals) is proportional to the concentration gradient; this relationship assumes that flow is laminar.

In addition to the spread (or sometimes contraction) of the slick under control of surface tension and viscosity, a variety of oil types in test and accidental spills, including the *Ixtoc I* spill, have been observed to spread into a complicated texture of thick "pancakes" and thin sheen. Formation of mousse is common. Windrows produced by surface convergences are further complications. Thicker patches have been observed to move downwind at a faster rate, leaving a thinner trailing sheen behind. Published observations of actual thickness variation downdrift from *Ixtoc I* are rare and none are quantitative. Several months after the well blew out, observers on research vessels, small boats, and helicopters, characterized the oil in the plume as occupying five zones (Atwood, 1980):

1. A continuous light-brown water-oil emulsion on the surface occurred in the immediate vicinity of the flames at the wellhead and extended no more than a few hundred meters down the plume.

2. The sea surface was 30 to 50 percent covered by a light-brown emulsion in disoriented streaks. This zone started a few hundred meters down-plume from the burn and extended several kilometers, depending on wind stress. At times it was virtually absent.

3. The sea surface was 20 to 50 percent covered by light-brown emulsion oriented in streaks parallel to the wind direction, apparently in the convergence zones of the Langmuir surface circulation.* The streaks were surrounded by a sheen, their width varied from a few centimeters to a few meters, and their length varied from one to tens of meters. The dimensions depended on wind stress. This zone extended from a few hundred meters from the flames to several kilometers down the plume.

4. The light-brown emulsion darkened until the streaks were black, apparently from photooxidation, since the degree of darkening seemed to depend on light intensity. Langmuir streaks were blackened in the center and light brown at the edges. At times they coalesced into lines of blackened oil several kilometers long. At the brown edges were small balls of viscous emulsion (mousse). The softer balls would coalesce on contact; in some instances grapefruit-sized balls formed a raft 50 m in diameter. Most of these phenomena

---

*The combination of wind forces and wave forces produce currents that rotate cylindrically. The convergences between these currents, or windrows, are straight running in the direction of the wind, and their distance apart ranges from 20 to 50 m depending on wind speed (Faller, 1978).

occurred 10 to 40 km from the burn. A light to heavy sheen of surface oil was always present in this zone.

5. An extensive sheen (1 to 10 μm), visible because it changed the reflectivity of the water surface, covered more than 50 percent of the surface, usually in the form of Langmuir rows, and extended to the farthest extremity of the plume.

Other observers noted that there were typically three to six "stringers," slicks much longer than they are wide, 1 to 3 mm thick (Fraser and Reed, 1982). Observers from Petroleanos Mexicanos also noted the complicated nonuniform structure of the *Ixtoc I* slick (Petroleanos Mexicanos, 1980, translated from the Spanish):

> The oil that did not disperse moved along the surface of the sea following the resultant marine surface currents and the direction of the winds, forming bands and strips of variable lengths and widths and other capricious shapes.
>
> During the first few days after the spill, the slick formed a maximum length of 15 km and a maximum width of 2 km. [As a result of] the combination of the containing barrier, recovery of spilled oil, and the reduction in flow of oil from the well, the size of the slick was reduced even though it advanced toward the coast along a corridor well defined by the currents. It can be shown that the slick did not advance as a compact and continuous mass, but as a series of strips that from aerial observation seemed like a web.

Physical Processes of Dispersion Related to Water Motion

In addition to spreading of an oil slick and dispersal of oil into droplets, spilled oil is distributed and transported by motion of the water mass in which it resides. Such motion includes drift of the slick caused by wind, tides, and other forces, motion of dispersed oil with the water mass, and redistribution of oil with respect to the water as a result of turbulent diffusion and vertical shear.

Conventional concepts of physical oceanography are frequently used to describe these processes. The change with time of oil concentration at a specific location is the sum of three terms:

1. advective changes, which depend on flow of the water mass;

2. turbulent diffusion changes, which depend on concentration gradients of the oil within the water mass and on wind- and current-induced turbulence; and

3. source or sink changes, which depend on physical, chemical, and biological interactions. Solid particles on which oil has adsorbed can settle out as sediment, dispersed oil droplets can be removed by

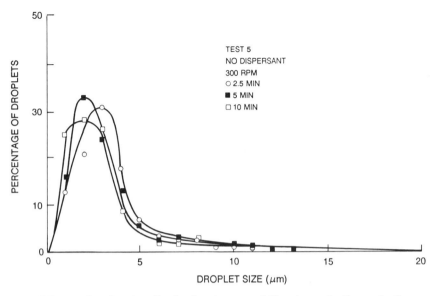

FIGURE 2-4 Droplet size distribution for test of Kuwait crude dispersal. Source: Jasper et al., 1978. Reprinted, with permission, from the American Society for Testing and Materials. © 1978 by ASTM.

recoalescence with the surface slick, or photochemical and biochemical degradation can occur.

Mechanical dispersion occurs primarily when waves break, which requires wind speeds greater than 10 kn. When wave action provides sufficient energy to overcome interfacial tension and create new oil-water interfacial area, an oil slick breaks into small droplets, usually less than 10 μm in diameter depending on slick thickness, that become suspended in the water column. Larger drops form in smaller numbers, and their distribution (Figures 2-4 and 2-5) falls off more steeply than exponentially (Bouwmeester and Wallace, 1986a,b; Franklin and Lloyd, 1986; Jasper et al., 1978; Lewis et al., 1985; Norton et al., 1978; Shaw and Reidy, 1979).

Advection

Bulk motion of water includes daily tidal currents; wind-induced currents (including the Langmuir surface circulation) that occur with daily sea breezes and longer-term time-scales (days to weeks); seasonal coastal currents; frontal eddies in boundary currents; and mesoscale eddies, which are common in the open ocean. Currents have been charted for centuries in waters frequented by ships. This accumulated knowledge is used, for example, by the U.S. Coast Guard, Navy Supervisor of Salvage, and the Canadian Coast Guard to predict drift so that boats and other objects lost at sea might be found.

Many currents are unpredictable, however, and advective flow is frequently complex and difficult to predict. For example, recent satellite-based studies of Gulf Stream eddies have shown the potential complexity of mesoscale circulation (Backus et al., 1981). This is also reflected in the subsurface distribution of dispersed oil in the

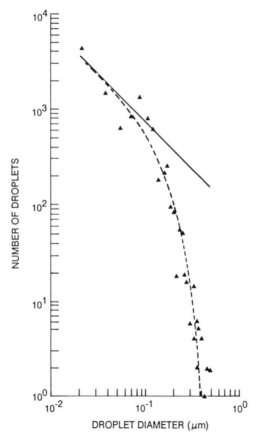

FIGURE 2-5 Droplet size distributions of No. 2 diesel oil in wind-wave tank with no dispersant. Initial (upstream) oil thickness = 0.15 mm. Wave height is condition II. Fetch: X = 10.6 m. Source: Boumeester and Wallace, 1986a.

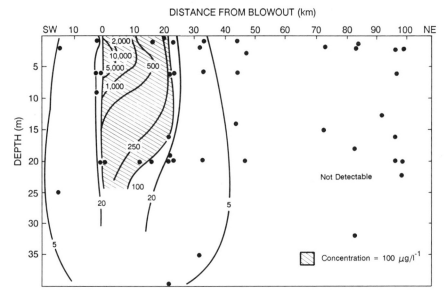

FIGURE 2-6   Oil concentrations (μg/liter) in the water column following the *Ixtoc I* blowout. Note the horizontal extent of the oil, which suggests the role of advection, while the downward penetration is indicative of the role of vertical turbulent diffusion. Source: Boehm and Fiest, 1982. Reprinted, with permission, from Environmental Science and Technology, Vol. 16, No. 2. © 1982 by American Chemical Society.

Ixtoc I plume (Figure 2-6). Observations of a section moving from the southwest to northeast suggested that oil was transported at least 40 km in subsurface water. Figure 2-6 shows substantial concentrations southwest of the blowout site ("upstream"). The upstream oil most likely was the result of advection prior to the date on which these measurements were made, when the current was in a different direction.

A second example of advective processes working at different time scales is shown in Figure 2-7. These data were derived from an experiment designed to test a model emphasizing physical processes, such as drift and spread, and weathering and vertical distribution (Johansen, 1984). The surface distribution of the oil slick followed the track of buoys quite well. In Johansen's model (discussed in more detail later) advection was computed as a "vectorial sum of the drift induced by the local wind, tidal currents, and an assumed stationary background current," which adequately summarizes the advective components.

Physical dispersion in many areas is dominated by unique local

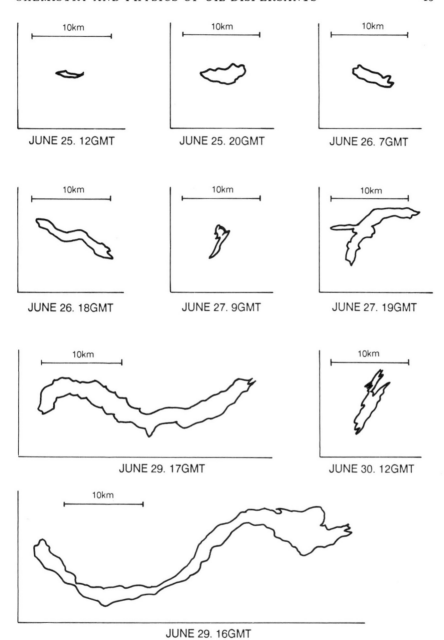

FIGURE 2-7 Time variation of an oil slick observed by remote sensing during the Halten Bank experiment. With time, the slick extended in line of advection and took on shapes determined by tidal currents, inert oscillations, and wind events. Source: Johansen, 1984.

TABLE 2-1 Typical Turbulent (Eddy) Diffusion Coefficients

Location	K (cm^2/sec)	References
	Horizontal Diffusion	
New York Bight	5,500	Okubo, 1971
Bering Sea	2,800	Coachman and Charnell, 1979
Harrison Bay	780	Wilson et al., 1981
	Vertical Diffusion	
Bering Sea	185	Cline et al., 1982
Beaufort Sea	25	Liu and Leendertse, private communication

SOURCE: Pelto et al., 1983.

currents. For example, currents near a coast are primarily determined by tidal flow, wind direction, and bathymetry, all of which are locally unique. Convergence zones created by internal waves can be a major factor in the onshore transport of slicks or tar lumps. Thus, in addition to models and theoretical studies, there is a need for knowledge of important physical processes specific to areas likely to be affected by oil spills.

Turbulent Diffusion

Turbulent diffusion coefficients are scale dependent since turbulent diffusion does not obey Fick's diffusion laws rigorously (some typical values are given in Table 2-1). However, the order of magnitude of values is instructive: horizontal turbulent diffusion tends to be more than 10 times faster than vertical. For oil dispersion, this means that surface spills tend to stay near the surface and to be advected horizontally, rather than to diffuse downward and enter a less observable regime.

Prediction of vertical movement may be highly simplified by using a "diffusion floor" (Mackay et al., 1982). Dispersant-treated oil tends to be under greater infuence from vertical diffusion than untreated oil (Chapman, 1985), and this type of physical transport can be addressed using a depth-dependent diffusion coeffcient.

The classic study of turbulent ocean diffusion was made by Okubo (1971), and recent theoretical and observational studies that give new information on the topic are Cline et al. (1982), Wilson et al. (1981), and Coachman and Charnell (1979).

Oil Concentration Under Slicks

Oil dispersed into the water column can come into contact with marine organisms that would not otherwise be affected by an oil spill. It is important to know the concentration of oil and its distribution in the water column in order to assess the environmental impacts of using a dispersant to treat an oil spill.

Vertical distribution of oil components in the water column depends on many physical factors, chiefly sea state, which breaks a slick into droplets and during storms can mix the upper layers of water to a depth of 10 m or more. (This vertical distribution has been studied in a number of field experiments described in Chapter 4.) Under untreated slicks, oil concentrations typically are a few parts per million to less than 0.1 ppm, diminishing with depth and with increasing time. For example, in one experiment water sampled beneath two untreated control slicks 2 to 6 hr after the spill gave oil concentrations ranging between 3 to 5 ppm at 1 m and 0.03 and 0.63 ppm at 2 m (Lichtenthaler and Daling, 1985).

In addition to redistribution of physically dispersed oil, the more soluble components of oil, such as benzene, toluene, xylene, and naphthalene, which are also more acutely toxic than other components, dissolve in the water and do not resurface if the droplets coalesce. Methods for analysis of oil in water are not consistent among researchers. For example, some researchers measure volatile hydrocarbons (C_1 to C_{10}). Others extract the water with a solvent and measure high molecular weight hydrocarbon by fluorescence, gas chromatography, or by weight.

The toxicity of a hydrocarbon component to marine organisms depends on the aqueous solubility of the compound, how concentrated it is in the portion of the water where the organism lives, the type of organism, the length of time it is exposed, and many other specific factors (see Chapter 3).

Evaporative Loss of Volatile Hydrocarbons

During the first 24 to 48 hr of an oil spill, evaporation is the single most important weathering process affecting mass transfer and removal of toxic lower molecular weight components from the slick. Evaporative loss is controlled by the composition, surface area, and physical properties of the oil and by wind velocity, air and sea temperatures, sea state, and solar radiation. The thinness of a slick and the small diameter of dispersed oil droplets allows the volatile hydrocarbons (C_1 to C_{10}) to evaporate quickly or go into solution. Evaporation greatly predominates over solution. For example, Harrison et al. (1975) demonstrated that evaporation is 100 times faster than solution for aromatics, and 10,000 times faster for alkanes. Biological toxicity of the remaining surface oil or droplets dispersed into the water column should thereby be greatly reduced (Anderson et al., 1974; McAuliffe, 1971, 1974; Wells and Sprague, 1976). From a physicochemical model, estimated evaporative loss of volatile hydrocarbons for different crude oils ranges from about 20 to 50 percent in 12 hr (Mackay et al., 1980a; Nadeau and Mackay, 1978; van Oudenhoven et al., 1983). The lower percentage occurs with more viscous oils.

Field measurements have demonstrated rapid loss and low concentrations of C_1 to C_{10} hydrocarbons in oil samples from surface slicks, and in water under untreated and chemically dispersed crude oil slicks. McAuliffe (1977) and Johnson et al. (1978) showed complete loss from a crude oil slick of low molecular weight aromatic hydrocarbons in 8 hr, for example, benzene and toluene within 1 hr, dimethylbenzenes by 5 hr, and trimethylbenzenes by 8 hr.

Evaporative losses of specific compounds are difficult to predict theoretically and thus limit theoretical modeling of weathering. Henry's law constants are required for hundreds of individual components whose rate of loss is usually assumed to be controlled only by individual mole fraction (which cannot be determined) in the crude or distillate product and pure component vapor pressure. This assumption can lead to some simplification, since pseudocomponents (or distillate cuts) of various boiling point ranges have been used in modeling evaporative behavior, and good agreement between observed and predicted behavior has been obtained (Page et al., 1983, 1984; Payne and McNabb, 1984).

Dissolved hydrocarbons, like the slick, are most concentrated near the oil spill release point, particularly if it is beneath the water surface. The regions in which the seawater is likely to be most toxic are therefore localized. As the slick spreads and evaporates, low molecular weight compounds are lost and the risk of subsurface exposure to these diminishes. Acute toxicity of oil on or near the surface oil therefore tends to be diminished by evaporative weathering (Gordon et al., 1976; McAuliffe, 1986; Wells and Sprague, 1976).

Photochemical Processes

Oxygen-containing products, resulting from exposure of oil to air and sunlight, are likely to have some surface activity (Klein and Pilpel, 1974). Photochemical oxidation of most oil on water occurs slowly, so the concentration of oxidation products and their effects on dispersibility are likely to be small in the first few days after an oil spill (Burwood and Spears, 1974; Hansen, 1975, 1977; Payne and McNabb, 1984; Payne and Phillips, 1985; Wheeler, 1978). However, for some oils (typically, waxy crudes), photolysis may have a significant effect on chemical dispersibility and on the formation and stability of water-in-oil emulsions after only a few hours of exposure (Daling, 1988; Daling and Brandvik, 1988).

Although some oxygenated products have been isolated from samples taken at large oil spills, most predictions are based on small-scale laboratory experiments, and little or no fieldwork has been done on this process (Overton et al., 1979, 1980; Payne and McNabb, 1984; Payne and Phillips, 1985).

Mousse Formation

The formation of stable water-in-oil emulsions appears to depend on the simultaneous presence of asphaltenes and paraffins (Bridie et al., 1980; Payne and Phillips, 1985). Although their experiments were not aimed at establishing the limits of either component, Bridie et al. noted the following:

- The original sample (Kuwait 200+ fraction, with 6.6 percent asphaltene and 9.8 percent paraffin wax) readily formed mousse.
- When waxes and asphaltenes were removed, no mousse could be formed.
- When either waxes or asphaltenes were added back to the basic oil singly, no mousse could be formed.
- When both waxes and asphaltenes were added back to the basic oil, mousses were easy to form.
- When only 10 percent of the original asphaltene content was added together with the waxes to the basic oil, mousses were easy to form.

Water-in-oil emulsions can be destabilized by adding surfactants that displace the indigenous surfactants from the interface (Canevari, 1982). One product performed well as an emulsion preventer at product-to-oil ratios as low as 1:5,000. The product has a high oil-to-water partition coefficient (10,000) compared to most dispersants (10), and is a better dispersant than Corexit 9527 at ratios between 1:400 and 1:5,000 (Buist and Ross, 1986, 1987). Similarly, dispersant formulation effectiveness may depend on interaction with the indigenous surfactants at the interface. Such an interaction appears to affect dispersant performance even more than the physical properties of crude oil (Canevari, 1985).

BEHAVIOR OF OIL-DISPERSANT MIXTURES

Criteria for Effective Dispersal

The chemical nature of dispersants and the physical and chemical processes that affect untreated oil have been reviewed, and this chapter now turns to the physical and chemical criteria for effective chemical dispersal. (The various factors that influence success or failure of a dispersant response operation—equipment design, dispersant regulation, dosage, remote sensing, application strategy, and logistics—are discussed in Chapter 5.)

Some of the processes discussed earlier, particularly advection and turbulent diffusion, apply to chemically dispersed oil as well as untreated oil. However, chemical dispersants can cause major changes in slick-spreading characteristics; droplet formation, stabilization, coalescence, and resurfacing; and adherence of oil to solid surfaces, suspended particulate matter, and sediments. Four basic criteria must be met for effective chemical dispersal of oil to occur:

1. Dispersant must be sprayed onto the slick. This obvious criterion can be a major problem in practice. Dispersants are usually sprayed from boats or aircraft onto a floating oil layer.

2. Dispersant must mix with oil or move to the oil-water interface. Dissolution of dispersant in bulk oil is not necessary as long as the surfactant molecules adsorb at the interface. The ideal dispersant application system produces drops small enough to just penetrate the slick; not so small that they remain at the oil-air interface or blow off-target with the wind, and not so large that they penetrate the slick and are lost in the water column (see Chapter 5; API Task Force, 1986).

3. The dispersant must attain the proper concentration at the interface; ideally, that is the concentration that causes a maximum reduction in interfacial tension (Rewick et al., 1981). The ideal dosage (quantity per unit area) is an amount sufficient to cause dispersion but not so high that dispersant is wasted. Although uniform dosage is the goal, actual dosage is often neither uniform nor precisely known (Chapter 5; API Task Force, 1986; Exxon Chemical Company, 1985).

If wind blows falling droplets outside the slick boundary, or if previous droplets cause herding—the movement of the slick into narrow bands interspersed with clear water—or dispersion, a falling droplet will be ineffective. In principle, thick areas require more dispersant and thin areas require less, but there is no practical way to vary the application rate to achieve a constant dispersant-to-oil ratio. Spray equipment is set to provide a uniform quantity per unit area. Dispersants may be applied "neat" (undiluted) or diluted by water or a hydrocarbon solvent. In practice, average dosage required is determined from manufacturer's recommendations (e.g., Exxon Chemical Company, 1985) and experience.

4. The oil must disperse into droplets. Energy is required to increase the oil-water interfacial area. The lower the interfacial tension, the less energy is required. Indeed, under optimum conditions (very low interfacial concentration for some dispersants), oil-water interfacial tension can be reduced to less than 1 dyn/cm, and almost any minor agitation will suffice. The energy can come from wind, waves, or mechanical stirring, although normal wave energy is usually adequate.

Relation of Oil Composition to Dispersibility

Earlier, the question whether dispersant formulation could be varied to match particular oil types was answered "no." Here, a related question is addressed: What oil composition factors influence the ease with which oil can be dispersed?

Crude oils naturally contain some surface-active compounds, which are believed to contribute to mousse formation and can interact unpredictably with a dispersant's surface-active components (Payne and Phillips, 1985). However, all oils are not equally dispersible (Canevari, 1987). This is not surprising given the many variations in oil composition and physical properties.

Available data about oil properties include volume yield during

TABLE 2-2 Labofina Effectiveness Results, Standard 1-Min Delay Before Sampling

Crude Oil	Emulsion Tendency	Dispersant[a]			
		A	B	C	D
Kuwait	Extremely strong	25	12	8	21
La Rosa	Extremely strong	24	--	--	20
North Slope	Strong	26	--	--	30
Guanipa	Strong	18	--	--	35
Loudon	Strong	23	4	5	21
Murban	Intermediate	17	10	10	21
Southern Louisiana	Relatively weak	20	11	4	15
Ekofisk	Weak	26	12	7	34
Saharan Blend	Weak	19	--	--	38
Goose Creek	Weak	31	--	--	29
C_{14}	None	46	39	24	50

[a]Identity of dispersant was not revealed in the publication.

SOURCE: Canevari, 1987.

distillation to temperatures that correspond usually to commercial product fractions, API gravity, viscosity, sulfur, nitrogen, nickel, vanadium, aromatics, naphthenes, smoke point, pour point, and aniline point (Bobra and Chung, 1986; Payne and McNabb, 1984). Of these data, the only ones that have been correlated with dispersibility are viscosity and pour point. Information on the influence of oil composition on dispersibility is poorly understood and is insufficient to allow a rigorous prediction of the dispersibility of a particular oil. Testing is usually needed.

Differences in performance of different dispersants with the same oil also occur, despite similarities in some dispersant compositions. An example from laboratory experiments with four dispersants on various crude oils is shown in Table 2-2 (Labofina test, discussed later in this chapter). Tetradecane (C_{14}) was used as a standard having no tendency to form water-in-oil emulsions. The four dispersants performed on C_{14} in the order D > A > B > C, which was generally preserved for the 10 crude oils tested. However, differences varied greatly. For example: for Ekofisk crude oil, dispersant C's effectiveness was rated 7, and that for D was 34; for C_{14}, C was 24 and D was 50; for Guanipa crude, D was 35 and A was 18; for Kuwait crude, A was more effective at 25 than D at 21 (Canevari, 1987).

Another example is the comparison of two crude oils adjusted to the same viscosity. Normally LaRosa crude (73 cSt viscosity at 16°C) is more difficult to disperse than Murban crude (6 cSt). When LaRosa crude was diluted with pure isoparaffin oil to a viscosity of 6 cSt, its dispersion efficiency (Mackay-Nadeau-Steelman [MNS] test, dispersant-oil ratio 0.003:1) was 50 percent compared to 78 percent for Murban crude of the same viscosity (Canevari, 1985).

Laboratory studies (NRC, 1985) have shown that metalloporphyrins, which are naturally occurring components of crude oils, with some surface-active properties, favor formation of water-in-oil emulsions (mousse). Canevari (1985) predicted that these trace components would tend to inhibit dispersion (oil-in-water emulsion formation) rather than enhance it. The literature on enhanced oil recovery indicates that other oil components might also affect dispersant effectiveness (Healy et al., 1976). For example:

• *Aromaticity.* Solubility in the oil phase of the oleophilic portion of surfactant molecules would be expected to increase with the concentration of aromatic constituents in the oil (Bancroft, 1913). On the other hand, aromatics facilitate dissolution of dispersant in oil (Mackay, private communication). Experiments by Mackay et al. (1986) indicate that addition of toluene to Alberta crude oil in some cases increased effectiveness as measured by MNS and rotating flask tests.

• *Naphthenic acids.* These are carboxylic acids of cyclic hydrocarbons. They are not strongly surface active but could interact with added surfactants. U.S. West Coast crudes typically contain much higher concentrations of naphthenic acids than Gulf Coast crudes (acid numbers are 3.5 to 4.0 in California versus 0.4 to 1.3 in the Gulf).

• *Paraffin.* The pour point of an oil is strongly correlated with its paraffin (wax) content. As spilled oil weathers and gradually loses its lower molecular weight components by evaporation, the higher molecular weight paraffins increase in relative amount and also become less soluble in the remaining oil, in some cases forming a separate solid phase. The result is an increase in pour point. If the pour point of the weathered oil becomes higher than the ambient water temperature, the oil will become solid or semisolid and be nondispersible.

• *Asphaltenes.* Wax and asphaltenes are stabilizing agents for water-in-oil emulsions that, with their very high viscosity, are extremely difficult to disperse. Hence asphaltenes can be important to determining dispersibility. Unfortunately, they are not a factor commonly determined in oil analysis. Lindblom (private communication) suggests that the 650+°F fraction might be taken as a surrogate in predicting whether mousse formation is likely. If that fraction is greater than 40 percent, the oil may tend to emulsify and will be difficult to disperse.

Effect of Oil Viscosity, Time, and Other Parameters on Dispersion

In this section, the influence of viscosity and other important physicochemical parameters will be considered in the context of optimizing dispersion. The effectiveness of oil dispersants as determined by laboratory tests is strongly dependent on oil viscosity. Viscosity has two effects: it retards dispersant migration to the oil-water interface, and it increases the energy required to shear off a drop from the slick. Forces applied to a viscous slick from water motion tend to be transmitted through the slick rather than being dissipated in the slick and causing dispersion (Mackay, private communication).

Dispersants are most effective for oil viscosities less than about 2,000 cSt, and almost no dispersion occurs over 10,000 cSt (see Figure 2-8) (Cormack et al., 1986/87; Lee et al., 1981; Morris, 1981). This viscosity limit was chosen by the United Kingdom as a reference point against which to consider the number of oils likely to be treatable by dispersants at sea (Cormack et al., 1986/87). Other researchers use 2,000 or 5,000 cSt as the limit (ITOPF, 1982).

Daling (1988) points out that the upper viscosity limit for chemical treatment of oils and water-in-oil emulsions is specific for different oils. It is therefore not possible to use a general viscosity limit, particularly on water-in-oil emulsions, where the dispersant has to break the emulsion into oil and free water at the surface before dispersion of the oil into the water column can take place.

Criticality of Timely Response

According to Cormack et al. (1986/87) most crude oils are treatable when freshly spilled, but because viscosity increases rapidly with weathering, time is critical. This time relationship is shown in Figure 2-9, which indicates increased viscosity of different crude oils, floating on the sea and constrained in 4-m diameter floating rings (Cormack et al., 1986/87; Martinelli and Cormack, 1979). In large accidental

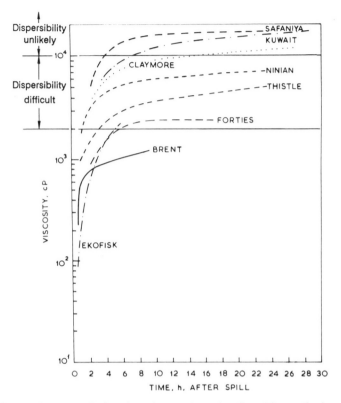

FIGURE 2-8 Increase of viscosity of several crude oils with weathering. Source: Cormack et al., 1986/87.

spills, oil is often 24 hr old or older when dispersant is first applied, and may have formed mousse as well.

The final viscosity values likely to be reached after weathering of oil at sea for 12 to 24 hr must be considered to determine whether and how long after a spill dispersants can be usefully applied (van Oudenhoven et al., 1983).

According to Cormack, all crude oils are initially amenable to dispersion except those crude oils with high initial viscosities, those that would be solid at seawater temperatures, and petroleum products normally carried in heated cargo tanks. All light fuel oils are amenable.

Lindblom (private communication) states that an oil is dispersible if it will spread on water. If the oil will not spread (i.e.,

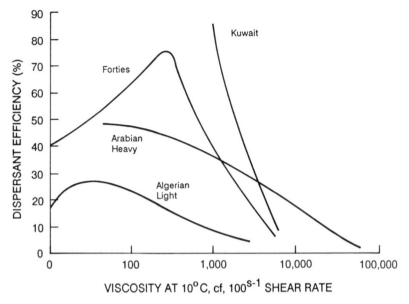

FIGURE 2-9   Effect of viscosity on dispersant efficiency using the Fina revolving flask test at the Warren Spring Laboratory. Source: Morris, 1981.

if the water temperature is below the pour point of the oil), dispersant drops will simply roll off of the oil layer.

The dispersant formulation used also may influence the dependence of dispersibility on oil viscosity. For example, hydrocarbon-solvent based dispersants appear to work better on mousse and on high-viscosity oils than do dispersants with water, glycol, or glycol-ether solvents, probably because their dissolution in the oil keeps oil and dispersant in contact longer (COAATF, 1986; Fraser, private communication).

Contact Time of Dispersant With Oil

Contact time between the dispersant and oil is an important factor in effectiveness of dispersal. If the dispersant is water soluble it can be diluted in the water before the proper interfacial structure can be developed to stabilize droplets. The Labofina test as practiced at Warren Spring Laboratory uses a contact time of 1 to 1.5 min before the onset of agitation. For oil or mousse with a viscosity of 10,000 cSt, the standard WSL Labofina test gives almost zero effectiveness, but with a contact time of several minutes the effectiveness could be as high as 40 percent (Cormack et al., 1987). With shipboard applications systems, estimated contact time may be as short as 2 sec.

Effect of Dispersant on Slick Dynamics

Because chemical dispersants lower interfacial tension between oil and water, they greatly alter the dynamics of slick spreading and the development of slick structures described earlier.

One of the most obvious and rapid effects, when surfactant is first applied to an oil slick, is herding (Chau et al., 1986; Delvigne, 1985; Mackay et al., 1986). The surfactant lowers the surface tension of the water thereby causing the oil slick to contract in a few seconds. This herding soon subsides and is not important after a few minutes.

On a longer time scale, oil slicks treated with dispersant appear to spread more rapidly than untreated slicks. Near St. John, New-foundland in 1981, investigators found that a treated slick initially spread slower than untreated ones, spread more rapidly after 2 hr 30 min, and after 3 hr 45 min was 33 percent larger (Goodman and MacNeill, 1984). Three sets of slicks laid down during a test near Halifax in 1983 also exhibited this effect—the treated slick spread slower at first (COAATF, 1986). The transition times and excess area varied, however—both slicks were the same size after 4 hr 15 min for the Corexit 9527 treated slick, 1 hr 45 min for Corexit 9550, and 2 hr for BPMA 700. The Corexit 9550 slick was 4 times larger than the untreated slick after 2 hr 30 min.

Application of dispersant to a mousse patch from the *Ixtoc I* blowout was described in the cruise report of the ship *Longhorn*. On August 18, 1979, the ship encountered a large area of reddish to chocolate-colored viscous oil, containing much debris. A strong odor, likened to the smell of "an old gas station," permeated the area. The mousse was so viscous it made a "slurping" sound against the side of the ship. When the oil surface parted, schools of small silvery fish and some sharks could be seen swimming below.

> Shortly after the mousse was found, a DC-6 aircraft appeared and sprayed the patch, presumably with a dispersant; after the spraying had ceased, the *Longhorn* steamed back into the mousse and found the area of the patch appeared smaller, as if the oil had been "herded" by the chemical. Its texture was now more liquid, but the layer was thicker. (Hooper, 1981:186)

These phenomena are rarely addressed by modelers, and not

much progress has been made in developing mathematical expressions for their effects (Zagorski and Mackay, 1981). Payne and McNabb (1984) have written:

> It is not possible at this time to model spilled oil behavior beyond first order estimations of total area potentially covered for a defined range of slick thickness.

The variables affecting modeling of dispersed oil behavior are even more complex; these are addressed later in this chapter.

Behavior of Droplets and Resurfacing

Addition of a chemical dispersant reduces the energy required to break a slick into droplets. If the interfacial tension is sufficiently low, chemically aided dispersion can occur in the absence of breaking waves. In addition, under similar conditions the number of small drops tends to be larger for oil dispersed chemically than mechanically (Jasper et al., 1978).

Droplet size greatly affects the physical transport of oil. Because oil drops are generally less dense than water, larger droplets rise to rejoin the slick, smaller ones rise more slowly, and the smallest ones tend to remain suspended in the water column as the result of turbulent diffusion. The suspended droplets are then transported horizontally by subsurface currents and can diffuse deeper into the water column. Oil droplets can be removed from the water column by combining with sediment and other abiotic particles, or becoming bound to or ingested by biota, such as plankton.

Size distribution of dispersed oil droplets in the water is an important measure of dispersant effectiveness (Mackay et al., 1986). A spectrum of oil drop diameters can be characterized by a variety of terms: most-frequent droplet size or mean droplet size are used, as is volume mean diameter (vmd), the droplet size that contains average volume. The latter measure is quite different from mean droplet diameter or size since the volume of a sphere increases with the cube of the diameter. For example, a dispersion may have a mean particle diameter of 5 μm, but a volume mean diameter of 70 μm (Byford et al., 1984).

Depending on the interfacial tension produced by the dispersant and the energy with which the oil is dispersed, the vmd may range from 1 to 200 μm. For example, the vmd of dispersed oil droplets (at dispersant-oil ratios of 1:16 to 1:100) were 14 to 225 μm in the MNS test, and 17 to 84 μm in the Labofina test (Lewis et al., 1985). Ideally, the dispersed oil droplets should be as small, if possible having a vmd as low as 1 to 5 μm, in order to avoid resurfacing.*

A laboratory study of the effect of temperature and energy on droplet size distribution (Byford et al., 1984) found that the vmd of oil droplets was most strongly influenced by dispersant formulation and energy input, but temperature (0° to 20°C) had only a minor effect. Dispersant-oil ratio had a great effect on vmd for a low-performance dispersant, but not for a high-performance dispersant. Density had a greater influence on vmd than viscosity, at constant temperature.

Delvigne (1987) showed that droplet size distributions became smaller with increasing energy input (d varied as $E^{-0.5}$) as well as duration. Oil type, weathering state, and temperature all affected viscosity, and droplet size was approximately proportional to the 0.34 power of viscosity. Droplet size distributions were similar whether they were drawn from submerged oil or a surface layer, and they were independent of salinity.

To characterize the resurfacing of the larger droplets for a given observation time and turbulent regime, Mackay et al. (1986) used in their model the concept of a critical oil droplet diameter, above which most oil will resurface and below which most oil remains dispersed. The higher the mixing energy, the larger this critical droplet diameter will be.

With time, the depth of dispersed oil particles increases (Mackay et al., 1986). In practice, this depth depends on oceanographic conditions, especially turbulent diffusion and mesoscale currents. It is not necessary that all oil droplets remain in the water column indefinitely for the dispersant to work.

It is apparent that for effective dispersal, oil droplets formed in the water should be as small as possible. The smaller they are, the longer they will stay in the water column without resurfacing (Gatellier et al., 1973; Mackay et al., 1980b, 1986; Nichols and Parker, 1985). Byford et al. (1984) summarized the situation by proposing that dispersed oil droplet size be regarded as a major factor in judging dispersant effectiveness.

---

*The ideal vmd is an estimate based on observations; one review of this report suggested that a 20 μm vmd is acceptable.

Oil Concentration Under Dispersed Slicks

Dispersal of an oil slick initially increases subsurface concentration, which may be rapidly diminished by physical transport processes. The larger oil-water interfacial area of chemically dispersed oil may cause dissolution to be increased, and evaporation, which depends on oil-air surface, to diminish.

Because the most water-soluble components of oil tend to contribute most to acute toxicity of dispersed oils (Abernethy et al., 1986; Anderson et al., 1974; Wells and Sprague, 1976), there is concern that the biological impact of chemically dispersed oil on subsurface organisms might be greater than for mechanically dispersed oil. This effect would be mitigated, in open water, by turbulent diffusion, which transports dispersed oil away from the surface and greatly dilutes its concentration. For a very large spill near shore, or for a smaller spill in a confined area, turbulent transport may not be adequate to disperse oil rapidly.

Some field studies have been designed to quantify the effects of chemical dispersion on physical transport processes and to measure the exposure conditions that result. The concentrations of dispersed oil measured in these tests should be compared with toxic thresholds estimated for the organisms and crude oils reviewed in Chapter 3. For example, during tests off southern California, La Rosa crude was dispersed by aerial application of Corexit 9527. An estimated 50 percent of the spilled oil was dispersed. In samples collected 20 to 25 min after dispersant application, the highest concentrations were 2 to 3 ppm at 1 to 3 m and 0.5 ppm at 6 m (McAuliffe et al., 1980). Using the concept of ppm-hr,* the estimated exposure was 0.67 to 1.25 ppm-hr at 1 to 3 m depth and 0.17 to 0.21 ppm-hr at 6 m. After 100 min, the concentration at 1 to 3 m had not diminished, so that the exposure factor became 3.3 to 5 ppm-hr.

Under a slick of Murban crude oil in the same test series estimated to be 90 to 95 percent dispersed 23 min after spraying, oil concentrations were 18 ppm at 1 m and 10 ppm at 3 m. After 50 to 57 min, concentrations had diminished to 3 to 4 ppm at 1 m.

---

*The ppm-hr concept is an attempt to relate laboratory bioassay test data (in which oil concentration is usually held constant) to exposure conditions in nature, where the concentration of dispersed oil varies with time. The ppm-hr concept assumes that toxic effects are linear, both with concentration and time. Thus, both for variable concentration and for constant concentration, the integral of concentration times time is assumed to be comparable. There are situations, particularly involving extremes of concentration or time, where linearity should not be assumed.

Exposure at 1 to 3 m ranged from 2.5 to 6.9 ppm-hr, and at 6 to 9 m it was 0.83 to 0.95 ppm-hr.

For the best dispersed (approximately 80 percent) Prudhoe Bay crude oil slick off southern California, McAuliffe et al. (1981) measured higher total oil concentrations 15 min after dispersant application by aircraft: 30 to 50 ppm at 1 m, 10 ppm at 3 m. After 220 min concentrations at 1 to 9 m had decreased to 0.5 to 2.3 ppm. By graphical integration, assuming exponential decrease of concentration with time, the exposure factors were 47 ppm-hr at 1 m depth and 12 ppm-hr at 9 m.

Cormack and Nichols (1977) measured concentrations of Ekofisk crude oil chemically dispersed within the first 2 min after spraying from a boat: 16 to 48 ppm at 1 m. The oil concentration decreased to 5 to 18 ppm within 5 to 10 min after spraying, and to 1 to 2 ppm after 100 min. By graphical integration, the exposure factor in this case was 2 to 6 ppm-hr for the first 100 min.

Aerially sprayed Statfjord crude oil slicks were sampled after 30 to 50 min. The highest oil concentrations were 25 to 40 ppm at 0.5 to 2 m depth; at 3 m only 4 ppm were found (Lichtenthaler and Daling, 1985). Water samples collected 165 to 225 min after spraying contained up to 7 ppm at 0.5 to 1 m and 0.2 to 1.4 ppm oil at 3 m.

Field experiments that measured oil concentration beneath naturally dispersed (untreated) slicks gave concentrations varying from less than 0.1 to 5 ppm, and exposures of 1 to 35 ppm-hr; the higher concentrations and greater exposure factors were found within the top meter of water. In the tests reviewed above, concentrations of chemically dispersed oil (up to 40 ppm) and exposure factors (up to 60 ppm-hr) are higher than those in undispersed cases.

Evaporation and Weathering of Dispersed Oil

Subsurface sampling of dispersed oil is seldom done because of logistical and measurement problems, and because concentrations decrease quickly to levels too low to be reliably analyzed. As a result, many questions about such interrelated processes as evaporation and weathering remain unanswered. For example, McAuliffe (1977) stated that "chemically dispersed oil appeared to weather in a manner similar to oil naturally dispersed under slicks." In contrast, Page et al. (1985) observed enhanced evaporation when oil was chemically dispersed.

Interaction of Dispersed Oil With Suspended
Particulate Matter and Sediment

Because a primary fate of spilled oil may be sedimentation, especially in estuarine and nearshore environments, it is important to know whether chemical dispersal increases or decreases transport of hydrocarbons to the sediment and whether sedimented oil is more available to benthic biota. In this section, physicochemical evidence for enhanced or reduced sediment transport is examined.

Mackay and Hossain (1982) conducted a laboratory study of mechanically dispersed oil and suspended particulate matter (SPM) interactions and sedimentation of mechanically and chemically dispersed crude oil (Alberta, Murban, and Lago Medio crude and Corexit 9527 and BP1100WD dispersants). They found that in the presence of sediment, dispersed oil was removed from the water column, and the settling velocity of oiled particulate matter was estimated to be as rapid as 1 m/hr. From 5 to 30 percent of the oil was incorporated in the settled sediment. Addition of dispersant decreased the fraction settled: for example, 30 percent with no dispersant, 10 to 15 percent at a dispersant-oil ratio of 1:10, and 6 percent at a dispersant-oil ratio of 1:5.

The most important factor appeared to be the degree of sorption of dissolved oil by organic matter in the sediment compared to the lower uptake of the mineral component (Karichoff et al., 1979). Association of oil and sediment particles changed the buoyancy of both; hence excess sediment tended to sink the oil, and excess oil (e.g., twice the volume of sediment) tended to keep fine sediment particles suspended. Smaller oil droplets (induced by chemical dispersion) were less susceptible to sedimentation than larger droplets.

Even in the absence of sediment, oil droplets do not remain suspended indefinitely. Like other particles in the size range 1 to 10 μm, they are scavenged by zooplankton, which absorb what nutrition they can and package the remainder into rapidly sedimenting fecal pellets (NRC, 1985; Sleeter and Butler, 1982). Oil droplets can also be degraded by bacteria or fungi, which are comparable in size and capable of metabolizing hydrocarbons if they are given adequate nutrients. This group of processes is generally known as "biologically mediated transport."

The agglomeration of both sediment particles and oil droplets is affected in a complex way by their interactions investigated by Harris and Wells (1979), Mackay and Hossain (1982), and Little et al. (1986 and earlier papers). Mackay and Hossain concluded that "prediction of the environmental behavior requires a detailed knowledge of the prevailing sediment depositional regime." Nevertheless, their results suggested that chemical dispersion of oil leads to reduced interaction with suspended particulate matter or sedimentation.

OIL FATE AND DISPERSION MODELS

Model Types

An ideal model simulates the following processes mathematically:

- advection of slick and water masses;
- evaporation of oil components;
- dissolution of oil components in water;
- dispersion of oil droplets in water;
- oxidation of oil components (particularly photooxidation);
- emulsification (mousse formation);
- biodegradation of oil components; and
- sedimentation (including biologically mediated transport).

All of these processes are time-dependent and must be described by dynamic models. State-of-the-art models include some, but not all, of these processes at varying degrees of sophistication; but field or laboratory experiments designed to calibrate or test models usually focus on only one process. Comprehensive models tend to be created in response to a need, such as the following (Mackay, 1986 and private communication):

- real-time spill trajectory—to facilitate countermeasures by on-scene commander (requires real-time environmental data, but can be used as a "game" for training);
- environmental impact assessment—to provide scenario for likely impact of oil-related developments such as offshore exploration and production or deep-water port;
- site-specific biological assessment—to provide an assessment of likely ecological effects;
- ecosystem—to provide long-term overall assessment of impacts and hazards on lakes, estuaries, bays, or open ocean.

Chemical dispersion is only infrequently addressed in these models; typically the only dispersion modeled is produced by wind and surface turbulence.

There is extensive literature on slick movement and oil fate models (Table 2-3). The report of Huang and Monastero (1982) is

TABLE 2-3 Oil Spill Models Considering Dispersion

Oil Spill Model	Author
DOOSIM	Johansen, 1987
Applied Science Associates	Spaulding, 1987
Three-dimensional shear diffusion	Elliott et al., 1986
University of Toronto	Mackay, 1986
SINTEF	Johansen, 1985
University of Toronto	Mackay et al., 1982
A. D. Little	Aravamudan et al., 1981
SLIKFORCAST	Audunson et al., 1980
University of Toronto	Mackay et al., 1980b
University of Rhode Island	Cornillon and Spaulding, 1978
SEADOCK	Garver and Williams, 1978
SLIKTRAKB	Blaikley et al., 1977
USC/API	Kolpack et al., 1977

SOURCES: Johansen, 1985a,b; Mackay et al., 1986.

the most complete review to 1982; Spaulding (1986) updates it. Only Mackay et al. (1980b, 1982, 1986) and Johansen (1985a,b) consider chemical dispersion.

Oil-weathering mathematical models developed by Payne et al. (1983, 1984) are based on measured physical properties and generate material balances for both specific and pseudocompounds (distillation cuts) in crude oil. They apply to open-ocean, estuary, lagoon, and land spills. Weathering processes included in the model are evaporation, dispersion into the water column, dissolution, water-in-oil emulsification, and slick spreading. Good agreement has been obtained between predicted and observed weathering behavior. The material balance and weathered-oil composition predictions generated as functions of time are useful for contingency planning, for assessing potential damage from spills, and in preparing environmental impact statements for outer continental shelf drilling activities.

Although oil is composed of hundreds of compounds, each with a distinct solubility (oil-water partition coefficient), volatility, reactivity, and diffusivity, most models treat "oil" as a single chemical species. This simplification is particularly inadequate when it is necessary to distinguish between components (such as benzene and toluene) that partition from oil into water and remain in the subsurface water mass and insoluble hydrocarbons that comprise the separate phase oil droplets that tend to rejoin the slick or attach to sedimenting particles.

Nonuniform Slick Thickness

Because slicks spread and have a nonuniform structure, some researchers (Mackay et al., 1986) have used a conceptual model that changes abruptly in moving from a small thick area to a sheen over the rest of the slick. This is important because uniform dispersant application based on such a model will result in overtreatment of the thin area by an estimated factor of more than 10 or more and undertreatment of the thick portions by a factor of 10 or more.

Advection and Diffusion

In modeling, it is frequently assumed that advection dominates the horizontal field, and turbulent diffusion dominates the vertical field and they are independent of each other. This greatly simplifies the differential equations and allows the two transport processes to be treated separately in numerical models (Mackay, 1984).

Resurfacing

The most common supposition in modeling the natural dispersion process is that breaking waves cause the oil layer to be propelled into the water column, thus forming a "shower" of oil droplets (Figure 2-10). This depends on wind speed and on oil density and viscosity. Most of the oil particles rise again to the slick and coalesce there, but smaller droplets diffuse downward and are retained by sedimentation or biologically mediated transport (Mackay et al., 1980b; Sleeter and Butler, 1982).

Breaking Waves

Dispersion rate is likely to be a function of slick thickness, oil-water interfacial tension, sea state, and fraction of the sea covered by breaking waves. It is well known that oil causes turbulence damping on a water surface; its presence thus presumably reduces the incidence of breaking waves and diminishes the fraction of sea they cover. It is also believed that dispersion occurs even in the absence of breaking waves, possibly as a result of "folding" of the oil when very

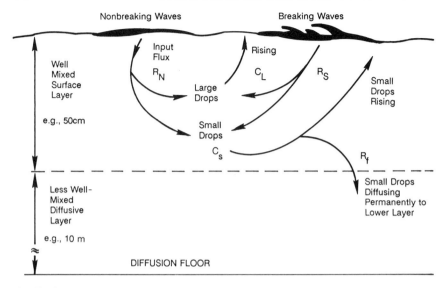

FIGURE 2-10   Schematic diagram of dispersion processes.   Source: Mackay et al., 1980b.

sharp, high-amplitude, short-wavelength waves pass through the oil layer and near-breaking conditions exist for a brief time.

Buist (1979) has done the most detailed mathematical treatment of this process, along with experimental wind-wave tank measurements that resulted in equations with adjustable parameters.

Integrated Approaches

Bringing all the variables affecting oil dispersion together in a single mathematical model is a formidable challenge. Two groups have made substantial progress, however: Donald Mackay and co-workers at the University of Toronto, and Oistein Johansen of the Continental Shelf Institute (IKU), Trondheim, Norway.

Mackay's Model

Mackay's early models are primarily one-dimensional, emphasizing mass balance, and do not attempt to predict a detailed three-dimensional distribution of dispersed oil. The equations developed by Mackay et al. (1980b, 1986) predict an increasing dispersion rate as a slick becomes thinner. Thinner slicks damp turbulence less effectively on the water surface, and fewer breaking waves are affected.

The droplets formed from a thin slick are expected to be smaller and the rate of dispersion faster than for thicker slicks (Figure 2-10).

In a simple diffusion model to estimate the oil concentration beneath a dispersed spill, Mackay et al. (1982), using data from McAuliffe et al. (1980), predicted that concentration decreases exponentially with increasing depth and time—at any time, it is directly proportional to oil volume. Thus a larger spill would be expected (after dispersal) to yield higher concentrations of oil in water than would a small spill. A limitation on this calculation is the time allowed for spreading of oil on water; the calculations by Mackay et al. implicitly assume dispersant application and dispersion very shortly after the oil has been spilled, and before it has (for a large spill) spread very far.

In the most recent model, Mackay et al. (1986) divide oil on the water surface into thick and thin slicks, the proportion of oil in each and the amount of dispersant sprayed on each are part of the input data. The influence of chemical dispersants is included in the form of an effectiveness factor, X; that is, the amount of oil dispersed is X times the amount of dispersant applied. A transition in dispersant effectiveness is recognized—from a "performance-limited" regime at low-dispersant concentrations to an "access-limited" regime of thin slicks in which herding and the relatively small amount of oil contacted by each dispersant drop make the amount of oil dispersed independent of the amount of dispersant added.

Dispersal is assumed to be a quadratic function of wind speed; the coefficient decreases by a factor of 2 as the temperature decreases from 25 to 0°C. The size of dispersed droplets is assumed to be distributed according to the Weibull function; those larger than a critical diameter will resurface (following Stokes' law, modified for eddy diffusion effects) and expand the thin slick. In addition, eddy diffusion can transport the smaller drops to the surface, where they rejoin the slick, or to deeper water, where they are lost. Horizontal transport of oil dispersed in the water column produces a linear increase in plume diameter with time, as based on the dye patch data of Okubo (1971).

The model developed by Mackay et al. (1980b) does not discriminate between dissolved and dispersed oil and does not include evaporation or any other process that might differ for components of the oil. It treats the sea as semi-infinite, without allowing for shoreline, bottom topography, or stratification. Aspects of this model include:

- expression for oil drop diameter as a function of the effectiveness factor, X;
- temperature and turbulence dependence of X;
- validity of the resurfacing expressions;
- validity of the horizontal diffusion expressions;
- use of a "diffusion floor" to simplify modeling of vertical advection; and
- calibration of the model with data on experimental spills.

Johansen's Model

Johansen's (1984) work illustrates the complexity of advection-diffusion processes and provides a new approach to including them in a model of dispersed oil. This model considers particles to be in one of three states: at the surface, entrained in the water column, or evaporated. The drift of a particle is determined by its state, and the transition from one state to another is determined by a random-number generator and a set of probability parameters:

- probability for entrainment—wind dependent;
- probability for resurfacing—dependent on density difference, droplet size distribution, time submerged, and wind force;
- probability for evaporation—computed from established evaporation models.

The advective movement of the spill is assumed to be dependent on the vectorial sum of the drift induced by the local wind, tidal currents, and an assumed stationary background current. These can be estimated by empirical correlations or from observations.

This integrated model was extended to contrast the behavior of chemically treated with untreated oil (Johansen, 1985b). This is done by using two different sets of entrainment and resurfacing parameters, as well as two different droplet size distributions. The advantage of chemically dispersed oils, from the physical viewpoint, is that the oil is dispersed into the water column rather than remaining as a surface slick. Chemical dispersion thus tends to enhance the probability of entrainment and, since chemically dispersed oil droplets tend to be smaller and recombine less easily than physically dispersed untreated oil globules, the probability for surfacing is reduced.

The most recent dispersion model, Dispersion of Oil on Sea Simulation (DOOSIM), is a two-layer drift model using a random-walk algorithm for spreading calculations (Johansen, 1987). A stochastic model is used for the mass budget, but an empirical model is used for weathering. The model produces a color display with a generalized map, a detailed map, and a continual presentation of mass balance.

Model Validation

As more information is obtained in the laboratory and in field tests, models can be more thoroughly calibrated and their assumptions tested. Good models help discipline thinking about complex processes and aid in designing and planning further work. The models discussed above have been tested against data obtained in the field under realistic conditions. To make such a test meaningful requires carefully planned measurements on a controlled spill. Ideally, the field experiment should be designed to test a specific hypothesis. For example, Mackay et al. (1982) used a one-dimensional model to fit some data from the field studies by McAuliffe et al. (1980).

Summary

Both untreated and chemically dispersed oil are transported by advective and diffusive processes. Oil left untreated on calm water tends to stay at or near the surface and thus tends to be controlled by wind-related surface drift. Untreated oil is more influenced by vertical turbulent diffusion as the seas become rougher, but it remains strongly influenced by surface currents. In contrast, dispersed oil enters the water column even in calm weather, is more influenced by vertical diffusion and vertical shear, and is less affected by horizontal advection. Resurfacing of oil in the water column is an important process that greatly complicates attempts to model, conceptually or numerically, the distribution of dispersed oil.

These considerations emphasize the need to understand more precisely the role of vertical turbulent diffusion and the vertical distribution of dispersed oils. Areas of needed research include near-surface wave, current, and oil dynamics; oil-ice interaction; algorithms for the time-dependent process of photooxidation; improved environmental data; and model validation with spill data. Trends include the development of comprehensive models, systems, development of portable stand-alone models, and the move toward interactive ("expert") system models (Spaulding, 1986).

LABORATORY STUDIES OF EFFECTIVENESS

Purpose of Laboratory Testing

The general objectives of laboratory testing of dispersants include the following:

- testing a variety of dispersants to rank their relative effectiveness (e.g., Doe and Wells, 1978; Mackay and Szeto, 1981; Mackay et al., 1984; Martinelli, 1984; Rewick et al., 1981, 1984); and
- testing effectiveness of dispersants under carefully controlled conditions to assess the role of oil type, weathering state, dispersant-oil ratio, mixing energy, salinity, temperature, and application methods (e.g., Byford et al., 1983; Lehtinen and Vesala, 1984; Mackay et al., 1984; Payne et al., 1985; U.S. EPA, 1984).

Laboratory tests are also used to screen dispersant types prior to more expensive field testing (Meeks, 1981; Nichols and Parker, 1985). They provide data for contingency planning; for stockpiling specific dispersants for particular environments, oil types, or deployment methods (Byford et al., 1983; U.S. EPA, 1984); and ultimately for deciding whether or not to use a particular dispersant (Mackay and Wells, 1983).

Mathematical models for dispersal of oil can be partially validated in the laboratory (Mackay, 1985). Appropriate concentrations for toxicity testing can also be determined in laboratory tests (Anderson et al., 1985; Bocard et al., 1984; Mackay and Wells, 1983; Wells et al., 1984b).

There are three generic types of laboratory tests in use as of 1987:

1. Tank tests with water volumes of 6 to 150 liters, including test vessels agitated using circulated seawater, and tests that employ breaking or nonbreaking waves to generate more realistic turbulent mixing energy. Examples are the EPA test (U.S. EPA, 1984), MNS test (Mackay et al., 1984), and the French Institute of Petroleum (IFB) test.

2. Shake-flask or rotating flask tests that are conducted on a 1-liter scale. Examples include the WSL Labofina test (Martinelli, 1984).

3. Interfacial tension tests that measure properties of the treated oil instead of degree of dispersal in a system with given energy input (Rewick et al., 1984).

The most common of these tests are compared later in this chapter.

All tests reviewed establish an oil slick and then apply dispersant in a defined dispersant-oil ratio. Dispersant may be applied by spraying neat or mixed with seawater, by pouring into a ring on the water surface that contains the oil slick, by adding dispersant to seawater, or by premixing the dispersant with the oil. Mixing energy is applied by a high-speed propeller, by rotating a separatory funnel containing the oil-dispersant mixture (Labofina), by a spray hose and circulation pump (EPA), or by a high-velocity air stream (MNS).

Dispersant effectiveness is determined by one of the following four criteria:

1. The amount of oil dispersed in the water. This can be measured by visual observation, or by solvent extraction and spectrophotometric analysis. The amount of dispersed oil may be determined under dynamic conditions (e.g., MNS and EPA tests) or after mixing has terminated (e.g., Labofina and MNS tests). It is also desirable to measure the amount of oil remaining in the surface slick to obtain a mass balance, but standard methods for doing so have not yet been developed (Nichols and Parker, 1985).

2. Dispersed oil droplet size (Byford et al., 1984; Lewis et al., 1985). This is another important criterion since larger droplets resurface some time after dispersal in the water column. The volume mean diameter in the MNS test was 14 to 226 μm depending on the dispersant-to-oil ratio (DOR) and the dispersant used (Byford et al., 1984; Lewis et al., 1985); in the Labofina test it was less than 154 μm.

3. Dispersed droplet stability as a function of time and turbulence in both static and dynamic systems (Mackay et al., 1984; U.S. EPA, 1984).

4. Interfacial tension (Mackay and Hossain, 1982; Rewick et al., 1981, 1984) has been used to rank dispersant formulations, but the static character of the measurement makes correlation with dispersal under turbulent conditions unrealistic.

Critical Factors

In all tests, oil-water ratio, dispersant-oil ratio, dispersant application method, mixing energy application, and methods of sampling and analysis were found to be critical factors in determining the precision of results. Oil-water ratio is most important for the relatively hydrophilic dispersant formulations (i.e., those with relatively high HLB) since greater dispersion occurs with higher concentration of surfactant; this will be the case if the volume of water is smaller for the same volume of oil and surfactant. The nearly infinite capacity of the open ocean for diluting hydrophilic dispersant is not normally accounted for in laboratory tests. Typical oil-to-water ratios are 0.02:1 for the Labofina test, 0.0017:1 for the MNS test, and 0.00077:1 for the EPA test.

At high oil-water ratios, collisions (and possible coalescence) of dispersed droplets are more frequent; this too is an unrealistic simulation of the open ocean.

Dispersant-oil ratio is especially important below 0.2:1, where a steep dependence of effectiveness on dispersant-oil ratio is observed (Rewick et al., 1981).

The method of applying the dispersant to the oil is a key factor in an effectiveness test. Types of application used include:

- dispersant premixed with water;
- dispersant premixed with oil;
- neat (undiluted) dispersant, poured on slick; and
- neat dispersant sprayed on slick (this is the only test that has direct field relevance).

For dispersant sprayed on the slick, droplet diameters in the 200 to 700 μm range are desirable. These diameters are similar to those produced by dispersant spray systems used in practice, which is a fortunate match of circumstances (Chapter 5). Dispersal tends to be more efficient with smaller drop sizes. Other factors affecting results include amount (volume) of dispersant in the slick surface, oil slick thickness, and drop-to-drop distance in the sprayed slick (Mackay et al., 1984). These are areas that continue to be important in laboratory research.

How mixing energy is applied in the laboratory is also a major factor. Recognizing that premixing dispersant with oil or water does not realistically represent field conditions, various methods have been employed to mix dispersant with oil: swirling flasks, water jetting, and air streams, for example. The mixing energy can be affected by the materials under study. In the MNS test, wave amplitude is reduced by No. 6 fuel oil and other viscous oils resulting in less dispersal (Mackay and Szeto, 1981; Mackay and Wells, 1983; Mackay et al., 1984). Oil-water ratio appears to be at least as important as viscosity and mixing energy.

Salinity and temperature are environmental factors that affect the results of all effectiveness tests. Salinity affects the hydrophilic-lipophilic balance, and salting-out effects diminish water solubility of ethoxylated surfactants. Lower temperatures tend to increase viscosity of both oil and dispersant as well as changing solubility of various components. In some cases the effect of temperature can be so great that an oil dispersible at 15°C may not be dispersible at 5°C (Lehtinen and Vesala, 1984; Wells and Harris, 1979).

Sampling and analysis are the last factors to be considered. Because oil drops resurface, the most reproducible results are obtained by sampling while mixing is proceeding, or at predetermined times immediately after mixing is stopped (Wells et al., 1984a). Contamination of surfaces with oil is frequently a cause for major errors.

Need for Standard Testing Oils

A number of investigators (Canevari, 1985; Mackay et al., 1986; Fingas, private communication) have expressed the need for a surrogate or standard oil for dispersant testing, which would improve reproducibility of product testing and provide international intercalibration of methods and products. Canevari (1985) recommended that tetradecane be used. However, Mackay et al. (1986) have shown that long-chain paraffins are a primary inhibitor of effectiveness. Short-chain alkanes also reduce dispersant effectiveness, but not as dramatically, and aromatic compounds increase effectiveness. In view of these findings, a single compound as surrogate oil (e.g., tetradecane) would not be representative of the dispersion properties of real oils.

A multicomponent mixture containing alkanes and aromatics might be more representative. In earlier work Mackay and Leinonen (1977) constructed such a synthetic oil, consisting of 10 components, to test evaporation rate modeling. Their mixture anticipated chemical and natural dispersion. The surrogate oil consisted of the normal isomers of butane, hexane, octane, decane, dodecane, and hexadecane, as well as benzene, toluene, naphthalene, phenathrene, and an "inert" component.

From 1980 to 1982 the American Petroleum Institute in cooperation with the U.S. Environmental Protection Agency set aside a number of oils for oil spill testing. They include Prudhoe Bay and Arabian light crudes and No. 6 and No. 2 fuel oils. The U.S. EPA has analytical data on all of the stored oils. The actual samples—19

55-gal (208-liter) drums of each of six oils—are stored* at EPA's Oil and Hazardous Materials Simulated Environmental Test Tank (OHMSETT) facility at Leonardo, New Jersey (Kolde, private communication). At the same time, Canada's Environmental Protection Service set aside a reference oil (Environment Canada, 1984; Fingas, private communication). These oils have been available upon request to investigators in Canada, the United States, and other countries.

Advantages and Disadvantages of Testing Methods

To be useful, a laboratory test should be fairly easy and quick to perform. This is satisfied by shake-flask tests (Abbott, 1983), but not by the more sophisticated tests requiring specialized apparatus and trained operators. On the other hand, these more complex tests are designed to mimic actual field conditions more closely.

A test should also be repeatable (within the same laboratory), reproducible (from one laboratory to another), and precise (variation coefficient of less than 20 percent). This is not always easy to achieve. Most significantly, the test should show a good correlation with real dispersant operations at sea. Although serious attempts have been made to mimic sea surface turbulence, so far no test satisfies this criterion.

For example, the French Institute of Petroleum test, which uses a beating hoop just under the surface and changes the water continuously, was compared with the WSL Labofina test, using a variety of oil types and dispersants (Bocard et al., 1984; Gillot and Charlier, 1986). No correlation on the basis of hydrophile-lipophile balance was observed, but rank order of effectiveness was often the same for both tests. Two primary differences are continuous washout of soluble materials in the IFP test, and greater mixing energy of the Labofina test (500 W/m^3 versus 1.5 W/m^3), although the longer run time of the IFP test (1 to 5 hr) brings total energy input into the same range.

The IFP test process, by its continuous removal of water containing dispersed oil, tends to simulate at-sea flow conditions while other test processes use a fixed volume of water. This feature of the IFP process can be a disadvantage since the test system is more

---

*Test oil stored at OHMSETT is kept under positive pressure and normal NO_2 atmosphere, and stored at the ambient temperature of the facility. Oil is checked for indications of aging or deterioration before testing (Tennyson, private communication).

complex to set up and transport. The test has been used in France and Norway, but few researchers are familiar with the process. In addition, the control of turbulence generated by the IFP agitation device is significantly affected by the level of the water surface. As a result, reproducibility of tests between laboratories is uncertain.

Labofina Test

The WSL Labofina test uses a standard separatory funnel in which the test fluids are mixed by a mechanical rotator (Martinelli, 1984). It has one major advantage: it is simple and fast—16 tests per day can be conducted. It is as reproducible as other tests (variation coefficient of 10 to 14 percent). The Labofina test shows a general bias toward lower effectiveness ratings for many oils owing to the relatively high oil-to-water ratio in this test (Daling, 1988). However, this test tends to give a relatively high efficiency rating for high-density oils, which do not rise rapidly after agitation has stopped (Daling, 1988).

Byford and Green (1984) report good agreement between the MNS and Labofina tests, when used to identify optimum surfactant combinations. The parameter that caused the greatest effect on the Labofina test was the shape of the conical separatory funnel. Precision of timing was also important: the stopcock orifice diameter affected results since a narrower orifice lengthened the time to collect a 50-ml aliquot of dispersed oil for analysis. Oil adhering to the flask walls can also produce major errors.

The major disadvantages of the Labofina test are that it uses an unrealistically high oil-water ratio, and mixing does not simulate the turbulence of an ocean surface. The samples are collected under static conditions, and the results depend on precisely when the samples are collected after mixing stops.

Mackay-Nadeau-Steelman Test

The MNS test uses a 20-liter closed glass, temperature-controlled tank, with a stream of air blowing tangentially on the water surface to generate reproducible waves and turbulence (Mackay and Szeto, 1981; Mackay et al., 1978, 1984; U.S. EPA, 1984; Wells and Harris, 1979). It reproduces turbulence that closely simulates actual mixing conditions. In this dynamic measurement, effectiveness can be assessed as a function of time, and oil film thickness can be independently controlled. Airflow rate is critical, however, and satisfactory waves cannot be generated unless the apparatus is precisely level (Byford and Green, 1984). Reproducibility is good (variation coefficient of 10 to 15 percent), and six tests can be completed per day by one operator.

The disadvantages of the MNS test are that its mixing energy, while reproducible, is difficult to quantify, and wave dampening by the materials under study can affect the results. It also gives anomalous results at 0°C, because of viscous shearing (Fingas et al., 1987). The airflow rate and direction must be critically adjusted, the apparatus is complex, and—since it is hand-made—tends to be expensive.

The MNS and WSL tests were compared in a study involving 13 dispersants and two oil types (Daling, 1986). Low-performance products were rated similarly by both tests, but the ranking of more effective products was generally different. The MNS test seems to be more reliable for low-viscosity oils, while the WSL test gives information for the higher-viscosity range. A recent comparison of effectiveness for 10 oils using the MNS, Labofina, and swirling flask methods demonstrated a poor correlation between results (Fingas et al., 1987).

U.S. Environmental Protection Agency Test

The EPA test uses a 130-liter stainless steel tank,* that, with its larger water volume, allows lower oil-water ratios (U.S. EPA, 1984). Use of a spray simulates application in the field from a boat. In this test also, effectiveness can be measured as a function of time, and samples can be obtained under dynamic conditions. Reproducibility is good for one operator using a consistent procedure (variation coefficient of less than 10 percent).

Critical factors include rate of application of dispersant to the oil surface, height above the surface from which it was poured, exact height of the jet spray nozzle, and need for care to avoid splashing oil onto the tank walls. Disadvantages of the test are that mixing energy (from the water jet) is not only difficult to quantify, it is difficult to reproduce. The test is sensitive to technique (Payne et al., 1985); some operators produce more reproducible results than others. High-shear turbulence is generated by the circulation pump,

---

*One reviewer recommends using a glass test container since it has hydrophilic properties; use of glass would remove oil adherence problems that are common to stainless steel containers.

and this affects results. The apparatus requires a skilled operator, is difficult to clean, and generates a large volume of contaminated water that must be disposed of. Only two tests per day are feasible.

Flume or Wave-Tank Tests

Flume or wave-tank tests do the best job of simulating turbulence from breaking or nonbreaking waves and currents—even in ice-covered waters (Brown et al., 1987; Delvigne, 1985; To et al., 1987). A flume, with its large volume, permits low oil-water ratios and greatly reduces wall effects. In a flume, the resurfacing of dispersed oil droplets can be studied, and droplet size distributions in the water column can be measured in the course of the test.

The disadvantages of a flume are obvious: it is expensive, complicated, and not portable. Large volumes of water are required, and the contaminated water must be heated and disposed of.

Bocard et al. (1984) and Bocard and Castaing (1986), noting the dissolution processes in at-sea trials, suggested that laboratory tests should incorporate a flow-through seawater system because closed vessels cannot duplicate the dilution process that occurs in the field. It would be desirable to separate the rate of dispersion from the advective loss of dispersed droplets. The IFP dilution test was designed to do this. Four grams of oil are agitated on the surface of water in a 4-liter reactor; the water is replaced continuously at 0.5 liter/hr, and the percentage of oil washed out is measured with time (Bardot et al., 1984; Demarquest et al., 1985).

Delvigne (1985) concluded that one shortcoming of most laboratory tests is that evaporation, photooxidation, emulsification, and nonhomogeneity of oil layer thickness cannot be modeled. Therefore, in the Delft flume experiments, he evaluated the following parameters:

- evaporation (0, 12, and 30 percent) in a pan outside the flume;
- photooxidation by ultraviolet exposure for 0.2 and 10 hr;
- emulsification by premixing the oil and water in laboratory beakers to a total water content of 70 percent; and
- layer thickness of 0.1, 0.5, and 2.5 mm.

Visual observations during the flume experiments showed that dispersant droplets barely incorporating oil had moved into the water column after falling or slowly sinking through the oil layer. The oil layer contracted into small slicks with open areas (herding) immediately after the dispersant droplets hit the oil and water surface.

Evaporation did not seem to affect dispersed oil concentrations with either naturally or chemically dispersed systems. It appeared, however, that oil droplet size may have decreased slightly with some evaporation.

Photochemical oxidation increased naturally dispersed oil concentrations, with no change in the chemically dispersed concentration. In both cases, however, droplet size distribution shifted to smaller volumes, presumably due to the formation of surface-active compounds in the oil slick, which lowered the oil-water interfacial surface tension (Payne and Phillips, 1985).

Emulsification (i.e., premixing the oil and water) decreased dispersed oil concentration in naturally dispersed experiments with Statfjord crude but had no effect in the chemically dispersed tests (Delvigne, 1985). Oil droplet size increased with emulsification in the naturally dispersed slick, but the oil droplet size did not increase despite emulsification in the case of chemical dispersion.

Layer thickness effects on both dispersed oil concentrations and oil droplet sizes were minor. Delvigne concluded that his flume experiments showed that the lack of dispersant effectiveness in field tests cannot be explained completely by the various parameters manipulated in his study.

Summary

Different laboratory tests give consistent results in discriminating broadly between high- and low-performance dispersants. Each test has special advantages and disadvantages; all give reproducible results, although some mimic field conditions better than others. Each test appears to measure different physical and chemical phenomena in the sense that the weight assigned to, or simulation of, processes and effects, such as oil-dispersant mixing, turbulence, drop size, distribution, and resurfacing tendency, are quite different. It is thus not surprising that they rank the most effective dispersants differently, corresponding to different performance criteria.

There is a consensus that it is impossible to simulate, in a laboratory system measured in tens of centimeters, turbulence characteristics that exist at the oceanic air-water interface, with its turbulent eddies and breaking waves. It is clearly necessary to introduce some turbulence in laboratory tests to promote and maintain dispersion, but an entirely satisfactory method of accomplishing this has not yet been devised. Even sophisticated, large-scale, experimental water tank systems cannot claim to simulate closely the ocean surface turbulence (Bouwmeester and Wallace, 1986a; Brown et al., 1987). No attempt is usually made in laboratory tests to simulate photooxidation. Only in the IFP tests is diffusive dilution simulated.

A recent evaluation (Anderson et al., 1985) has compared a number of effectiveness and toxicity tests, and evaluated them. There is no strong correlation between laboratory and field tests (see Chapter 4). A simple strategy for screening dispersants is to apply a reliable test such as the rotating flask test or the MNS test (see CONCAWE, 1986).

NEED FOR RESEARCH

There has been a regrettable lack of input into dispersant research on the basic interactions of dispersants with oil by professional surfactant scientists, with the obvious exception of the notable contribution by those who have formulated the products. Unfortunately, little of the commercially funded research has appeared in print. Had there been a complimentary program of research in this area, the state of knowledge could have been greatly advanced.

As a manifestation of the absence of research, no references have been found to recent papers in the peer-reviewed surfactant literature, for example, the *Journal of Colloid and Interfacial Science* or *Journal of Dispersion Science and Technology.* A considerable literature exists in journals and texts, such as that by Eicke and Parfitt (1987), on the behavior of surfactants in hydrocarbon and water systems. The thrust of most oil spill related studies has apparently been to determine if existing commercially available products work in the laboratory and at sea when applied by conventional methods. This would have been appropriate if it had been found that dispersants work in an efficient, predictable manner. In reality, the performance is often in doubt (see Chapter 4). It can be argued that more effort should have been devoted to determining why dispersants do, or do not, work on different oils under different application conditions. This research did not occur. As a result, there is an inadequate understanding of dispersant phenomena.

A program of research is needed to elucidate the mechanisms by which droplets of dispersant contact an oil film, mix and penetrate into it, how the surfactant forms various phases with the oil and migrates to the oil-water interface, and the microscopic process by which emulsions actually form. This appears to be a complex transient process and will be difficult to observe, but a knowledge of these phenomena is fundamental to determining why dispersants work, and at times why they do not work. For example, it appears that oil composition (as distinct from physical properties) affects dispersibility, but the reason for this is not known. It is still not too late to start a program to fill this lack of understanding.

3
Toxicological Testing of Dispersants and Dispersed Oil

This chapter describes what is known about biological—especially toxicological—effects of dispersants and dispersed oils from laboratory studies; reviews the evidence on oil-induced damage to organisms and how it is modified by dispersant use; and notes the applications and limitations of this knowledge (Figure 3-1). Experience with oil spill dispersants over the years has resulted in less toxic formulations. However, some questions pertaining to the effects of dispersants with and without oil remain, and they are addressed throughout this chapter.

OVERVIEW OF TOXICOLOGICAL TESTING

Toxicity, the potential of a material to cause adverse effects in a living organism, is a relative measure (see Glossary). Estimates of toxicity depend on many experimental physicochemical and biological factors. In addition, there are many different testing methods and variations in the products tested. A related problem has been the uncertain applicability of toxicity data from one species in one body of water to another species or area. For example, are species' sensitivities to dispersed oils in New England waters applicable to Texan waters? This question, which is of concern to regulators and industry, is addressed in this chapter.

81

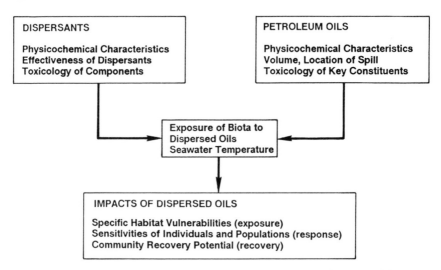

FIGURE 3-1   Factors to consider in the assessment of biological effects of dispersed oil in marine environments.

The objectives of toxicity testing of dispersants and dispersed oils in the laboratory are:

• to provide data on relative acute toxicities of effective products to commonly used test species under standardized conditions so that dispersant users have a basis for selecting effective and acceptably low toxicity products;

• to assure that dispersants do not significantly increase the acute and chronic toxicities of dispersed petroleum hydrocarbons; and

• to determine factors that modify dispersant toxicity, or enhance or ameliorate oil toxicity under natural conditions.

Different types of toxicity tests can satisfy these objectives. Tests are chosen to detect potentially harmful products both rapidly and reliably. They are not intended to be ecologically realistic or to predict effects in the field.

In measuring toxicity effects of oils, exposure comparisons may be made using the integral of concentration multiplied by time of exposure to 50 percent mortality LC_{50} (Anderson et al., 1980); the results of exposure tests are usually expressed as mg/liter per day or per hour. Since concentration may also be stated in approximate terms as parts-per-million (ppm) or parts-per-billion (ppb), and the exposure period as hours or days, some data on dispersed oil presented in this chapter will be stated as ppm-hour (ppm-hr) or ppm 1-day or 2-days. This allows a comparison to be made between different exposures used by different investigators who use the same analytical techniques. The exposure-time expression also allows an exposure to be expressed as concentrations change rapidly in the field. This concept probably holds during time periods from 1 hr to 4 days for oil and dispersed oil exposures. The use of ppm-hr assumes that organisms will respond in the same manner to a toxicant if exposed, for example, to 20 ppm for 1 hr or to 1 ppm for 20 hr. The concept is approximately valid for some of the data shown in this chapter.

There are obvious limits to this concept. If the time is short and the concentration high, the organism may be killed immediately. If the time is long and the concentration correspondingly lower, many organisms can tolerate, adapt to, or metabolize hydrocarbons and dispersants and survive and recover without apparent adverse effects. This concept has long been used in radiation exposures. It was proposed and used by Anderson et al. (1980, 1984), and McAuliffe (1986, 1987a) used the concept to compare laboratory bioassays that actually measured the dissolved hydrocarbons in the water-soluble fraction and chemically dispersed oil exposures with those measured in the field.

Toxicological Testing Methods

Considerable attention has been paid, especially by regulatory agencies, to

- the choice of suitable exposure regimes (static, continuous flow);
- test species;
- acute versus chronic testing;
- influence of modifying factors; and
- standardized testing protocols.

International workshops on these issues have been held by the United Nations Food and Agriculture Organisation (FAO) and the United Nations Environmental Programme (UNEP). Work from the United Kingdom has included Shelton (1969), Perkins (1972), and Beynon and Cowell (1974). Work from the United States has included Tarzwell (1969, 1970), Zillioux et al. (1973), and Becker et al. (1973). Canadian work has included Mackay et al. (1981), Wells (1984), and workshops leading to the Canadian Dispersant

Guidelines, 2d edition, Environment Canada (1984). FAO has been represented by White (1976) and UNEP by Thompson (1980 and private communication).

It is difficult to compare disperant formulations or sensitivities of different species, unless such work is conducted comprehensively in qualified laboratories (Doe and Wells, 1978; Wells, 1984; Wilson, private communication). Furthermore, information obtained using rigorously controlled and standardized testing protocols is desirable for reliable interpretation of toxicological information. Major components and trace contaminants should be known and exposures verified by analyzing the water in which the organisms are exposed.

Fish, arthropods (usually decapod crustaceans), mollusks (pelecypods), annelids (polychaetes), and algae have been the favored test species. Some researchers have also studied sensitive life stages; behavioral, biochemical, and developmental responses; and multispecies interactions, either acute or chronic. Testing of current formulations can be acute (i.e., short term), single-species, lethal, or sublethal; it is usually done in static rather than flowing systems, and at ambient temperatures. Some testing includes standard samples or reference toxicants.

Dispersant toxicity thresholds are most often reported as nominal concentrations—total amount of dispersant or oil divided by the total volume of water in the experimental chamber—rather than measured concentrations of materials to which organisms are actually exposed. This can lead to major errors in some cases. For some water-immiscible formulations at high concentrations, dispersant in the bioassay chambers can separate into a floating and dispersed upper-surface layer, several millimeters thick, and a dissolved subsurface fraction during the tests. For example, BP1100X in static tests separates like this immediately. Expressing the LC_{50} or EC_{50} on the basis of nominal concentration then gives a higher (and incorrect) value than if the water-soluble fraction were analyzed and used as the basis. Thus the toxicity of a water-soluble material may be underestimated. The same problem arises because of the immiscibility of water and dispersed oil (as discussed later in this chapter). For some dispersant formulations, this is an important but generally unrecognized source of error for toxicity estimates.

Dispersant Screening Procedures for Toxicity: Considerations

The qualities of a good laboratory screening test are that it is easy to perform and control, and it is reliable, reproducible, and adequately sensitive. Because its purpose is to determine the relative toxicity of one dispersant versus other previously tested dispersants, practicality and known sensitivity are weighed against ecological realism. Screening tests are usually conducted with a single species, but do not yet attempt to simulate interactions of two or more species, that is, community responses (Cairns, 1983; Mount, 1985). Screening tests can include

- various test species and life stages;
- response parameters other than mortality;
- various test materials;
- different exposure modes;
- varying length of exposure; and
- various pass-fail criteria.

Laboratory tests are poor simulations of natural conditions because they are conducted under standard controlled conditions. Generally, this means exposing animals in the laboratory to more or less constant concentrations for 2 to 4 days, while in the ocean initial concentrations of dispersants and dispersed oil would be diluted progressively and generally rapidly.

Because the effectiveness and toxicity of a dispersant may be positively correlated, screening tests should consider both criteria in sequence (Bratbak et al., 1982; Doe and Wells, 1978; Mackay and Wells, 1983; Nes and Norland, 1983; Norton et al., 1978; Swedmark et al., 1973). Both criteria have already been considered together when evaluating dispersants for government agencies (Anderson et al., 1985; Araujo et al., 1987; Environment Canada, 1984). Screening tests should accurately evaluate and accommodate the possibly greater acute toxicity of more effective dispersants.

For improved accuracy and utility of hazard assessments, future screening toxicity tests should consider the above factors, the physical chemistry of the dispersant solutions, and the responses of the test organisms during short exposures.

Dispersant Screening Procedures in Canada, the United States, and Other Countries

Until 1982, most countries used a combination of dispersant and dispersed oil tests, oil and dispersed oil tests, or tests with all treatments (Wells, 1982a). The primary concern was to evaluate the toxicity of oils upon dispersal. Such an approach was particularly supported by the United Kingdom's sea and beach test.* However, the inclusion of oils in dispersant tests is experimentally complex because it introduces a new set of variables associated with the oil and is subject to errors in interpretation because of immiscibility (Wells, 1982a; Wells et al., 1984a,b). Yet most countries, out of concern about dispersed oil effects and joint toxicity of oil and dispersant constituents, have included both oil and dispersant in their tests.

Some countries screen dispersants only (e.g., Australia, Canada, several Asian countries). At least 10 countries employ a toxicity screening test for dispersants or dispersed oil: Australia, Canada (linked to effectiveness test), France, Hong Kong (modified U.K. sea test), Japan, Norway (modified U.K. sea test), Singapore (modified 1970s Canadian test; Environment Canada, 1973), South Africa (modified U.K. sea test), United Kingdom, and the United States. Brazil (Araujo et al., 1987), Nigeria, the Philippines, and Sweden are also developing testing approaches (Schalin, 1987). Screening methods and status are listed in Tables 3-1, 3-2, and 3-3; of particular note are procedures for Australia, Canada, South Africa, the United Kingdom, and the United States (Table 3-1; reviewed in detail by Moldan and Chapman, 1983; Thompson, 1985; and Wells, 1982a). A number of other countries are thought to be doing tests.

The most frequent combination of test materials are dispersant, oil, and dispersed oil, that is, dispersant and oil mixture (Table 3-3). Most tests use seawater, and lethality is the usual toxicity response. Many different test species are used, with little uniformity among countries. Both indigenous and standard species have been selected, such as local shrimp and *Artemia*, and in most countries local species are used as the standard (rainbow trout, *Salmo gairdneri*, in Canada; brown shrimp, *Crangon crangon*, in the United Kingdom; and mummichog, *Fundulus heteroclitus*, in the United States). Most countries have pass-fail criteria, but they vary. When dispersed oil is tested in the laboratory, oil composition is variable and differs from place to place, the water-soluble fraction is normally not separated, and hydrocarbon exposures are normally not measured. Hence, the same dispersant submitted to different countries for approval may be subjected to quite different toxicity screening methods and pass-fail criteria.

---

*The United Kingdom screens for effectiveness first, and dispersants that pass go onto the toxicity-testing phase. The work is conducted in two laboratories and is a phased approach rather than a linked approach.

TABLE 3-1 Worldwide Survey of Dispersant Toxicity Screening

Geographic Location	Author/Year	Report/Result
United States	Battelle Memorial Institute, 1970	Early test procedures developed for American Petroleum Institute (API)
	Blacklaw et al., 1971	U.S. Toxicity Test Procedure
	California State Water Resources Control Board, 1971	Early test procedures developed for California
	Becker et al., 1973	Regional U.S. survey for bioassey species for tests
	McCarthy et al., 1973	U.S. EPA standard toxicity tests
	U.S. Department of the Navy, 1973	U.S. Military Dispersant Specifications
	Exxon Chemical Company, 1980	U.S. policies on dispersant use
	Cashion, 1982	Draft ASTM Method No.6, dispersed oil
	Smith and Pavia, 1983	Dispersant use guidelines, California
	Lindstedt-Siva et al., 1984	Use guidelines, ecological considerations, coast
	Pavia and Smith, 1984	Use guidelines, California
	U.S. Environmental Protection Agency (EPA), 1984	U.S. revised standard dispersant toxicity test
	API, 1985	API Dispersant Use Guidelines
	Pavia and Onstad, 1985	Use guidelines, California
Canada	Abbott, 1972	Ontario guidelines, Ministry of the Environment
	Environment Canada, 1973	Canadian Dispersant Acceptability Guidelines
	Doe and Harris, 1976	Selection of suitable species for toxicity tests
	Environment Canada, 1976	Standard Listing of Acceptable Dispersants, Department of Energy (DOE)
	Harris and Doe, 1977	Toxicity methods for screening dispersants, DOE
	Wells, 1982a	Summary/review, toxicity testing worldwide for regulatory control
	Abbott, 1984	Discussion paper, Canadian Dispersant Acceptability Guidelines
	Environment Canada, 1984	Dispersant Acceptability Guidelines, 2d ed.
	Harris et al., 1986	Regulatory considerations, acute toxicity test spp.
	Trudel and Ross, 1987	Dispersant use decision-making methods
European Nations		
Europe	Wilson et al., 1973, 1974	Review, toxicity tests
Finland	Kerminen et al., 1971	Policy on toxicity, fish, early reports
France	Division Qualite des Eaux, 1979	Toxicity protocol
	Auger and Croquette, 1980	Acceptability list, use guidelines
	CTGREF, Division Qualite des Eaux, 1981	Toxicity testing protocol
Norway	Norwegian Ministry of Environment, 1980	Regulations on dispersant composition and use
	Westerngaard, 1983	Dispersant policy
Sweden	Lehtinen et al., 1985	Study for criteria for guidelines

TABLE 3-1 (Continued)

Geographic Location	Author/Year	Report/Result
United Kingdom	Moore, 1968	Early paper, U.K. dispersal experience in ports
	Jeffery and Nichols, 1974	List of approved dispersants and rationale
	Blackman et al., 1978	Procedure for U.K. screening, sea and beach tests
	Lloyd, 1980	U.K. role of toxicity tests, registration and notification
	Norton and Franklin, 1980	Methods, dispersant toxicity, sea and beach tests
	Norton, private communication	U.K. methods, dispersant toxicology
	Wilson, 1981	Toxicity tests, rationale for choice
	Franklin and Lloyd, 1982	Toxicity, 25 oil-dispersant mixtures, sea and beach tests
	Lloyd, private communication	U.K., MAFF toxicity approach
	Wilson, 1984	U.K. policies on dispersant use, risk analysis
Other Countries		
Australia	Henry, 1971	Policy, dispersant use, early report
	Thompson, 1985	Program, effect and toxicity of dispersants
	Thompson and McEnnally, 1985	Resource atlas for spill countermeasures
Bahrain	Linden, 1981	Fisheries, use recommendations
Hong Kong	Thompson and Wu, 1981	Toxicity testing/screening methods
Japan	Ministry of Transportation, 1974	Testing standards, toxicity
	White, private communication	Testing method
Singapore	Port of Singapore Authority, 1976	Toxicity testing methods
South Africa	McGibbon, private communication	Dispersant testing program
	Moldan and Chapman, 1983	Review, toxicity methods
International Agencies		
FAO	White, 1976	Course, methods for oils and dispersants
IPECA	International Petroleum Industry Environmental Conservation Association, 1986	Statement of environmental concerns, fate and effects
IMO/UNEP	International Maritime Organization, 1982	Guidelines, environmental considerations
	Hayward, 1984	Bonn Agreement, toxicity test methods
IMO	IMO Organization, 1986	Hong Kong, acceptance list

KEY: ASTM--American Society for Testing and Materials; FAO--Food and Agriculture Organisation of the United Nations; MAFF--Ministry of Agriculture, Fisheries, and Food; and UNEP--United Nations Environment Programme.

TABLE 3-2 Summary of Aquatic Toxicity Test Procedures for Oil Spill Dispersants, 1969-1984

Country	Species and Life Stage	Basic Description of Toxicity Test[a] (1) Materials; (2) Mode; (3) Duration; (4) Fresh Water/Salt Water; (5) Response Parameters; (6) Pass-Fail Criteria	References
United States[b]	Two adult fish (flat-head minnow, Pimephales promelas, mummichog, Fundulus heteroclitus); one invertebrate (larval; brine shrimp (Artemia sp.)	(1) Dispersant; dispersant plus No. 2 fuel oil (D+O, 1:10); reference toxicant (dodecyl sodium sulphate [DSS]) (2) Static (3) Fish = 4 days; shrimp = 2 days (4) Fish = fresh and salt water; shrimp = salt water (5) Lethality (6) Not given; $LC_{50}s$ estimated, only	(1) McCarthy et al., 1973 Other references: (1) Tarzwell, 1969 (2) Tracy et al., 1969 (3) LaRoche et al., 1970 (4) Hazell et al., 1971 (5) Tarzwell, 1971
	Five fish (golden shiner, rainbow trout, striped bass, stickle-back, California killifish); brine shrimp (Artemia sp.)	(1) No. 6 fuel oil plus dispersant (D:O of 1:5); dispersant (2) Static (3) 4 days (4) Salt water (5) Mortality (6) Not given	Battelle Memorial Institute, 1970
	Brine shrimp (Artemia sp.)	(1) Dispersant (2) Static (3) 2 days (4) Salt water (5) Mortality (6) Not given	U.S. Department of the Navy, 1973
	Many suggested species; Preferred species is fish, especially Fundulus heteroclitus	(1) Dispersant; dispersant plus oil (undefined type and undefined D:O ratio); 2 reference toxicants: DSS and phenol	(1) Cashion (private communication) based on ASTM standard practice, 1982

TABLE 3-2 (Continued)

Country	Species and Life Stage	Basic Description of Toxicity Test[a] (1) Materials; (2) Mode; (3) Duration (4) Fresh Water/Salt Water; (5) Response Parameters; (6) Pass-Fail Criteria	References
United States[b] (continued)		(2) Static; static with replacement; flow-through (preferred) (3) 4 days or less (4) Fresh or salt water (5) Primarily lethality; other effects being considered (6) Not given; $LC_{50}s + EC_{50}s$ are estimated	U.S. EPA, 1984
	Fundulus heteroclitus (juveniles); Artemia sp. (larvae)	(1) Dispersant; No. 2 fuel oil alone; dispersant plus oil, 1:10 mixture; reference toxicant, DSS (2) Static (3) 4 days (Fundulus); 2 days (Artemia) (4) Salt water (20 ± 1 ppt) (5) Mortality (6) Not given; $LC_{50}s$ are reported	
Canada	Rainbow trout (Salmo gairdneri) juveniles	(1) Dispersant; dispersant plus No. 2 fuel oil (D:O, 1:1); reference toxicant (2) Stirred, static (3) 4 days (4) Fresh water (5) Lethality (6) Dispersant 4-day $LC_{50} > 1{,}000$ ppl (nominal concentration) dispersant plus oil 4-day $LC_{50} > 100$ ppm (nominal concentration)	(1) Environment Canada, 1973 (2) Harris and Doe, 1977 (3) Doe and Wells, 1978

Country	Species	Test conditions	References
Canada (continued)	Rainbow trout (juveniles) and threespine stickleback (Gasterosteus aculeatus) adult; Salmo gairdneri (juveniles)	Same as above, fresh water (trout) and salt water (stickleback); hydrocarbon analyses conducted on oil dispersions	(1) Wells and Harris, 1980
		(1) Dispersant; reference toxicant, DSS (2) Static, aerated (3) 4 days (4) Fresh water (5) Mortality (6) LC_{50} (median lethal time) > 4 days, at 100 ppm or X ppm (derived from effectiveness test)	Environment Canada, 1984
European Nations			
France	Eel (Anguilla anguilla) adult; brown shrimp (Crangon crangon) adult; mussel (Mytilus edulis) adult	(1) Dispersant; dispersant plus reference oil, BAL 150 (2) Stirred, static (3) Exposure of 24 hr; 3-day recovery in uncontaminated seawater (4) Salt water (5) Mortalities after 3 days in recovery tanks (6) Not given	(1) CTGREF, 1981; Division Qualité des Eaux, 1979
Norway	Brown shrimp and common limpets Alga (Chlamydomonas reinhardtii)	Sea test; beach test (see test for salt water under United Kingdom) (1) Dispersant plus Ekofisk crude oil (2) Static (3) 40 hr (4) Fresh water (5) Mortality and growth (6) $LC_{50} \geq 200$ ppm (?)	(1) Norwegian Ministry of Environment, 1980 (2) Blackman et al, 1977, 1978 (1) Norland et al., 1978 (2) Heldal et al., 1978; Heldal, private communication
United Kingdom	Brown shrimp (Crangon crangon) adults	Sea test (1) Crude oil (fresh Kuwait); crude oil plus dispersant (D:O, 1:1); oil = 1,000 ppm (nominal) (2) Stirred, static (3) Exposure 100 min; 24 hr recovery (4) Salt water (5) Mortality at 24 hr (6) Mortalities in oil vs. oil plus dispersant using t-test	(1) Blackman et al., 1977, 1978 (2) Norton et al., 1978

TABLE 3-2 (Continued)

Country	Species and Life Stage	Basic Description of Toxicity Test[a] (1) Materials; (2) Mode; (3) Duration (4) Fresh Water/Salt Water; (5) Response Parameters; (6) Pass-Fail Criteria	References
United Kingdom (continued)	Common limpet (Patella vulgata)	Beach test (1) Dispersant; crude oil (fresh Kuwait) (2) Exposure by spraying; air; then flowing seawater (3) Exposure; 6 hr in air; 72 hr recovery in seawater (4) Salt water (5) Mortalities at 72 hr (6) Mortalities in oil vs. oil and dispersant, using t-test	(1) Blackman et al., 1977, 1978 (2) Norton et al., 1978 Other references: (1) Norton and Franklin 1980 (2) Wilson, 1974 (3) Norton, private communication
Other Countries Australia		Toxicity test methods under development; meanwhile, substantial evidence of low toxicity in use should be submitted with application for treatment quality	Piranig, private communication
	Yellow-eyed mullet, or other organisms	(1) Gippsland and Kuwait crudes; dispersants plus oils (effective ratios) (2) Flow-through (3) Indefinite (4) Salt water (5) Lethality (6) Not given	Department of Transport, Australia[b]

Country	Species	Test	References
Brazil	Brine shrimp (Artemia sp.); other appropriate salt water sp. sought	Using U.S. EPA (1973) procedure with Artemia sp., in experimental studies dispersants	(1) Goldstein and Araujo, private communication (2) McCarthy et al., 1973
Hong Kong	Rabbitfish (Siganus canaliculatus); sea urchin (Anthocidaris crassispina)	Tests are the beach test and modified U.K. sea test (1) Gas oil (i.e., lighter fraction of marine diesel oil); gas oil plus dispersant (D:O, 1:1) (2) Stirred, static (3) Exposure of 100 min; 24 hr recovery in uncontaminated salt water (4) Salt water (5) Percent mortality at 24 hr at 100 ppb (fish) and 330 to 3,300 ppb (urchin) (6) Comparison with oil alone (all nominal concentrations)	(1) Thomson and Wu, 1981 (2) Wu, 1981
Japan	Killifish (Oryzias lapipes); diatom (Skeletonema costatum)	Fish test (1) Dispersant (2) -- (3) -- (4) Salt water (5) Lethality (6) 4-day LC_{50} > 3,000 ppm	(1) Ministry of Transportation of Japan, 1974 (2) Cashion, private communication
Republic of South Africa	Sand shrimp (Palaemon pacificus)	(1) O (250 ppm); D+O (250 ppm or each); (2) Stirred, static (3) Exposure of 100 min; 24 hr clean seawater (4) Salt water (5) Lethality (6) O+D mixture toxicity ≤ O alone	(1) Chapman, 1980 (2) Chapman, 1981, in Mackay et al., 1981 (3) Chapman, private communication (4) Moldan and Chapman, 1983

94

TABLE 3-2 (Continued)

Country	Species and Life Stage	Basic Description of Toxicity Test[a] (1) Materials; (2) Mode; (3) Duration (4) Fresh Water/Salt Water; (5) Response Parameters; (6) Pass-Fail Criteria	References
Singapore	Glass fish (Chanda gymnocephalus)	(1) Dispersant and standard marine fuel oil (MFO V1100/1200); D:O = 1:1 (2) Static (3) 4 days (4) Salt water (5) Lethality (6) 4-day LC$_{50}$ ≥ 100 ppm Toxicity test is a modification of the Canadian (Environment Canada, 1973) test	(1) Port of Singapore Authority, 1976

NOTE: Static tests following from Blackman et al., 1978; continued flow experimental, chronic tests. Nominal concentration is based on added amounts or reagents; no measurements of hydrocarbons occur.

[a]EPA has determined that its experience with dispersants and other chemicals in oil spills is not yet sufficient to support preparation of a schedule to permit routine testing (U.S. EPA, 1984).
[b]This is a proposed test in 1978, outlined in an Exxon report, Chemicals for Oilspill Control, distributed by Exxon Corporation, April 1980.

SOURCE: Modified from Wells (1982a).

TABLE 3-3 Summary of Regulatory Aquatic Toxicity Tests for Dispersants, 1986

Country	Species			Materials[a]			Type of Water[b]			Toxicity Response	Pass-Fail Criteria Given
	Plant	Inverte- brate	Verte- brate	D	D+OO	RT	FW	SW	Both		
United States	x	x	x	x	x	x			x	Lethality	No
Canada (1984)		x	x	x		x	x			Lethality	Yes
European Nations											
France		x	x	x	x			x		Lethality	No
Norway	x	x		x	x	x			x	Lethality, growth	Yes
United Kingdom		x		x	x			x		Lethality	Yes
Other Countries											
Australia		x		x		x		x		Lethality	Yes
Brazil		x		x		x		x		Lethality	Yes
Hong Kong		x	x	ND[c]	x			x		Lethality	Yes
Japan	x				x		x		ND	ND	
Singapore			x	x				x		Lethality	Yes
South Africa		x		x	x			x		Lethality	Yes

[a] Dispersant (D), dispersant and oil (D+O), oil (O), reference toxicant (RT).
[b] Fresh water (FW), salt water (SW).
[c] ND--no data.

SOURCE: Modified from Wells (1982a).

The lack of a standardized approach displays a lack of consensus about screening test objectives. For example:

• The United Kingdom attempts realism in both its sea and beach tests, one with oil exposure in the water column and one simulating exposure to a beach application of dispersant, respectively. However, the sea test ignores the experimental complexity of preparing and controlling the oil preparation, and the beach test assumes that dispersants will be used directly on rocky shorelines.

• The U.S. test covers all treatments but simply lists the toxicity data without interpreting it for use by the on-scene coordinator (OSC).

• Canada's test is linked closely to the effectiveness test through a formal decision framework, but it only screens the dispersant using a single-species freshwater fish test.

• The state of São Paulo in Brazil is adopting an approach similar to Canada's, ultimately screening the dispersant with a valuable indigenous shrimp.

To date, only Canada and Brazil link effectiveness and toxicity screening tests in a formal decision-making framework, although the United States is considering such a linkage based on Anderson et al. (1985). Yet, as pointed out by Thompson (1985), it is important for all countries to recognize "contemporary problems arising from the development of effective third-generation dispersants and more accurate methods for determining the toxicity of oils" in the design of their toxicity screening tests.

There are significant advantages to screening only the dispersant formulation, but an international consensus on the key treatments to test has not yet been reached. A consensus on methods would allow reliable comparisons of data from one country to another.

TOXICITY OF DISPERSANTS

This section summarizes the aquatic toxicology of dispersant components and commercially available dispersants. Data are reported for early formulations, as well as second- and third-generation dispersants that generally are less acutely toxic than earlier products.

Toxicity is a relative measure that is influenced by many factors, particularly concentration, duration of exposure, and type of organism. Most of these experiments use concentrations and exposure durations that substantially exceed expected field exposures. Nevertheless, the following factors are important to understand:

- the risks of misapplication of dispersants;
- the environmental fate and effects of dispersant materials added to ocean waters to treat oil spills;
- the range of responses in different species and to different formulations and to different environments; and
- which components contribute most to toxicity in order to improve formulations.

Acute Toxicity of Components

Knowledge about the toxicity of the primary components of dispersants would assist in evaluating dispersant toxicology and the toxicities of dispersed oils. All surfactants are toxic at high concentrations. Many surfactants have unique toxicological properties, are usually but not always nonspecific or physical toxicants, can cause narcosis, and can disrupt membranes physically and functionally. A number of factors control the toxicity of surfactants to aquatic organisms, among them, ethoxylate chain length, the presence of esters versus ethers, and hydrophilic-lipophilic balance (HLB). Rates of uptake and penetration into an organism's tissues are highly dependent on species (Abel, 1984; Wells, 1974).

Acute toxicity data of some surfactants used in current formulations (circa early 1980s) are presented in Table 3-4 (Wells et al., 1985). New and more effective formulations may have different solvents and different combinations or types of surfactants. Toxicity in Table 3-4 is expressed as a 1- or 2-day EC_{50} for two crustaceans, *Artemia* sp. and *Daphnia magna*. Results of these laboratory tests show that the anionic surfactants are generally more toxic than the nonionic surfactants or esters, toxicities being influenced by alkyl chain length, degree of dispersion, and HLB.

Studies on surfactants cover a wide range of organisms because of concern for effects on membranes, reproductive stages, bacteria, behavior (especially chemoreception), and other subtle sublethal changes in exposed organisms (Abel, 1974; Moore et al., 1986).

Solvents were the most toxic components of some early dispersants, due to high concentrations of aromatic hydrocarbons in the petroleum fractions employed (Nelson-Smith, 1972; Smith, 1968). Several types of solvents are now used in most formulations (see Chapter 2 and Appendix A), and they are far less toxic (Caneveri, 1986). Nagell et al. (1974) and Wells et al. (1985) have shown toxicities to decrease in the order: aromatic hydrocarbon > saturated

TABLE 3-4   Acute Toxicity of Surfactants From Dispersant Formulations to Two Planktonic Crustaceans, Artemia sp. and Daphnia magna, at $20^{\circ}C$

Description		Behavior in Water[a,b]	
Chemical Family	General Description	Salt Water	Fresh Water
Anionic surfactant	Oxyalkylated (6.5 m)C_{12}-C_{13} alcohol	S	S
Surfactant salt	Sulfate salt of oxyalkylated (3 m) C_{12}-C_{15} alcohol (60% in water and ethanol)	S	S
Sulfosuccinate, anionic surfactant	Alkyl (C_8) sulfosuccinate salt, 70% solution	D	S
Amine surfactant	Alkyl (C_{13}) sulfosuccinate salt, 70% solution	D	D
Anionic surfactant	Oxide ester of DDBSA in amine salts	D	D
Aromatic surfactant	Oxyalkylated (9 m) alkyl phenol	D	D
Anionic surfactant, salt	Sodium xylene sulfonate, 40% in water	S	S
Nonionic surfactant	Sorbitan monooleate	D	D
Surfactant ester	Oxyalkylated (20 m) sorbitan monooleate	S	S
Surfactant ester	Oxyalkylated (20 m) sorbitan trioleate	D	D

[a]These data are expressed as ppm, which is approximate since variability in calculating and measuring field concentrations and toxicity thresholds used in analyses are much greater than the difference in the relationship between ppm and mg/liter. Data in Table 3-4 were originally expressed in mg/liter.
[b]Behavior in water code: S--soluble, D--dispersible.

1-Day EC$_{50}$s (ppm)[c,d,e]		2-Day EC$_{50}$s(ppm)	
Artemia sp.	Daphnia sp.	Artemia sp.	Daphnia sp.
11.3 (n.c.)	14.0 (11.1-17.7,1.5)	8.2 (5.6-11.9,1.8)	6.0 (4.0-9.0,2.1)
23.0 (14.4-36.7,2.1) [13.8*][8.6-22.0]	23.5 (19.5-28.3,1.5) [14.1][11.7-17.0]	13.5 (8.3-21.9,2.2) [8.1][5.0-13.2]	6.0 (3.8-9.4,2.2) [3.6][2.3-5.6]
14.8 (10.1-21.6,2.4) [10.4] [7.1-15.1]	32.0 (21.2-48.2,1.6) [22.4] [14.8-33.7]	18.5 (13.2-25.9,2.2) [12.9][9.2-18.1]	21.0 (15.2-28.9,2.8) [14.7][10.6-20.3]
> 1,000 [> 700]	180 (93-350,3.2) [126][65-245]	n.c. [< 39]	< 56
60.0 (49.6-72.5,1.5)	3.8 (2.0-7.2,6.1)	32.0 (23.4-43.7,2.0)	< 1
46.5 (36.1-60.0,1.8)	20.0 (16.5-24.2,1.39)	68.0 (45.5-101.7,4.4)	6.3(4.2-9.5,2.0)
> 3,000 [> 1,200]	430 (239-773,2.27) [172][96-309]	> 1,000 [> 400]	100-320 [40-128]
> 1,000	300 (181-498,3.3)	> 1,000	90 (39-209,3.3)
> 1,000	220 (100-482,3.9)	--	< 100
> 1,000	205 (86-487,4.6)	--	75 (36-157,4.7)

[c]n.c.: Not calculated.
[d]Values in parentheses are: (95 percent confidence limits, slope function).
[e]Values in brackets have been adjusted to represent 100 percent of parent surfactant.

SOURCE: Wells et al., 1985.

hydrocarbons > glycol ethers > alcohols. Third-generation concentrate dispersants tend to use less toxic solvents, such as glycol ethers.

Acute Toxicity of Formulations

More than 100 studies have been conducted on acute lethal toxicity of dispersants alone, more than half of them on currently used second-generation dispersants (Doe and Wells, 1978; Doe et al., 1978; Dye and Frydenborg, 1980; Nelson-Smith, 1972, 1980, 1985; Pastorak et al., 1985; Sprague et al., 1982; Wells, 1984). Such an extensive data base of varying quality invites periodic critical analysis by dispersant, organism type and stage, and method of exposure before definitive statements can be made about the acute toxicity of any one formulation. This has been done for Corexit 9527 (Wells, 1984).

Table 3-5 lists toxicity data, expressed as LC_{50}s for a wide variety of species and dispersants. A wide range of values is reported, including the following:

• In Wells (1984) 4-day LC_{50}s were 150 to greater than 10,000 ppm.

• Eleven dispersants tested with rainbow trout showed 4-day LC_{50}s of 260 to greater than 10,000 ppm (Doe and Wells, 1978).

• Early EPA data showed 2-day LC_{50}s for *Artemia* sp. of 1.2 to 100,000 ppm for 15 products (Dye and Frydenborg, 1980).

• In a study with freshwater phytoplankton and several dispersants, estimated 2-day LC_{50}s were 1 to 575 ppm (Heldal et al., 1978).

• Studies by Kobayashi (1981) with sea urchin embryonic stages gave threshold concentrations of 0.32 to 320 ppm.

• A range of 0.1 to 20,000 ppm for a wide range of species and stages were compiled in Pastorak et al. (1985).

From Table 3-5 it can be seen that the "second-generation" conventional dispersants, such as BP1100X and Corexit 7664, are generally much less toxic than earlier formulations (BP1002, early Finasol formulations).

One useful way to present dispersant toxicity data is by testing several products on one species (Table 3-6). Table 3-6 illustrates that the majority of products had LC_{50}s and EC_{50}s greater than 100 ppm to a planktonic crustacean, that is, a marine copepod.

Some formulations with high toxicities to certain species still exist. Most research studies have examined only Corexit, BP, and

TABLE 3-5 Acute Lethal Toxicity of Some Oil Spill Dispersants to Marine Organisms, a Selection of Current Data by Taxonomic Group

Organisms	Species/Stage	Dispersant	Exposure Time (hr)	Threshold Concentration Expressed as LC$_{50}$ (ppm)	References
CRUSTACEANS					
Brine shrimp	Artemia sp.	Corexit 9527	48	52-104	Wells et al., 1982
Brine shrimp	Artemia sp.	Cold Clean	96	28	McAuliffe et al., 1975
Brine shrimp	Artemia sp.	Corexit 7664	96	130,000	McAuliffe et al., 1975
Isopod	Gnorimosphaeroma oregonensis	Corexit 9527	96	> 1,000	Duval et al., 1982
Amphipod	Anonyx laticoxae	Corexit 9527	96	> 140	Foy, 1982
Amphipod	A. nugax	Corexit 9527	96	97-111	Foy, 1982
Amphipod	Boeckosimus edwardsi	Corexit 9527	96	> 80	Sekerah and Foy, 1978
Amphipod	Onisimus litoralis	Corexit 9527	96	> 175	Foy, 1982
Amphipod	Onisimus litoralis	Corexit 9527	96	80-160	Foy, 1982
Amphipod	Gammarus oceanicus, juvenile	Corexit 9527	96	> 70	Sekerah and Foy, 1978
Amphipod	Gammarus setosus	Corexit 9527	96	> 80	Foy, 1982
Amphipod	Gammarus spp.	Corexit 9527	96	38-803	Foy, 1982
Amphipod	Gammarus spp.	BP1100X	96	>10,000	Ladner and Hagström, 1975
Amphipod	Gammarus spp.	BP1100X	96	> 10,000	Nagell et al., 1974
Amphipod	Gammarus spp.	BP1100X	96	300	Nagell et al., 1974
Amphipod	Gammarus spp.	Water-based dispersants	96	200	Ladner and Hagström, 1975
Amphipod	Gammarus spp.	Petroleum-based dispersants	96	> 10,000	Ladner and Hagström, 1975
Amphipod	Gammarus spp.	Petroleum-based dispersants	96	200 ± 130	Ladner and Hagström, 1975
Copepod	Pseudocalanus minutus	Corexit 9527	48	8-12	Wells et al., 1982
Copepod	Pseudocalanus minutus	Corexit 9527	96	5-25	Wells et al., 1982

TABLE 3-5 (Continued)

Organisms	Species/Stage	Dispersant	Exposure Time (hr)	Threshold Concentration Expressed as LC$_{50}$ (ppm)	References
Mysid	Neomysis integer	Water-based dispersants	96	> 4,500	Ladner and Hagström, 1975
		Petroleum-based dispersants	96	~150	
Mysid	Neomysis integer	Corexit 7664	96	> 4,500	Nagell et al., 1974
Mysid	N. integer	Corexit 7664	96	> 4,500	Ladner and Hagström, 1975
Mysid	N. integer	BP1100X	96	150-200	Ladner and Hagström, 1975
Mysid	N. integer	BP1100X	96	150	Nagell et al., 1974
Grass shrimp	Palaemonetes pugio	Corexit 7664	48	2,700	McManus and Connell, 1972
	P. pugio	Corexit 7664	96	> 10^4 (27°C), nontoxic (17°C)	Ozelsel, 1981
		Atlantic-Pacific		1,000 (27°C), 1,800 (17°C)	
		Gold Crew	96	150(27°C),380(17°C)	
		Nokomis-3	96	140 (27°C), 250 (17°C)	
Grass shrimp	P. pugio	Corexit 7664	96	> 10,000	Ozelsel, 1981
Grass shrimp	P. pugio	Gold Crew	96	150 (27°C), 380 (17°C)	Ozelsel, 1981
Grass shrimp	P. pugio	Conco K	96	500 (27°C), 700 (17°C)	Ozelsel, 1981
Grass shrimp	P. pugio	Cold Clean	96	170 (27°C), 850(17°C)	Ozelsel, 1983
Grass shrimp	P. pugio	Cold Clean	96	3,700	Ozelsel, 1983
Grass shrimp	P. pugio	Corexit 9527	96	640 (27°C), 840 (17°C)	McAuliffe et al., 1975 Ozelsel, 1983

Common name	Species	Test material	Time (h)	Concentration	Reference
Shrimp	Crangon crangon	Shell LT	48	1,000	Perkins et al., 1973
Shrimp	C. crangon	BP1100X	48	> 1,000	Perkins et al., 1973
Brown shrimp	C. crangon	10 conventional dispersants, 7 concentrated dispersants (unnamed)	48	3,300 > 10,000	Norton et al., 1978
			48	2,800 > 10,000	Norton et al., 1978
Prawn	Leander adspersus	Corexit 7664	96	> 1,000	Swedmark et al., 1973
Prawn	L. adspersus	BP1100X	96	> 688	Swedmark et al., 1973
Lobster	Homarus americanus, fourth stage	BP1100X	96	18,200–42,800	Wells and Doe, 1976
				17,000–41,000	Doe and Wells, 1978
Shore crab	Carcinus maenas	BP1100X	48	20,000	Perkins et al., 1973
Shore crab	C. maenas	Shell LT	48	20,000	Perkins et al., 1973
Hermit crab	Eupagurus bernhardus	Shell LT	48	> 10,000	Perkins et al., 1973
Hermit crab	E. bernhardus	BP1100X	48	> 10,000	Perkins et al., 1973
Hermit crab	E. bernhardus	Corexit 7664	48	10,000	McManus and Connell, 1972

OTHER INVERTEBRATES

Common name	Species	Test material	Time (h)	Concentration	Reference
Stony coral	Madracis mirabilis	Shell dispersant LTX	24	162	Elgershuizen and de Kruijk, 1976
Oligochaete	Marionina subterranea	Corexit 7664	96	> 1,000	Giere, 1980
		Finasol OSR-2			
		Finasol OSR-5			
Intertidal limpet	Patella vulgata	BP1100X	96	- 3,700	Nuwayhid et al., 1980
		BP1100WC	96	- 270	Nuwayhid et al., 1980

MOLLUSKS

Common name	Species	Test material	Time (h)	Concentration	Reference
Cockle	Cardium edule	BP1100X	96	> 688	Swedmark et al., 1973
Cockle	C. edule	Corexit 7664	96	1,000	Swedmark et al., 1973
Cockle	C. glaucum	Corexit 7664	96	950–1450	Ladner and Hagström, 1975
Cockle	C. glaucum	Corexit 7664	96	1,500	Nagell et al., 1974
Cockle	C. glaucum	BP1100X	96	≥ 2,000	Nagell et al., 1974

TABLE 3-5 (Continued)

Organisms	Species/Stage	Dispersant	Exposure Time (hr)	Threshold Concentration Expressed as LC$_{50}$ (ppm)	References
Cockle	C. glaucum	BP1100X	96	> 2,000	Ladner and Hagström, 1975
Mussel	Mytilus edulis	BP1100X	96	> 2,000	Ladner and Hagström, 1975
Mussel	M. edulis	BP1100X	96	≥ 2,000	Nagell et al., 1974
Mussel	M. edulis	BP1100X	96	> 688	Swedmark et al., 1973
Mussel	M. edulis	Corexit 7664	96	> 1,000	Swedmark et al., 1973
Mussel	M. edulis	Corexit 7664	96	1,400	Nagell et al., 1974
Mussel	M. edulis	Corexit 7664	96	900-1,300	Ladner and Hagström, 1975
Scallop	Pecten opercularis	Corexit 7664	96	250	Swedmark et al., 1973
Scallop	P. opercularis	BP1100X	96	> 688	Swedmark et al., 1973
Scallop	Argopecten irradians	Corexit 9527	96	200 (20°C); 1,800 (10°C)[b]; 2,500 (2°C)	Ordzie and Garofalo, 1981
Clam	Brachidontes variablis	Corexit 7664	96	1-5	Avolizi and Nuwayhid, 1974
Clam	Protothaca staminea	Corexit 9527	96	- 100	Hartwick et al., 1982
Clam	Donax trunculus	Corexit 7664	96	0.1-1	Avolizi and Nuwayhid, 1974
Oyster	Ostrea edulis	Shell LT	48	> 10,000	Perkins et al., 1973
Oyster	O. edulis	BP1100X	48	2,500	Perkins et al., 1973
ECHINODERMS					
Starfish	Asterias rubens	Shell LT	48	> 6,000	Perkins et al., 1973
Starfish	A. rubens	BP1100X	48	3,000-6,000	Perkins et al., 1973
Fish					
Plaice	Pleuronectes platessa	BP1100X	48	7,100	Perkins et al., 1973

Common name	Species	Dispersant	Time	Concentration	Reference
Plaice	_P. platessa_	Shell LT	48	> 10,000	Perkins et al., 1973
Sole	_Solea solea_	Shell LT	48	> 10,000	Perkins et al., 1973
Fish larvae	_P. platessa_	Corexit 7664	96	400	Wilson, 1977
	S. solea				
Baltic herring	_Clupea harengus membras larvae_	BP1100X	48	50-300	Linden, 1974
Spot	_Leiostomus xanthurus eggs_	Corexit 9527	48	40-100	Slade, 1982
Rainbow trout	_Salmo gairdneri_	BP1100X	96	2,100-2,900	Wells and Doe, 1976
Rainbow trout	_S. gairdneri_	Corexit 9527	96	96-293	Wells and Doe, 1976
Rainbow trout	_S. gairdneri_	OSE-20	96	29-93	Wells and Doe, 1976
Rainbow trout	_S. gairdneri_	Corexit 7664	96	490-857	Doe and Wells, 1978
Cod	_Gadus morhua_	BP1100X	96	> 688	Swedmark et al., 1973
Cod	_G. morhua_	Corexit 7664	96	130	Swedmark et al., 1973
Killifish	_Fundulus similis_	Corexit 7664	96	13,400	McAuliffe et al., 1975
Killifish	_F. similis_	Cold Clean	96	110	McAuliffe et al., 1975
Killifish, adult	_F. heteroclitus_	AP oil dispersant (GFC Chemical Co.)	48 72	100 50-100	Butler et al., 1982 Butler et al., 1982
Mummichog	_F. heteroclitus_	Cold Clean	96	130	McAuliffe et al., 1975
Tidewater silverside	_Menidia beryllina_	Cold Clean	96	75	McAuliffe et al., 1975
Stickleback	_Gasterosteus aculeatus_	Water-based dispersants	96	950 ± 250	Ladner and Hagström, 1975
		Petroleum-based diaspersants	96	> 10,000	Ladner and Hagström, 1975
Stickleback	_G. aculeatus_	Corexit 7664	96	700-1,200	Nagell et al., 1974
Stickleback	_G. aculeatus_	Corexit 7664	96	700-1,200	Ladner and Hagström, 1975
Stickleback	_G. aculeatus_	BP1100X	96	> 10,000	Ladner and Hagström, 1975
Stickleback	_G. aculeatus_	BP1100X	96	> 10,000	Nagell et al., 1974
Stickleback	_G. aculeatus_	Cold Clean	96	70	McAuliffe et al., 1975
Stickleback	_G. aculeatus_	Corexit 7664	96	2,600	McAuliffe et al., 1975

TABLE 3-5 (Continued)

Organisms	Species/Stage	Dispersant	Exposure Time (hr)	Threshold Concentration Expressed as LC_{50} (ppm)	References
Dace	Phoximus phoximus	Corexit 7664	96	1,200–1,600	Ladner and Hagström, 1975
Dace	P. phoximus	Water-based dispersants		1,400 ± 200	Ladner and Hagström, 1975
Coho salmon	Oncorhyncus kisutch	BP1100X		1,700	McKeown, 1981
Coho salmon	O. kisutch	BP1100X		1,700 (15% salinity)	McKeown, 1981
Fourhorn sculpin	Myoxocephalus quadricornis	Corexit 9527	96	< 40	Foy, 1982
MACROPHYTES					
Turtlegrass	Thalassia testudinum	Corexit 9527	96	200	Baca and Getter, 1984

SOURCE: Adapted from Wells (1984).

TABLE 3-6 Acute Toxicity Threshold Concentrations for
Several Dispersants and Marine Calanoid Copepods (Primarily
Pseudocalanus minutus) at $10°C$

Concentration Range[c] (ppm)	1-Day[a,b]	4-Day[a,b]
LC_{50}[c]		
1-10	DSS	DSS, C-9527
$10-10^2$	C-9527	
10^2-10^3	BP1100WD	BP1100WD, C-7664
10^3-10^4	C-8867, BP1100X, C-7664	C-8667, C-9550, BP1100X
10^4	C-9550	
EC_{50}[d]		
1-10	DSS	DSS, C-9527
$10-10^2$	C-9527	BP1100WD
10^2-10^3	BP1000WD	C-8667, C-7664
10^3-10^4	C-8667, BP1100X, C-7664	C-9550, BP1100X
$>10^4$	C-9550	

[a]C is Exxon's Corexit; BP is British Petroleum.
[b]Dodecyl sodium sulfate (DSS) is the reference toxicant.
[c]Median lethal concentration (LC).
[d]Median effective (lethality plus incapacitation)
concentration (EC).

SOURCE: Wells, 1985.

Finasol products. A few studies have related dispersant toxicity to its
detailed chemical composition (Ladner and Hagström, 1975; Nagell
et al., 1974; Wells et al., 1985). Lethal toxicities of dispersant formulations vary greatly with product, testing conditions, species, and life
history stage (e.g., Thorhaug and Marcus [1987a,b] on seagrasses).

It is particularly useful to know the differences in toxicities by
species for dispersant products. A single product can have a wide
range of acute toxicity. This is clearly shown with Corexit 9527 (Table 3-7), which has been used in much recent biological and toxicological research on dispersants and dispersed oils (Nelson-Smith, 1985;
Peakall et al., 1987; Sergy, 1987; Wells, 1984; Wells et al., 1984a).
Table 3-7 shows that for some algae, protozoa, fish, copepods, and
mollusks the dispersant is not very toxic—LC_{50} concentrations are

TABLE 3-7 Aquatic Toxicity of Corexit 9527 to Invertebrates and Fish

Species	Type of Threshold[a]	Threshold Concentrations (mg/liter)				References[c]
		>100	10-100	1-10	<1	
Fresh Water						
Algae (Chlamydomonas)	2-day LC$_{50}$	575				Heldal et al., 1978; Vorlond et al., 1978
Zebra fish (Brachydanio)	2-day LC$_{50}$	550				Esso, 1978
Ciliate protozoa	IEC	100-350[b]				Rogerson and Berger, 1981
Trout (Salmo gairdneri)	4-day LC$_{50}$	140-233				Doe and Wells, 1978; Wells and Harris, 1980
Salt Water						
Scallops (Argopecten irradians)	6-hr LC$_{50}$	1,900, 2,500				Ordzie and Garofalo, 1981
Calanoid copepods	1-day LC$_{50}$	100-1,000				Lönning and Falk-Petersen, 1976
Amphipods (several sp.)	4-day LC$_{50}$	104->170				Foy, 1982
Clam (Protothaca staminea)	LT$_{50}$ at 100 mg/liter		3-5 (days)			Hartwick et al., 1979
Barnacle (Balanus) larvae	6-day EC$_{50}$		10-100			Lönning and Falk-Petersen, 1976
Shrimp (Artemia) larvae	2-day EC$_{50}$		42-72			Wells et al., 1982
Shrimp (Artemia) larvae	2-day LC$_{50}$		40			Dye and Frydenborg, 1988
Copepods (Acartia)	2-day LC$_{50}$		22-35			Wells et al., unpublished

Organism	Endpoint			Reference
Sea urchins (4 sp.) eggs	4-day EC$_{50}$	10-100		Lönning and Hagström, 1976
Plaice (Pleuronectes) eggs	10-day EC$_{50}$	10-100		Lönning and Falk-Petersen, 1978
Cod (Gadus morhua) eggs	21-day EC$_{50}$	10-100		Lönning and Falk-Petersen, 1978
Arctic fish (Myoxocephalus) embryo	4-day LC$_{50}$	<40		Foy, 1982
Stickleback (Gasterosteus)	4-day LC$_{50}$	28		Wells and Harris, 1980
Natural arctic phytoplankton	EC stimulation		10	Hsaio et al., 1978
Coelenterate larvae	4-day EC$_{50}$		1-10	Lönning and Falk-Petersen, 1978
Copepods (Pseudocalanus minutus)	2-day LC$_{50}$		7.1-11.0	Wells et al., unpublished
Clams (Protothaca staminea)	LT$_{50}$ at 10 mg/liter		>10 (days)	Hartwick et al., 1979
Sea urchins (4 sp.) larvae	EC$_{50}$ (several days)		1-10	Lönning and Hagström, 1976
Sea urchin (Paracentrotus) sperm	10-min EC$_{50}$		0.03-0.05	Hagström and Lönning, 1977
	30-min EC$_{50}$		0.003-0.0003	Hagström and Lönning, 1977
Cod (Gadus morhua) eggs	21-day EC$_{50}$		<1	Lönning and Falk-Petersen, 1978

[a]LC$_{50}$ = median lethal concentration; EC$_{50}$ = median effective concentration; IEC = incipient effective concentration; LT$_{50}$ = median lethal time.
[b]Range of several threshold values.
[c]Based on literature published to 1982.

SOURCE: Wells, 1984.

above 100 ppm for exposures of 6 hr to 4 days. However, many more reported toxicity thresholds are below 100 ppm, and some life stages are extremely sensitive. A 10-min EC_{50} for sea urchin sperm is 0.03 to 0.05 ppm, and is the only laboratory-derived threshold concentration so far that is likely to be still encountered in the field hours after dispersant use.

Patterns of sensitivity do not readily emerge. The threshold concentrations for eggs, embryos, and larvae are widely spread, showing EC_{50} values from 0.0003 to 1,000 ppm. Crustaceans exhibit wide ranges of toxic thresholds. Freshwater species appear to be less sensitive than marine species.

For Corexit 7664 (Table 3-8), Ladner and Hagström's (1975) LC_{50}s show that there may be increasing sensitivities from crustaceans to mollusks to fish, although there are exceptions. Seventy-five percent of toxicity thresholds for this dispersant are above 1,000 ppm. A similar evaluation for BP1100X shows the greater sensitivity of some crustaceans, and approximately 80 percent of all reported threshold concentrations for BP1100X are over 1,000 ppm. Table 3-9 compares the three dispersants in terms of the number of tests in each concentration range. Although there is considerable overlap, Corexit 9527 is generally more toxic than the other two. Such evaluations (Tables 3-7 to 3-9) could be usefully compiled for all extant dispersant formulations and local species of interest as an aid to on-scene coordinators at spill sites (Trudel, private communication).

Factors Influencing Acute Toxicity

A number of physicochemical and biological factors influence the toxicity of a dispersant formulation (Wells, 1984). These factors are important to understand because estimates of toxicity are relative—not absolute—numbers, and they change depending on environmental conditions and biological populations being exposed.

Physicochemical Factors

Surfactant molecular structure and ionic state were considered by Portmann and Connor (1968), Bellan et al. (1969), George (1971), Wildish (1972), Abel (1974), Nagell et al. (1974), Macek and Krzeminski (1975), Tokuda (1977a,b), Tokuda and Arasaki (1977), Wilson (1977), Ernst and Arditti (1980), and Wells et al. (1985).

Solvent type and aromatic content were considered by Shelton (1969), Nagell et al. (1974), Ladner and Hagström (1975), Wilson

TABLE 3-8 Aquatic Toxicity of Two Dispersants to Invertebrates and Fish

Species	Type of Threshold (LC$_{50}$)	Threshold Concentrations (ppm)			References
		$> 10^4$	10^3-10^4	10^2-10^3	
		Corexit 7664			
Hermit crab (Diogenes sp.)	2 days	10,000			McManus and Connell, 1972
Amphipod (Gammarus sp.)	4 days	>10,000			Ladner and Hagström, 1975
Grass shrimp (Palaemonetes pugio)	4 days	>10,000 (27°C), nontoxic (17°C)			Ozelsel, 1981
Brine shrimp (Artemia sp.)	4 days	130,000			McAuliffe et al., 1975
Brine shrimp (Artemia sp.)	2 days	99,500			U.S. EPA, 1984
Killifish (Fundulus similis)	4 days	13,400			McAuliffe et al, 1975
Oligochaete (Marionina subterranea)	4 days		>1,000		Giere, 1980
Cockles (Cardium edule)	4 days		1,000		Swedmark et al., 1973
Cockles (C. glaucum)	4 days		950-1,450		Ladner and Hagström, 1975
(C. glaucum)	4 days		1,500		Nagell et al., 1974
Mussel (Mytilus edulis)	4 days		>1,000		Swedmark et al., 1973
	4 days		1,400		Nagell et al., 1974
	4 days		900-1,300		Ladner and Hagström, 1975
Mussel (Brachidontes variablis)	4 days		2,000		Avolizi and Nuwayhid, 1974
Grass shrimp (Palaemonetes pugio)	2 days		2,700		McManus and Connell, 1972
Mysid (Neomysis integer)	4 days		>4,500		Nagell et al., 1974;
Prawn (Leander adspersus)	4 days		>1,000		Ladner and Hagström, 1975 Swedmark et al., 1973

TABLE 3-8 (Continued)

Species	Type of Threshold (LC$_{50}$)	> 10^4	10^3-10^4	10^2-10^3	References
			Corexit 7664 (Continued)		
Copepod (Pseudocalanus minutus)	2 days		1,200->10^4		Wells, unpublished data
Brine shrimp (Artemia sp.)	2 days		>5,600		Wells, unpublished data
Brown shrimp (Crangon crangon)	2 days		7,500-10^4		Portmann, 1970
Dace (Phoxinus phoxinus)	4 days		1,200-1,600		Ladner and Hagström, 1975
Killifish (Fundulus heteroclitus)	4 days		1,150		U.S. EPA, 1984
Stickleback (Gasterosteus aculeatus)	4 days		2,600		McAuliffe et al., 1975
Clam (Donax trunculus)	4 days			300	Avolizi and Nuwayhid, 1974
Scallop (Pecten opercularis)	4 days			250	Swedmark et al., 1973
Amphipods (4 genera, 8 species)	4 days			> 70->175	Foy, 1982
Copepods (Pseudocalanus minutus)	4 days			520-800	Wells, unpublished data
Cod (Gadus morhua)	4 days			130	Swedmark et al., 1973
Stickleback (Gasterosteus aculeatus)	4 days			720-1,200	Nagell et al., 1974
Rainbow trout (Salmo gairdneri)	4 days			490-851	Doe and Wells, 1978
Flounder larvae (Pleuronecta platessa) (Solea solea)	4 days			400	Wilson, 1977

BP1100X

Organism	Duration			Reference
Lobster (Homarus americanus, fourth stage)	4 days	18,200–42,800		Wells and Doe, 1976
Shore crab (Carcinus maenas)	2 days	20,000		Perkins et al., 1973
Hermit crab (Eupagurus bernhardus)	2 days	>10,000		Perkins et al., 1973
Stickleback (Gasterosteus aculeatus)	4 days	>10,000		Ladner and Hagström, 1975; Nagell et al., 1974
Oyster (Ostrea edulis)	2 days		2,500	Perkins et al., 1973
Cockle (Cardium glaucum)	4 days		>2,000	Nagell et al., 1974
Mussel (Mytilus edulis)	4 days		>2,000	Ladner and Hagström, 1975
Copepod (Pseudocalanus minutus)	4 days		2,350	Wells, unpublished data
	4 days		1,700	Wells, unpublished data
Shrimp (Crangon crangon)	2 days		>1,000	Perkins et al., 1973
Starfish (Asterias rubens)	2 days		3,000–6,000	Perkins et al., 1973
Plaice (Pleuronectes platessa)	2 days		7,100	Perkins et al., 1973
Coho salmon (Oncorhyncus kisutch)	4 days		1,700	McKeown, 1981
Rainbow trout (Salmo gairdneri)	4 days		2,100–2,900	Wells and Doe, 1976
Cockle (Cardium edule)	4 days	>688		Swedmark et al., 1973
Mussel (Mytilus edulis)	4 days	>688		Swedmark et al., 1973
Scallop (Pecten opercularis)	4 days	>688		Swedmark et al., 1973
Amphipod (Gammarus sp.)	4 days	300		Nagell et al., 1974
Mysid (Neomysis integer)	4 days	200		Ladner and Hagström, 1975
Prawn (Leander adspersus)	4 days	150–200		Ladner and Hagström, 1975
Baltic herring (Clupea harengus membras larvae)	2 days	>688		Swedmark et al., 1973
		50–300		Linden (1974)
Cod (Gadus morhua)	4 days	>688		Swedmark et al., 1973

TABLE 3-9 Reported Acute Toxicity Thresholds of Some Common Dispersants

Dispersant	N[a]	Threshold Concentrations (ppm)					
		$> 10^4$	10^3-10^4	10^2-10^3	10-10^2	1-10	< 1
Corexit 9527	24	0	1	6	10	5	3
Corexit 7664	32	7	17	8	0	0	0
BP1100X	26	4	17	5	0	0	0

[a]N = number of reported data sets, up to 1986.

(1976), Tokuda and Arasaki (1977), and Wilson (1977). For example, a study by Mommaerts-Billiet (1973) found that Finasol with a relatively toxic aromatic carrier lengthened the lag growth phase of the nanoplankter *Playtymonas tetrathele* to a greater extent than did the less toxic water-soluble Finasol. Diminished growth rates were also noted at high concentrations.

Parameters affecting the condition of the dispersant in water have been considered:

• concentration of a dispersant and duration of test (McManus and Connell, 1972; Wilson, 1976, 1977);
• temperature, salinity, oxygen, and so on (Abel, 1974; Wells et al., 1982; Wilson, 1977); and
• chemical stability of the dispersant and age of test solution (Wilson, 1977).

Biological Factors

Biological characteristics of the exposed organisms were divided into three groups: species (phylogeny), life history, and physiology. Sensitivities to anionic and nonionic surfactants varied widely: 4-day LC_{50}s ranged from 0.1 to 800 ppm (Abel, 1974; Czyzewska, 1976). Sensitivity to surfactants followed the general order: crustaceans < bivalve mollusks < teleost fishes (Butler et al., 1982; Eisler, 1975; Nagell et al., 1974; Ladner and Hagström, 1975; McManus and Connell, 1972; and Swedmark et al., 1971, 1973).

Sensitivity to water-based dispersants fell in the same order as sensitivity to surfactants—crustaceans < bivalves < fishes; but sensitivity to petroleum-based dispersant fell in the reverse order—fishes < bivalves < crustaceans (Ladner and Hagström, 1975; Nagell et al., 1974; Swedmark et al., 1971, 1973). Other comparisons of fishes, bivalves, and crustaceans were made by Macek and Krzeminski (1975), Lönning and Falk-Petersen (1978), and Wells et al. (1982).

Phylogeny

Phylogeny is an important factor. Dispersant ranking tests differ with phyla (Abel, 1974; Beynon and Cowell, 1974; Boney, 1968; Heldal et al., 1978; LaRoche et al., 1970; Wilson, 1974). Foliose lichens are more sensitive than crustose (Cullinane et al., 1975).

Life History Stage

The influence of life history stage varies for different species. Teleost fishes, despite their great commercial importance, have received only limited study. For example, fish eggs are often very susceptible to dispersant at time of fertilization. Developing embryos are less sensitive than fish larvae (Linden, 1974; Wilson, 1976). Sensitivity of both embryos and larvae vary with dispersant formulation (Linden, 1974; Lönning and Falk-Petersen, 1978). For some fish larvae, the difference in susceptibility between species is less than the difference between different ages of a single species (Wilson, 1977).

Young life stages of other organisms, such as echinoid sperm and larvae, and some species, such as copepods, appear to be particularly sensitive. Larval resistance of crustaceans increases with age, based on studies with surfactants only (Czyzewska, 1976). For polychaetes, the most sensitive stages are gravid animals (Foret, 1975). Other studies considering life history include Portmann (1969), Bellan et al. (1972), and Abel (1974).

Physiological Factors

Physiological factors include seasonal variation in susceptibiity to dispersants (Baker and Crapp, 1974; Braaten et al., 1972; Crapp, 1971a,b,c; Fingerman, 1980; Perkins et al., 1973). Previous exposure and acclimation makes a difference (Abel, 1974). The transition from yolk sac to feeding is especially susceptible (Lönning and Falk-Petersen, 1978; Wilson, 1977). Health and feeding state are also important (McManus and Connell, 1972; Wilson, 1977). Starvation of fish larvae has been found to increase their susceptibility (Wilson, 1977).

Of all the many possible factors, five have the primary influence on toxicity thresholds, by 1 to 3 orders of magnitude when changed: type of surfactant, type of solvent, water temperature, phylum, and stage of development (Wells, 1984). Few studies varying these factors have been conducted in a single laboratory where each variable can be controlled. Wilson's (1977) study with fish larvae was one; it considered the influences of aromatic content of the solvents, temperature, salinity, and species. Work with copepods and brine shrimp (Wells et al., 1982; Wells, unpublished data) has also shown the influences of temperature and species, toxicity thresholds increasing in magnitude with declining temperatures.

The influence of these factors becomes apparent when evaluating sets of data for any one dispersant (see Tables 3-8 and 3-9). Most products produce wide threshold ranges, even the reportedly "low toxicity" products such as Corexit 7664. Relatively few products approved for use in several countries fall into the range of high acute toxicity (4-day LC_{50} less than 10 ppm). Products that do give such high toxicities to local organisms should be used with caution, especially in nearshore environments (Tables 3-5 and 3-6).

Additional work on standardizing methods (Wells, 1981; Wells et al., 1984b) and on studying sublethal effects and their causes should clarify the major influencing factors, the degree of their influence, and the implications of such variable responses to the field effects of dispersed oils.

Temperature Influence on Toxicity of Dispersants

A wide range of studies (Ordzie and Garofalo, 1981; Wells, 1984; Wilson 1977) show that dispersants become less toxic with lowering temperatures. This is most accurately shown by comparing threshold concentrations at the different temperatures. This relationship holds for Corexit 9257 which is, or has been, used as a reference dispersant in many studies.

Wells reported 1-day LC_{50}s, median lethal concentrations, showing an order of magnitude lower toxicity at 15°C (51 to 96 ppm concentration) than at 25°C (greater than 560 ppm concentration) with *Artemia* sp. and Corexit 9257. For scallops, Ordzie and Garofalo (1981) found that as temperature increased, the concentration of Corexit 9527 required to kill 50 percent of the scallops decreased: 200 ppm at 20°C, 1,800 ppm at 10°C, and 2,500 ppm at 2°C. They also noted that dispersant concentrations that were not lethal to scallops during winter temperatures caused greater than 50 percent mortality at summer temperatures.

The reasons for the higher toxicity at higher temperatures may be a combination of increased uptake rates of chemicals, greater exposure due to increased activity of the organisms in the water, and combined factors such as higher temperatures and lower-oxygen levels. The exact mechanisms for the higher toxicity have not been elucidated.

In general, the same phenomena occurs with dispersed petroleum oils—generally higher toxicities at higher temperatures in laboratory exposures (Bobra and MacKay, 1984). The implications of these studies are that water temperature has a profound influence on the toxicity of dispersants. There are significantly higher sensitivities of organisms in warmer waters and in summer as compared to winter conditions. Individual dispersants should be screened at the range of expected environmental temperatures and threshold concentrations reported that could be used by on-site commanders in decisions regarding dispersant deployment.

Sites and Physiology of Toxic Action

Contact, uptake, internal storage, toxic action, detoxification, and depuration are all processes by which a marine organism responds to and may be affected by dispersants. Thus, understanding these processes within exposed organisms is crucial to understanding why species sensitivities vary, which sublethal effects are significant, and whether dispersant-oil combinations might be more harmful than either one by itself. It is also important to know whether toxicity is temporary or permanent.

Few studies have used well-characterized dispersants or known constituents, hence much understanding is tentative or based on detergent-surfactant literature. Early dispersant formulations and anionic and nonionic surfactants act "often physically and irreversibly, on the respiratory organs and other tissues of aquatic organisms, and reversibly, depending upon exposure time, on their nervous systems" (Wells, 1984).

A large number of studies describe the work of early investigators on the sites and physiology of toxic action of dispersants, much of the work being conducted at high concentrations in order to investigate the various responses (Wells, 1984). Some constituents of dispersants appear to cause disruptive effects to membranes and narcosis of the whole organism. No single action is implicated, but rather a total response of the surface membranes and tissues, particularly gills, to the exposure of surface-active agent. Behavioral responses include cessation of feeding, slowed swimming, disorientation, impaired locomotion, and paralysis often leading to death (if the exposures are high and long enough). For example, fish gills were damaged under 200 ppm exposures to Oilsperse 43 over 1 to 4 days (McKeown and March, 1978). Blood enzyme (cathepsin D and acid phosphatase) activities were increased in shrimp exposed to a high concentration (10 to 100 ppm) of a nonionic detergent, Solo, for 72 hr, presumably reflecting a change in the membrane permeability of lysosomal enzymes (Drewa et al., 1977). Asphyxiation due to the swelling of gill lamellae and changes in membrane permeability is the principal manifestation of toxicity in fish (Granmo and Kollberg, 1976). Reduced surface tension may also play a role with HeLa cells (a specialized cell line used in bioassays) and nonionic surfactants (Ernst and Arditti, 1980). Certainly, "asphyxiation by surfactants packing at gill surfaces appears to be one main physical toxic mechanism of surfactants" (Pastorak et al., 1985), but evidence of the exact mechanisms and the NOEL (no-observable-effect level) concentrations are not yet conclusive.

Surfactant molecules have both lipophilic and hydrophilic chemical groups, and surfactants of different HLB characteristics are usually blended in the solvents to ensure the separation of the oil droplets when dispersed. Crustacean surfaces and gills, which are largely hydrophobic, tend to be contacted by low HLB surfactants and hydrocarbons. Fish gills are coated with mucus and are less hydrophobic. Such differences may help explain the different sensitivities of crustaceans and fish to water immiscible and water-miscible dispersants (Nagell et al., 1974). Considering the differences between the external structure and composition of fish and crustaceans, the influence of the molting cycle on water uptake and loss in crustaceans, and the way by which ions and water are regulated or otherwise controlled in many species, variations in sensitivity are not surprising.

Sublethal Effects

Many studies, performed mostly in the 1970s, have examined sublethal effects. Sublethal responses such as reproduction, behavior, growth, metabolism, and respiration usually occur at levels well below lethal thresholds, and hence are the most sensitive biological

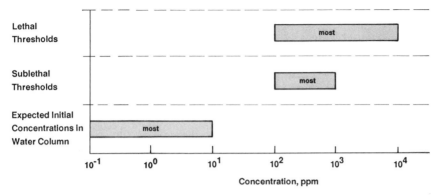

FIGURE 3-2   Comparison of lethal and sublethal threshold concentrations and expected concentrations in the water column for dispersants, as of 1982. Source: Adapted from Wells (1984).

responses. Normally, although not always, the laboratory exposure durations are 1 to 4 days longer than organisms that would be exposed in most dispersant use situations in open waters. Further, laboratory exposure concentrations of reported sublethal effects normally are 1 or 2 orders of magnitude above highest anticipated concentrations in field use (see Figure 3-2). Organisms from bacteria to algae and invertebrates to fish exhibited varied biological responses to dispersants in 50 to 100 sublethal laboratory studies (Wells, unpublished compilation; Nelson-Smith, 1985). As with acute toxicity, the range of threshold concentrations is extremely wide, from less than 1 ppm to 10^6 (undiluted dispersant) based on exposures from 10 min to 3 weeks (Table 3-10).

Some studies have described effects of dispersant applied directly to external surfaces or injected into organisms. This work is particularly relevant to interpreting dispersant effects on intertidal organisms, if dispersants are applied directly onto shorelines. Limpets dropped off rock surfaces in response to dispersants (Blackman et al., 1977). However, Fingerman (1980) observed that the regeneration of killifish caudal fins were unaffected when BP1000X was injected in amounts of 1:80 by weight for each fish.

The principal studies of sublethal effects of solvents have included hydrocarbons (such as naphthalene) and mixtures (such as kerosene) that contain them (Neff, 1979; Nelson-Smith, 1972; NRC, 1985); many responses have been measured. In contrast, few sublethal studies have been conducted using other components known to be in dispersant formulations.

TABLE 3-10 Some Sublethal Responses and Their Threshold Concentrations for Various Marine Organisms Exposed to Second-Generation Oil Spill Dispersants

Dispersants	Threshold Concentrations (ppm)	Exposure Times	Biological Responses	References
Corexit 9527	$10^{-3}-10^{-4}$	20-40 min	Fertilizing capabilities (sea urchin)	Hagström and Lönning, 1977
A-P dispersant	1-10	3 days	Schooling behavior (fish)	Butler et al., 1982
Corexit 7664	$1-10^2$	2 days	Inhibit growth rate (phytoplankton)	Portmann, 1972
Corexit 9527	$1-10^2$	2 days	Decreased metabolism (bacteria)	Griffiths et al., 1981
Seven dispersants	$1-10^2$	10 min	Induction of brachycardia (fish)	Kiceniuk et al., 1978
Eight dispersants	$10-10^2$	30 min	Inhibition of fertilization (sea urchins)	Sawada and Ohtsu, 1975[a]
A-P dispersant	$10-10^2$	36 hr	Declined precopulatory behavior (amphipod)	Butler et al., 1982
BP1100X, Slickgone LT2	$10-10^2$	2 days	Inhibition of food detection (shrimp)	Evans et al., 1977[a]
Gulf agent 1009	$10-10^2$	24 hr	Decreased swimming behavior (fish)	Wildish, 1974
BP1100X, Corexit 7664	$10-10^3$	96 hr	Decreased shell closure (bivalve)	Swedmark et al., 1973
Corexit 7664	10^2-10^3	10 min	Avoidance reactions (amphipod)	Portmann, 1972
BP1100X, Corexit 7664	10^2-10^3	96 hr	Disrupted equilibrium (fish)	Swedmark et al., 1973
BP1100X, Corexit 7664	$10-10^3$	96 hr	Decreased byssal activity (bivalve)	Swedmark et al., 1973

Corexit 8666	10^2–10^3	10 min	Induction of brachycardia (fish)	Kiceniuk et al., 1978
BP1100X, BP1100WD	10^2–10^3	24 hr	Gill damage (limpets)	Nuwayhid et al., 1980
Chemoserve, OSE 750	10^2–10^3	40 hr	Reduction of nitrate generation (beach microflora)	
84 Products	10^2–10^3	1–4 days	Decreased growth (phytoplankton)	Harty and McLachlan, 1982
Oilsperse 43	10^1–10^3	1–4 days	Gill damage, reduction in blood constituents (trout)	Tokuda and Arasaki, 1977[a]
Finasol OSR-2	10^3	3 hr	Slowed rate of fertilization (sea urchins)	McKeown and March, 1978
Corexit 7664	10^3–10^4	1.5 days	Inhibited embryonic development (sea urchins)	Lönning and Hagström, 1975
BP1100X	10^4–10^6	1–3 days	Delayed embryonic development (sea urchins)	Vashchenko, 1978[a]
Shell SD LTX	10^6	3 weeks	Weight decline, gonadal pathologies (small)	Kobayashi, 1981[a]
				Battershill and Bergquist, 1982

[a] As cited by Wells (1984).

SOURCE: Adapted from Wells (1984).

One example is Payne (1982) who showed that fish (*Salmo gaird-neri*), crabs (*Cancer irroratus*), and mollusks (*Chlamys islandicus*) "had the capacity for enzymatic hydrolysis of the complex fatty acid ester mixtures found as surfactants in second-generation dispersants."

Laboratory and fieldwork comparing oil with dispersed oil has tended to replace sublethal toxicity studies of dispersants alone, because it is believed that organisms under natural conditions would be exposed only briefly to very low concentrations of dispersant. However, this position requires examination for various marine habitats and organisms, using recognized hazard assessment approaches (Cairns et al., 1978).

Hazard Assessment of Dispersant Alone

Effects due to the dispersant solvent and surfactants in the water column may be surmised only from laboratory studies, as field studies have not examined this question. A commonly accepted approach for laboratory and field comparisons and predictions does not yet exist. Exposure in the water column depends on the concentration-time profile of components as they dilute and degrade through various processes, such as advection, volatilization, solubilization, diffusion, bacterial degradation, uptake by organisms, and detoxification.

Normally, a portion of the dispersant applied to an oil slick misses the oil and enters the water column directly, particularly if dispersant droplets are large and slick thickness and distribution varies (Chapters 2 and 5). Dispersant may also partition from the oil droplets into the water. This effect has been demonstrated by laboratory toxicity experiments with Corexit 9527 and mineral oil (Wells et al., 1982), and may be a common phenomenon (Fingas, private communication).

The above dispersant concentrations that were measured are:

- range of less than 0.2 to 1.0 ppm (McAuliffe et al., 1975);
- 1 to 10 ppm for a short period after application (Canevari and Lindblom, 1976); and
- up to 13 ppm at various depths and times (Bocard et al., 1984).

Concentrations in the water for uniform mixing to various depths and rates of application have been calculated:

- 1 to 12 ppm for shallow inshore areas (Griffiths et al., 1981);

- 1 ppm for 1,000 sec in top 3 m (Mackay and Wells, 1983); and
- 0.5 to 2.0 ppm in the top 2 m, after an application of 10 to 40 liters/ha (Mackay, private communication).

McAuliffe et al. (1981) calculated total dispersant exposure under the best dispersed oil slick to be 3.2 ppm-hr in southern California field tests (see Chapter 4). This calculation is based on an integration of the dispersed oil concentrations measured over 12 hr. In addition, it was assumed that the dispersant, which was sprayed at a concentration of 5 percent that of oil, was distributed in the water, as were the dispersed oil droplets. The above numbers are dependent on application rates, depths, sampling, and chemical analysis methods. The evidence is limited, but at presently recommended application rates with effective dispersants, concentrations as high as 10 ppm may be expected initially. In the open ocean, small-scale field tests (see Chapter 4) have indicated that the concentration of dispersant in water falls to less than 1 ppm within hours. Hence, even initial concentrations in the water column are below most, but not all, estimated lethal and sublethal concentrations (Figure 3-2) derived from "constant" exposure experiments. These experiments were conducted for a much longer period (24 to 96 hr). Therefore, the effects on the field were expected to be much less.

Some effects of dispersants on organisms in the sea may occur, considering the variety of organisms and biological processes in the upper-water column, their frequent concentration at oceanic boundaries (surface microlayers and convergence zones caused by surface currents, winds, and Langmuir circulation) and nutrient locations, and the demonstrated sensitivity of some single organisms (particularly reproductive and larval stages). However, such effects would probably be minor and short-lived due to dilution of materials and recruitment of organisms from unaffected areas.

In conclusion, major effects (other than on insulation capability of fur and feathers) should not occur in the near-surface waters due to dispersant alone, provided properly screened dispersants are used at recommended application rates.

TOXICITY OF DISPERSED OIL

This section addresses exposure assessment, comparative toxicity, and joint toxicity. Laboratory work is reviewed later in this chapter; field studies, which assess both toxicological and ecological effects of dispersed oil, are described in Chapter 4. Case histories of spills in which dispersants were used are reviewed in Appendix B.

The following discussions of marine ecotoxicology of chemically dispersed oils are based on considerable literature on dispersants, oils, and dispersed oils (e.g., Allen, 1984; Doe et al., 1978; Nelson-Smith, 1972, 1980, 1985; NRC, 1985; Sprague et al., 1982).

Exposure Assessment

Exposure assessment involves estimating the concentrations of toxicants to which the organisms will be exposed and the time of exposure. This assessment is the first step in the process required to estimate potential damage to marine organisms. Once the exposure to toxic materials is known, it can be combined with laboratory measures of toxicity to obtain a hazard assessment. Exposure assessment must take into account the several factors affecting oil concentration. Untreated oil produces a certain level of exposure to surface or near-surface organisms; treatment with chemical dispersants modifies this exposure, moving the oil from the surface slick into the water column as droplets with a significant lifetime. Chemically dispersed oil thus reaches a greater volume in which organisms can be affected, but at the same time it is being diluted so that those effects will be mitigated. Measured concentrations of oil in water reported at test oil spills (Chapter 4) have frequently been regarded as representative (McAuliffe et al., 1980, 1981; Nichols and Parker, 1985). When dispersants are used in confined areas with poor circulation, concentrations of dispersed oil in the water column will be higher than those found under open-water, experimental spills. In Chapter 4 recent field experiments designed to stimulate the latter situation will be discussed.

Factors Affecting Comparative Toxicity

Some of the physicochemical and biological factors influencing toxicity of dispersed oils and the magnitude of their effects are well known (MacKay and Wells, 1981; NRC, 1985; Sprague et al., 1982). Key biological factors to consider with chemically dispersed oil include phylum, life stage, physiological condition, and habitat. An ideal theoretical structure for understanding the influence of these factors would allow for

- extrapolation from one species or region to another;

- evaluation of joint toxicity of dispersant and oil in dispersions prepared under different circumstances; and
- evaluation of the linkages between dispersant effectiveness and toxicity of dispersant and dispersed oil.

The current limitations to fundamental understanding of such broad relationships, due to the limited data base, somewhat restrict the application of the laboratory-derived and field toxicity data bases and their use in predicting possible sensitivities of species exposed to dispersed oils under various field conditions.

Joint Toxicity

Joint toxicity, also referred to as joint action or mixture toxicity, occurs where two or more chemicals are exerting their effects simultaneously. Although the terminology is not standardized and is often used ambiguously, one model of joint toxicity describes the effects of mixtures of chemicals as additive, more than additive (synergistic), or less than additive (antagonistic) (Calamari and Alabaster, 1980; Marking, 1985; Rand and Petrocelli, 1985; Sprague, 1970). The combined effect of dispersant and oil could be a simple combination of effects each causes separately; but synergistic toxicity, which is greater than the sum of the two separate exposures, is a possibility that must be seriously considered.

As discussed earlier, joint toxicity cannot be assessed by a straightforward comparison of oil toxicity with dispersed oil toxicity. In most experiments using oil alone, the oil remains primarily at the surface of the experimental tank, and only a small fraction— the water-soluble fraction (WSF), which is sometimes called "water-accommodated," a term including fine droplets as well as truly dissolved components—dissolves or is dispersed in the water. When dispersant is added, the limited volume of the experimental system experiences a much higher concentration of dissolved plus dispersed oil components. Even if the WSF is compared with the dispersed oil, this situation exists. Thus, based on such exposures, many early experiments concluded that dispersed oil was much more toxic than oil alone.

In contrast, laboratory experiments comparing toxicity of the WSF of oil with the WSF of dispersed oil, generally have found the toxicity of the two indistinguishable.

This experience in accounting for the WSF suggests that meaningful evalution of joint toxicity must involve adequate chemical analysis of the oil in the water phase to allow comparisons of hydrocarbon toxicity on the same hydrocarbon scale, whether dispersants are present or not. This observation also reinforces the need for thorough, comprehensive experimental design in any experiments examining the toxicity of hydrocarbon versus hydrocarbon plus dispersant (Wells et al., 1984b).

Questions also arise concerning the most appropriate experimental procedures for studying dispersed oils, such as constant exposures versus declining concentrations. Simple, measurable toxicity parameters in individual organisms are needed to predict effects on population success (i.e., survival of individuals, reproduction and development, and recruitment).

Dispersant composition influences toxicity directly and also indirectly because a more effective dispersant mobilizes more oil into the water column (Mackay and Mascarenhas, 1979; Mackay and Wells, 1983; Nes and Norland, 1983; Wells, 1984; Wells et al., 1984a). This relationship must be considered in an accurate assessment of joint toxicity.

LABORATORY STUDIES WITH DISPERSED OIL

Since the *Torrey Canyon* spill in 1967, many studies of the effects of dispersed oils on marine organisms under laboratory conditions have measured toxicities and relative toxicities of various oils, oil with dispersants, and dispersants themselves. This section reviews studies employing dispersed oil or oil-dispersant mixtures on a wide range of organisms from phytoplankton to fish.

To evaluate hazards to marine organisms caused by dispersant use, the most important toxicity information needed is the comparison of chemically dispersed oil (particularly the effects of dispersant at concentrations normally used) with undispersed or physically dispersed oil, under conditions approximating those in the field. As discussed previously, the most appropriate laboratory measurement is the toxicity of the water-soluble fraction of oil or dispersed oil.

Unfortunately, about two-thirds of the literature published prior to 1987 does not give values for oil concentration in the water phase, but instead uses the total oil per unit volume, or nominal concentration. Approximately one-third of the many tests measured the dissolved hydrocarbons that cause immediate biological toxicity. As noted earlier, in systems where oil or dispersant forms a separate (floating layer) phase, basing toxicity on nominal concentrations leads to unrealistically high LC_{50} or EC_{50} values (i.e, underestimates of toxicity). A major portion of the toxic fraction remains in the floating layer and does not reach the test organism, resulting in erroneous estimates of exposure concentrations and toxicities.

If physically dispersed oil (where most of the oil resides in the surface layer) is compared with chemically dispersed oil (where much of the oil is accommodated in the water) using nominal concentrations, the chemically dispersed oil appears to have a higher toxicity. For example, based on studies using nominal concentrations, it has been hypothesized that under natural field conditions, toxicity of oil-dispersant mixtures to organisms would be greater than that of untreated oil (Nelson-Smith, 1972; Swedmark et al., 1973). As will be shown in the following sections, tests in which water-soluble fractions and water-accommodated fractions are measured and used as a basis for toxicity generally show no difference between physically dispersed and chemically dispersed oil.

As stated above, approximately one-third of the studies, varying with organism grouping, reviewed for this report measured the water-soluble fraction. Many studies used much higher concentrations of oil and dispersant than would likely be found under field conditions, except in highly enclosed bodies of water where the volume of oil may be large relative to the receiving waters. Some studies compared the toxicity of dispersed oil to dispersant alone, but not to oil alone. It is, therefore, difficult to estimate threshold concentrations accurately from these studies, and hence address the joint toxicity of dispersed oil.

Swedmark et al. (1973) and Doe and Wells (1978) proposed that the primary difference between untreated oil and dispersed oil under laboratory conditions was "that effective dispersants simply make more oil or its many components available to aquatic organisms," rather than causing greater-than-additive effects. Norton et al. (1978) likewise correlated the "coarsening [formation of larger particles] of the dispersion to a reduction in toxicity," that is, less oil was then available to the organisms.

Bobra et al. (1979) recognized the importance of separating the contributions of dissolved hydrocarbons, dispersed oil particles, and dispersants in the laboratory to identify which effect dominates toxicity based on the fate of materials under different aquatic conditions. This information was then incorporated into a hazard assessment of dispersed oil. Bobra et al. (1979, 1984) and Mackay and Wells (1983) have attempted to model these contributions of dispersant, dissolved oil, and particulate oil to acute toxicity at different oil-water ratios and volumes and at different weathering states, using crustacea as model organisms and standardized physicochemical and toxicity data for the remaining input. For example, when a fresh Norman Wells crude oil is dispersed with Corexit 9527, 85 percent of the toxicity (to *Daphnia magna*) was attributed to the dissolved hydrocarbon fraction, 14 percent to the suspended particles, and less than 1 percent to the dispersant. A similar system, but with the oil weathered by 42 percent, gave 87 percent of the toxicity in particles, 10 percent in the dissolved fraction, and 2 percent in the dispersant (Bobra and Mackay, 1984; Bobra et al., 1984). This kind of model is useful for assessing the source(s) of toxicity in oil dispersions to different organisms over the crucial time periods after a major spill.

It has been hypothesized that acute toxicity of chemically dispersed oils falls between that of the whole oil and its water-soluble fraction (Figure 3-3; Wells, 1985; Mackay and Wells, unpublished data). The exact location of the toxicity curve for each oil dispersion is determined uniquely by each laboratory or field spill situation, as well as the chemical components.

Mackay and Wells (1980, 1981, 1983) and Wells et al. (1984a,b) described the many factors known to influence the toxicity of dispersed oils (compare with discussion entitled "Factors Influencing Acute Toxicity," earlier in this chapter). These include

- the stability of the hydrocarbon mixture;
- the ratio of dissolved to particulate oil; and
- changes in concentration and composition of oil over exposure time.

Whether toxicity may be synergistically enhanced—that is, organisms will experience greater effects from hydrocarbons and dispersant surfactants together than would be predicted from experiments with either alone—is a more difficult hypothesis to test because it involves eliminating the large effects of individual components. Only a few investigators have seriously attempted to investigate this problem. One attempt was a detailed study of fish responses to dispersant, oil, and dispersed oil that included hydrocarbon measurements in the water (Wells and Harris, 1980). They concluded that the interaction of dispersant and oil was additive, and that "with an effective but low-toxicity dispersant, the acute toxicity of the chemically dispersed oil reflected primarily the toxicity of the oil-derived hydrocarbons." This study and others discussed below (e.g., Peakall et al., 1987)

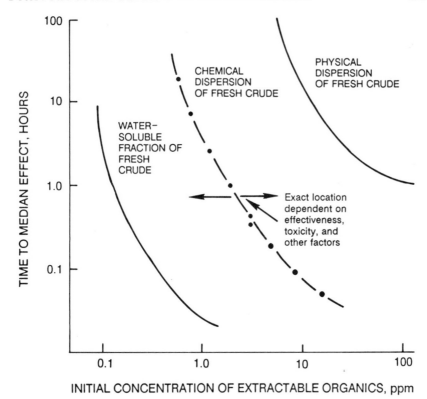

FIGURE 3-3 Hypothetical relationship of the toxicity curves for water-soluble fractions (WSF), chemical dispersions, and physical dispersions of fresh crude oils. Concentrations are of extractable organics measured by gas chromatography or fluorescence. Numbers on axes represent approximate values for marine planktonic crustaceans. Curves would be three-dimensional in practice. This figure formed part of the overall hypothesis tested in the study on the relationship between the effectiveness of dispersants and the toxicity of dispersants and dispersed oils (see Mackay and Wells, 1983). Source: Wells, 1985.

show that the apparently greater toxicity of chemically dispersed oil is generally a reflection of exposure, not a reflection of a greater inherent toxicity.

The following discussion is devoted to a detailed survey of the current literature covering phytoplankton, marine plants, zooplankton, crustaceans, and other marine organisms. This literature has not been recently reviewed (the last reviews were by Nelson-Smith, 1973; Sprague et al., 1982; and Pastorak et al., 1985). More important, a substantial portion of the studies have misinterpreted the toxicity of the dispersed oil because they used nominal concentrations rather than measured ones. Even though their numerical values for threshold concentrations are incorrect, their observations on type, duration, and recovery of responses are useful in understanding the toxicity of dispersed oil and have been noted, where relevant, in the following discussion.

Phytoplankton

Laboratory studies of the toxicity of dispersed oil to phytoplankton are summarized in Table 3-11. Four of the seven studies (Chan and Chiu, 1985; Hsaio et al., 1978; Lacaze and Villedon de Naide, 1976; Villedon de Naide, 1979) report that dispersed oil is more toxic than undispersed oil. However, these studies were incorrectly based on the use of nominal oil concentrations and are not considered further.

One investigator (Trudel, 1978) analyzed oil concentrations in water by infrared spectroscopy, and measured response by carbon fixation. His dose-response relationship was the same in oil and 1:1 oil-dispersant mixtures, and no change occurred in toxicity of the dispersed oil with the dispersant present.

Another study (Fabregas et al., 1984) measured the water soluble fraction of weathered crude from the wreck of the tanker *Urquiola,* and concluded that the toxicity of the dispersed oil was the same as that of the dispersant (Seaklin 101-NT).

There were other observations as well. An increase in light intensity increased the toxicity of dispersed Kuwait crude oil to phytoplankton. In the presence of Corexit 8666, toxicity increased by a factor of 5 in darkness and increased by a factor of 9 in light (Lacaze and Villedon de Naide, 1976). The weathered oil mixture, when illuminated and mixed with dispersant (1:1), was the most toxic. In another study, growth of the marine diatom *Skeletonema costatum,* under the influence of dispersed oil, was the same as for oil alone, but greater than that for dispersant alone (Tokuda, 1979). Both studies demonstrated similar toxicities of oil and chemically dispersed oil to phytoplankton.

Macroscopic Algae and Vascular Plants

As with phytoplankton, three of the five papers reviewed on macroscopic plants (Table 3-11) employ nominal concentrations and conclude, without convincing evidence, that dispersant-oil mixtures are more toxic than oil alone (Ganning and Billing, 1974; Hsiao et al., 1978; Thelin, 1981).

The other two studies (the last two entries in Table 3-11) employed gas chromatographic analysis of the water-soluble fraction; they consider seagrasses, which are discussed in Chapter 4.

Zooplankton

This section covers all groupings that have been studied except crustacea, which are discussed later.

Protozoa

Little work has been conducted with protozoa and dispersed oils. Goldacre (1968) was the first to describe the narcotic effects of hydrocarbons and some nonionic dispersants on the cell membrane of amoebae, but no oil-dispersant mixtures were evaluated.

Rogerson and Berger (1981) determined the toxicity of oil-dispersant mixtures to ciliates, *Tetrahymena pyriformis* and *Colpidium campylum*, on the basis of growth rate. Corexit 9527 concentrations above 1 ppm (nominal) were acutely toxic. The protozoa grew better in dispersed oil tests than in oil alone. This was attributed to the more rapid volatilization of the more toxic, aromatic fraction of the oil from the dispersed oil mixtures. After oil had weathered, the dispersant was apparently the primary toxicant.

Polychaetes

Polychaetes are known to be tolerant of oil and are often the first species to colonize the benthic community after an oil spill (NRC, 1985). The acute toxicity (1-day LC_{50}s) of Corexit 7664 to Spionid larvae was 889 ppm for the dispersant and 222 ppm for an Iraq crude oil-dispersant mixture (Latiff, 1969). Likewise, 48-hr LC_{50}s for Corexit 7664 with the polychaete *Ophryotrocha* were extremely high: 35,000 ppm for males, 30,000 ppm for females, and 12,000 ppm for larvae (Åkesson, 1975). With oil (not described) in a ratio 1:2, the toxicities became 580, 420, and 420 ppm, respectively. Even though the data from these experiments were analyzed using nominal concentrations, Åkesson concluded correctly that the oil-dispersant mixture was more toxic than the dispersant alone.

Mollusca

The only reported study on molluskan plankton deals with the gametes, embryos, and larvae of two oysters (*Crassostrea angulata*

TABLE 3-11 Laboratory Studies of Toxicity of Dispersed Oils to Plants

Organisms	Method of Analysis		Oil Type	Dispersant Type	Response
	Nominal	Measured			
PHYTOPLANKTON	x		Kuwait crude	Corexit 8666	Lethality (?)
Arctic species	x		Four crudes	Corexit 8666 (?)	Primary production
Four species	x		Kuwait crude	Nine dispersants	Photosynthesis
Chlorella salina	(?)	(?)	BP light Diesel	BP1100 X BL1100WD, Shell oil herder	Growth, chlorophyll a, photosynthesis, respiration
Natural population, Conception Bay, Newfoundland		IR	Lago Medio crude	Corexit 9527	Carbon fixation
Tetraselmis suecica		WSF	Weathered Urquiola crude	Seaklin 101-NT	Growth, chlorophyll a
MARINE DIATOM Skeletonema costatum			Oil products	Several	Growth
MARINE PLANTS Fucus vesiculosus (bladder wrack)	x		(?)	Corexit 7664, BP1100X	Dissolved oxygen uptake
Laminaria saccharina; Phyllophora truncata	x		Venezuelan and three other crudes	Corexit 8666	In situ primary production
Fucus serratus	x		Ekofisk, Statfjord crudes	Corexit 9527	Effects on zygotes and seedlings
Thalassia testudinum (seagrass)		GC	Prudhoe Bay crude	Corexit 9527	Lethality leaf coloration
Thalassia, Halodule, Syringodium		GC	Murban and Louisiana crudes	Corexit 9527	Mortality blade condition

NOTE: Uncertain information is noted by (?).

[a] O + D > O--oil and dispersant toxicity is greater than for oil alone.
[b] O + D < O--oil and dispersant toxicity is less than for oil alone.
[c] Water-soluble fraction.

Toxicity Comparison		Comments	References
O+D>O[a]	O+D<O[b]		
x		Illumination, enhanced photooxidation	Lacaze and Villedon de Naide, 1976
x			Hsaio et al., 1978
x			Villedon de Naide, 1979
O+D>O, high concentration	O+D<O, low concentration		Chan and Chiu, 1985
O+D=O			Trudel, 1978
		O+D=D (growth)	Fabregas et al., 1984
O+D=O		O+D>D	Tokuda, 1979
x		Threshold concentration>100 ppm	Ganning and Billing, 1974
x			Hsiao et al., 1978
x		D+Ekofisk> D+Statfjord	Thelin, 1981
	x	WSF[c] compared with dispersed oil D+O=D	Baca and Getter, 1984
x		WSF of dispersed oil had concentration 10 times oil alone	Thorhaug et al., 1986; Thorhaug and Marcus, 1985

and *C. gigas*) and the mussel *Mytilus galloprovincialis* (Renzoni, 1973). Only very high nominal concentrations of hydrocarbons in water, with and without dispersants, were toxic, producing responses with the eggs and embryos. Oil-dispersant mixtures (1 to 1,000 ppm) were toxic to fertilization. The toxicity of oil and oil-dispersant mixtures at high concentrations were similar, although the analytical method using nominal concentrations is incorrect.

Echinoderms

A number of studies have been conducted exposing sea urchin eggs, embryos, and larvae to oil-dispersant mixtures (Falk-Petersen, 1979; Falk-Petersen and Lönning, 1984; Lönning and Hagström, 1975, 1976), but their conclusions have been brought into question because toxicities were based on nominal concentrations and the water-soluble fraction of the oil was not analyzed. Experiments were conducted with Kuwait and Ekofisk crudes, a range of dispersants, especially Corexit 9527, and a number of sensitive sublethal embryological responses.

Ichthyoplankton

The high commercial value of fish, combined with the vulnerability of early life stages to oil, makes the toxicology with ichthyoplankton particularly important. Unfortunately, only one paper compares the toxicity of the water-soluble fraction of physically dispersed oil and chemically dispersed oil. That study found chemically dispersed oil less toxic than oil alone (Borseth et al., 1986). In other studies, nominal concentrations were used and, as expected, chemically dispersed oil was reported to be more toxic. Six of these papers are compared in Table 3-12.

Iran crude oil (1,000 ppm nominal concentration) and Corexit 7664 (100 ppm) produced narcosis or lethality in 1-day-old herring larvae (Kuhnhold, 1972). After 2 days in static laboratory conditions, the physically dispersed oil had lost its toxicity, but the chemically dispersed oil had retained or increased its toxicity.

Another early study considered the effects of Russian crude oil and dispersants on the eggs and larvae of northern pike, *Esox lucius* (Hakkila and Niemi, 1973). The study's main finding was a description of the comparative sensitivities of the life stages to oil dispersions. More than 60 ppm oil plus dispersant caused some egg mortality and 300 ppm caused complete mortality; this is high compared to other results. Larval tests gave 2-day LC_{50}s of 66 ppm (stage II) and 4.4 ppm (stage III). The toxicity of the oil-dispersant mixture was greater than the oil, but the same as the dispersant alone; and was attributed to the dispersant since it was a nonionic surfactant in an aromatic solvent.

Linden (1975, 1976) studied the effects of Venezuelan crude plus dispersant (BP1100X, Finasol SC, and Finasol OSR-2) on the ontogenetic development (i.e., embryonic movement and heart rates, morphology, and length of larvae) of the Baltic herring, *Clupea harengus*. Both acute and sublethal effects of the oil increased 2 to 3 orders of magnitude if the oils were dispersed by the least toxic dispersant (BP1100X), and 3 to 4 orders of magnitude with a highly toxic dispersant formulation, such as Finasol SC.

Two studies of oil-dispersant mixtures were conducted by Mori et al. (1983, 1984). They found that the tolerance of common sea bass eggs is greater than those of parrotfish and flounder (Mori et al., 1983), and that oil-dispersant mixtures were more toxic than oil alone for the young of sea bream and Japanese flounder and the larvae of stone flounder (Mori et al., 1984). The results of these experiments are questionable because the water-soluble fraction of the oil was not measured, and toxicities were based on nominal concentrations.

Borseth et al. (1986) conducted a comparison of in vivo and in situ exposure of plaice (*Pleuronectes platessa*) eggs to the water-soluble fraction of Statfjord A + B crude oil topped at 150°C and mixtures with Finasol OSR-5. To cause significant mortalities (98 percent) required the full strength WSF of the oil-dispersant mixture. In contrast, 10 percent mortality occurred with the maximum WSF concentration for the topped crude. At 1:1 dilutions of both WSFs, mortalities were equivalent.

In spite of these experimental studies, a dependable generalized hazard assessment with ichthyoplankton, and therefore to fisheries potential, cannot be made without considerably more research.

Crustaceans

More work with dispersed oil has been conducted on crustaceans than any other type of organism, due to their ecological and commercial importance and the relative ease with which many species can be studied. The sensitivity of many crustaceans, particularly in young and molting stages, to dissolved and physically dispersed

TABLE 3-12 Laboratory Studies of Toxicity of Dispersed Oils on Ichthyoplankton

Organisms	Method of Analysis		Oil Type	Dispersant Type
	Nominal	Measured		
Herring (Clupea harengus, larvae)	x		Iran crude	Corexit 7664
Northern pike (Esox lucius, eggs and larvae)	x		Russian crude	Several
Baltic herring (Clupea harengus, embryos and larvae)	x		Venezuela. crude	BP1100X, Finasol SC, Finasol OSR-2
Spot (Leiostomus xanthurus eggs)	x		Ixtoc I crude	Corexit 9527
Sea bass, parrotfish, Japanese flounder, stone flounder, sea bream (eggs)	x		(?)	NEOS AB 3000
Cod (Gadus morhua, eggs and larvae) Flounder (Platichthys fleus, larvae)	x		Ekofisk crude	9 including Corexit 9527
Plaice (Pleuronectes patessa, eggs)		Gravi-metric^d	Statfjord A+B, crude oil topped at 150°C	Finasol OSR-5

NOTE: Uncertain information is noted by (?).

^{a}O+D>O = toxicity of oil plus dispersant is greater than oil alone.
^{b}O+D<O = toxicity of oil plus dispersant is less than oil alone.
^{c}WSF = water-soluble fraction.
^{d}Presumed, based on abstract.

Biological Response	Toxicity Comparison		Comments	References
	O+D>O[a]	O+D<O[b]		
Lethality, narcosis	x		High concentration only 1-day larvae	Kuhnhold, 1972
Lethality	x		O+D=D	Hakkila and Niemi, 1973
Lethality, embryonic function, larval length	x		Dispersant toxicity contributing factor	Linden, 1975, 1976
Lethality	x		O+D>D	Slade, 1982
Lethality	x			Mori et al., 1983, 1984
Lethality	x		O+D>D	Falk-Petersen and Lönning, 1984
Lethality	x		Undiluted WSF[c] of O and O+D compared; O+D= O 1:1 dilutions of WSFs of O and O+D compared (100% WSF)	Borseth et al., 1986

hydrocarbons as well as to dispersant formulations has been well documented (NRC, 1985; Sprague et al., 1982). A number of studies are compared in Table 3-13, and additional details are given in the text.

Copepoda

Lipid metabolism, swimming behavior, respiration, and survival of the copepods *Acartia* sp. and *Cyclops* sp. were used as indicators of toxicity in the early work of Gyllenberg and Lundquist (1976). A number of potential sublethal effects were identified with the dispersal of oil by Finasol OSR-2 and Finasol SC, but comparisons were not reliable because only nominal concentrations were used.

Venezia and Fossato (1977) studied the acute and chronic effects of suspensions of Kuwait crude oil and Corexit 7664 on Harpacticoid copepods, *Tisbe bulbisetosa*. They prepared aqueous phases of dispersant and oil (1:5), which they analyzed by gas chromatography and fluorometry. Noting that the hydrocarbon concentration in the aqueous phase depends mainly on the amount of dispersant present, and less on the quantity of oil added, they rejected the hypothesis of a synergistic effect of hydrocarbons with this dispersant, assigning the effects to the higher oil concentrations in the oil-dispersant suspension. Even though the acute toxicity of the separated aqueous (1:5) phase of the dispersant and oil mixture was 37 ppm (9-day LC_{50}) and 9 ppm (20-day LC_{50}), exposure of the developing eggs and nauplii of the copepods to 39 to 45 ppm (or lower) of this mixture showed no significant effects on numbers of eggs or nauplii, or percent hatching.

Spooner and Corkett (1974) studied the effects of Kuwait 250°C residue oil combined with BP1100X on the feeding rate of *Calanus helgolandicus* females. Toxicities were similar between oil and dispersed oil. Exposure to 10 ppm total oil (analyzed spectroscopically) plus 2 ppm dispersant for 20 hr reduced the number of fecal pellets produced, whereas 2 ppm oil and 0.4 ppm dispersant had no effect. Feeding rates recovered in all treatments. Spooner and Corkett (1979) found clear effects of oil on four species of copepods at 10 ppm.

Wells et al. (1982) studied the mortality of Calanoid copepods (predominantly *Pseudocalanus minutus* with some *Acartia hudsonica*) in a white, low-toxicity, paraffinic mineral oil (Marcol 70) with Corexit 9527. Temporary loss of the toxic components of the dispersant into the oil was shown by mixing the oil and dispersant together at different speeds and testing their aqueous dispersions. When mixing stopped, the dispersant components, which had been dissolved in the oil or associated with the interface, eventually partitioned back into the water. The experiment demonstrated that dispersant could stay in the oil phase and not the water phase with mixing as done in the laboratory. Therefore, an incorrect conclusion may have been made about dispersant concentrations in oil-water-dispersant mixture dilutions in many bioassays, unless the experiments were conducted during continuous mixing of all materials.

Foy (1982) demonstrated that 96-hr LC_{50}s for *Calanus hyperboreus*, when based on hydrocarbon concentrations measured by fluorescence spectroscopy, were lower in Prudhoe Bay crude oil-water mixtures than in oil-Corexit 9527 mixtures, that is, 73 (51 to 103) ppm versus 196 (161 to 238) ppm, respectively. There was no evidence that oil dispersal with the dispersant led to a dispersion more toxic to copepods. The copepods were quite resistant to the Prudhoe Bay crude but suffered fouling with oil that may reduce survival in similar exposures under natural conditions.

Falk-Petersen et al. (1983) studied the effects of oil and dispersants on plankton. And in 1984, Falk-Petersen and Lönning (1984) summarized their many studies of the effects of oils and dispersants on embryos and larvae of sea urchins and fish, and copepods. The main finding of this body of work is that the fertilization and embrionic development of organisms may be affected by exposures at quite low concentrations, but this requires confirmation with chemical analysis. Their studies also demonstrated the changes in toxicity thresholds with organism and life stages.

Decapoda

Franklin and Lloyd (1986) used the standard U.K. Ministry of Agriculture, Fisheries, and Food (MAFF) *Crangon crangon* (brown shrimp) dispersant "sea test" to examine the relationship between oil droplet size and the acute toxicity of oil-dispersant mixtures with the use of a Coulter Counter. Toxicity increased with decreasing droplet size. The authors speculated that with a distribution of very small droplets, soluble toxic components were more rapidly transferred into aqueous solution.

Anderson et al. (1980, 1981, 1984, 1987) and Anderson (1986) have reported on the toxicity of dispersed and undispersed crude

Table 3-13 Laboratory Studies of Toxicity of Dispersed Oils to Crustacea

Organisms	Method of Analysis Nominal Measured		Oil Type	Dispersant Type	Response
COPEPODA					
Tisbe bulbisetosa		GC, fluor.	Kuwait crude	Corexit 7664	Lethality, reproduction development
Calanus helgolandicus		Spect.	Kuwait 250 degree residue	BP1100X	Feeding
C. hyperboreus		Fluor.	Prudhoe Bay crude	Corexit 9527	Lethality
Pseudocalanus minutus, Acartia sp.	x		Marcol 70 refined mineral oil	Corexit 9527	Lethality
Acartia sp., Cyclops sp.	x		Venezuelan crude	Finasol SC, Finasol OSR-2	Survival, lipid metabolism, respiration, swimming behavior
Field-caught sp., Metridia sp., Calanus finmarchicus	x		Ekofisk crude	Corexit 9527, Finasol OSR-5, Finasol OSR-7	Lethality
DECAPODA					
Crangon crangon (shrimp)	x		Iraq crude	Corexit 7664	Lethality, behavior
C. crangon	x		25 oils	BP1100X, Corexit 9527, and others	Lethality
C. crangon				Noramium DASO	Reference toxicant
Pandalus danae (shrimp)		IR, GC	Prudhoe Bay crude	Corexit 9527	Lethality
Homarus americanus larvae		GC	South Louisiana crude	Corexit 9527	Lethality, respiration, other subleth
Paragrapsus quadridentatus	x	Gravi- metric	Kuwait light	BP/AB	Lethality
Palaemon serratus (shrimp)	x		BAL 150 crude	BP1100WD, BP1100X, Finasols, others	Enzyme respiration
Carcinus maenas (shore crab)	x		Forties crude	BP1100WD	Cardiac action, oxygen uptak perfusion ind

Toxicity Comparison		Comments	References
O+D>O[a]	O+D<O[b]		
	x	Only aqueous phase of oil and dispersed oil tested	Venezia and Fossato, 1977
	O+D=O		Spooner and Corkett, 1974, 1979
	O+D=O		Foy, 1982
x		Toxicity due to dispersant	Wells et al., 1982
x		Oil+OSR-2<OSR-2, Oil+SC>SC	Gyllenberg and Lundquist, 1976
x		O+D>O, Finasols>Corexit	Falk-Petersen et al., 1983
x		O+D>D	Latiff, 1969
x	x	Dispersant type modifies availability of oil fractions	Franklin and Lloyd, 1982, 1986
			Bardot and Castaing, 1987
	O+D=O		Anderson et al., 1980, 1981, 1984, 1987; Anderson, 1986
	O+D=O		Capuzzo and Lancaster, 1982
	O+D>O nominal; O+D<O analyzed		Ahsanullah et al., 1982
x			Papineau, 1983; Papineau and LeGal, 1983; Papineau and Cheze, 1984
	O+D=O	WSF preparation shaken, not stirred; synergistic effect unlikely	Depledge, 1984

TABLE 3-13 (Continued)

Organisms	Method of Analysis Nominal	Measured	Oil Type	Dispersant Type	Response
OTHER CRUSTACEA					
Artemia sp. (shrimp larvae)	x		Tunisian crude	Finasol OSR-2	Lethality, respiration rate
Artemia sp. (larvae)	x	Fluor., GC	Lago Medio crude	Corexit 9527	Lethality
Onisimus affinis (arctic amphipod)		Fluor.	Pembina, Norman Wells, Atkinson Pt.	Corexit 8666	Metabolism, respiration
Anonyx nugax, Boeckosimus edwardsi, Gammarus setosus (arctic amphipods)		Fluor.	Prudhoe Bay	Corexit 9527	Lethality
Gnorimosphaeroma oregonensis (estuarine isopod)		Fluor.	Prudhoe Bay	Corexit 9527	Lethality, physiology behavior
Mysidopsis bahia		IR, GC	Prudhoe Bay	Several	Lethality

KEY: Fluor. = fluorescence; GC = gas chromatography; IR = Infrared; Spect. = spectroscopy
[a] O+D>O = toxicity of oil plus dispersant is greater than oil alone.
[b] O+D<O = toxicity of oil plus dispersant is less than oil alone.
[c] WSF = water-soluble fraction.

oil to the decapod shrimp *Pandalus danae*. Toxicity indexes (ppm-hr) were measured for physically dispersed and chemically dispersed oil, under constant and diluting exposures. All four were similar. Toxicity indexes for Prudhoe Bay and light Arabian crude were also similar.

Anderson et al. (1987) pointed out that toxicity indexes correlated with the presence of mono- and di-aromatics in the chemically dispersed oil and in the water-soluble fraction of physically dispersed oil. Toxicity was summarized as the lower limits at which effects would be observed: 48 ppm-hr for total hydrocarbons, 3 ppm-hr for aromatics. If aromatic compounds were removed, neither the WSF nor the dispersed oil were toxic.

Papineau (1983), Papineau and LeGal (1983), and Papineau and

Toxicity Comparison O+D>O[a]	O+D<O[b]	Comments	References
x		O+D<D pre-exposure induces higher tolerance	Verriopoulos et al., 1986; Verriopoulos and Moraitou-Apostolopoulou, 1982, 1983
	O+D=O	Dispersant was contributing toxic factor	Mackay and Wells, 1980, 1981; Wells et al., 1982
x		Oil type major factor influencing reversed respiratory rates	Percy, 1977; Wells and Percy, 1985
	x	Aromatics in WSF of oil preparation were more concentrated than dispersed oil	Foy, 1982
Lethal O+D>O; sublethal: O+D>O		Pattern of sublethal respiration: O+D=O; recovery also occurred, D+O>D	Duval et al., 1980, 1982 Anderson et al., 1985

Cheze (1984) studied sublethal effects of dispersants, oil, and oil dispersions on the gills of the shrimp *Palaemon serratus*. Inhibition of the gill sodium-potassium-magnesium-ATPase enzyme system and a change in its kinetic properties were caused by exposure to low oil concentrations, and these changes were suggested as a way of estimating "safe" concentrations. Effects on gill structure included cellular damage, membrane deterioration, and interference with blood circulation. No conclusions can be drawn about the comparative effects of oil and oil dispersants since only nominal concentrations were reported.

Depledge (1984) measured the changes in cardiac activity, oxygen uptake, and perfusion indices in the shore crab *Carcinus maenas* after exposures to sublethal concentrations of Forties crude water-soluble fraction. A 20 percent dilution of this WSF, BP1100WD

(10 percent solution), and mixtures (1:1) were tested. All treatments caused increases in cardiac activity and oxygen consumption and disruption of feeding behavior, but effects were reversible. Conclusions on comparative effects of oil and dispersed oil can only be considered as tentative because of the use of nominal concentrations.

Ahsanullah et al. (1982) described the acute lethal responses of the crab *Paragrapsus quadridentatus* to Kuwait light crude, an Australian dispersant (BP/AB), and an oil-dispersant mixture. Based on oil concentrations measured by hexane extraction, 4-day LC_{50}s were 63 to 70 and 69 to 106 ppm for the oil and oil-dispersant mixture, indicating no significant difference in acute toxicity.

Capuzzo and Lancaster (1982) studied the physiological effects of physically dispersed and chemically dispersed Southern Louisiana crude oil and ingestion of an oil-contaminated food source on larvae of lobsters, *Homarus americanus*. No enhanced toxicity was observed due to the presence of dispersant, but the sublethal effects occurred at concentrations expected in the water column during the first hours or days of a major spill. Larvae were exposed to dispersions in a continuous-flow system at 20°C for 4 days. Physically dispersed oil had droplet sizes of 20 to 50 μm (0.25 ppm) and chemically dispersed oil (Corexit 9527) had sizes of 10 to 20 μm (0.025 ppm). Concentrations were unusually low for this type of study and designed to resemble the exposure expected in the water column during the first days after a major spill. Survival of larvae was the same in all treatments (i.e., oil, oil plus dispersant, and controls). Exposure reduced metabolism, as shown by lowered respiration rates and reduced ratios of oxygen consumed to ammonia and nitrogen excreted. This was interpreted to be a result of increased dependence on protein catabolism, inhibited lipid utilization, and delayed molting. Larvae did not accumulate oil droplets in their digestive tracts, but those exposed to oil and dispersed oil had reduced metabolism during the exposure. Lobsters exposed to oil-contaminated food also showed reduced respiration.

Other Crustaceans

Percy (1977) reported on the effects of sublethal exposures to dispersed oils on the respiratory metabolism of an arctic marine amphipod, *Onisimus (Boekisimus) affinis*. The relative magnitude of the change in respiration rate was less in oil-dispersant mixtures than in oil alone, but in both cases, reversal of respiratory depression was influenced by oil type. Low concentrations (13 to 21 ppm initial measured concentration) significantly depressed respiration rates, whereas increased hydrocarbon concentrations (268 to 800 ppm initial measured concentration) reversed depression, and respiration rates reached and even exceeded the controls (Wells and Percy, 1985).

Duval et al. (1980, 1982) studied the lethal and sublethal physiological and behavioral effects of physically and chemically (Corexit 9527) dispersed oil on the estuarine isopod *Gnorimosphaeroma oregonensis* under computerized flow-through exposure to the water-soluble fraction. The extent of responses was greater in oil-dispersant mixtures, even at lower total oil concentrations. The organisms recovered in clean water. Corexit itself had a low toxicity (no mortality in 4 days at 1,000 ppm). The 48-hr LC_{50} for oil dispersant was 32 (13-78 c.l.*) ppm, compared to 70 (24-203 c.l.) ppm for physically dispersed oil, suggesting slightly more toxicity in the chemically dispersed oil. The toxicity curves are linear but nonparallel, suggesting that the mode of toxic action might be different in the two dispersion types. The sublethal studies (Duval et al., 1982) showed that oil alone gave patterns of sublethal responses as *G. oregonensis* similar to oil plus dispersant: respiration rates, carbon assimilation rates and efficiencies, naphthalene uptake, behavior (coordinated motor ability), molting, and reproduction (frequency of mating).

Mackay and Mascarenhas (1979), Bobra et al. (1979), and Mackay and Wells (1983) discussed the need to develop a mathematical analytical framework for expressing individual and combined toxicities of dispersant constituents and dispersed oil. A formula was developed to examine data for *Daphnia* sp., *Artemia* sp., and copepods (Abernethy et al., 1986; Bobra et al., 1984). This work will result in a computerized joint toxicity model for the mixture that takes into account dispersant, dispersant components, water-soluble fraction of the oil, specific hydrocarbons, and oil droplets or particles (Bobra et al., 1987; Wells, 1985; Mackay, private communication).

Mackay and Wells (1980, 1981) reported the effects of oil dispersions, water-soluble fractions, Corexit 9527, and chemically dispersed oil on *Artemia* nauplii in static exposures at 20°C. In the chemical dispersion, most of the oil was in particle (droplet) form, the stability of the dispersion was greater, and the water-soluble oil fraction

---

*Range of 95 percent confidence limits.

was more concentrated (shown by gas chromatography). At the toxic threshold for the dispersion, the dispersant itself was at its lethal threshold (2-day LC_{50}s were 24 to 45 ppm; Wells and Mackay, unpublished data). Oil physically dispersed, settled, and diluted produced few effects in *Artemia* at 100 percent dispersion: 2-day LC_{50}s were greater than 41 to 65 ppm, based on concentrations of water-accommodated oil (by fluorescence and gas chromatography) at time zero hour. Chemically dispersed oil, prepared in a similar manner (18 to 24 percent dispersion), was similarly toxic: 2-day LC_{50}s were 82 to 120 ppm oil (plus 37 to 55 ppm dispersant, since the dispersant-oil ratio was 1:2.5).

In another experiment with dispersed fresh and weathered Lago Medio crude and *Artemia* (Mackay and Wells, 1980, 1981), 50 percent mortalities were reached within 3 days in dispersions of 2.9 to 7.6 ppm (measured by fluorescence) or 1.0 to 2.1 ppm (gas chromatography). There were few differences between fresh and weathered dispersions. Acute toxicities decreased as the dispersant-oil ratio decreased from 1:10 to 1:50, confirming that the dispersant could contribute to acute toxicity.

Mollusks

Pelecypod and gastropod mollusks, with many species in the littoral and shallow sublittoral zones, are particularly susceptible to oiling. At least nine laboratory studies (Table 3-14) show the ranges of species, oils, and dispersants and responses evaluated in studies to determine if dispersants change oil or oil component toxicity. With bivalves, studies with measured concentrations showed equivalent toxicities or lowered toxicities between the dispersed oil and the oil (or its WSF) alone. Studies with nominal concentrations showed greater toxicities for the mixtures, but such conclusions are suspect because exposures were unknown for all compared treatments. Physiological, behavioral, and recovery experiments predominated with the bivalve research and illustrate differential sensitivities and often the capacity to recover from exposures.

For gastropods (Table 3-14), all studies used nominal concentrations, thus invalidating conclusions about the comparative effects of dispersed oils to oil alone, but showed that respiratory and behavioral responses were quite sensitive to oil or dispersant exposures.

In summary, laboratory research on mollusks shows that high concentrations of dispersed oils can be toxic, but that little defensible evidence (with measured concentrations) exists to suggest higher toxicities of oils and their components in the presence of dispersants.

Comparison of Laboratory Studies and Field Studies With Measured Hydrocarbons

This section reviews and compares laboratory bioassays that measured dissolved hydrocarbons (C_1 to C_{10}) in the water-soluble fraction from untreated and chemically dispersed crude oils with those measured in the field, and it also compares the C_1 to C_{10} hydrocarbon fraction bioassays or behavioral studies of untreated crude oil with dispersed oil bioassays using the same organisms.

Oil toxicity to organisms is thought to result principally from hydrocarbons that dissolve into water from crude oils or refined products (NRC, 1985). A large number of laboratory bioassays have been conducted with the WSF obtained by equilibrating oil and water, but many have not measured these dissolved hydrocarbons. Table 3-15 summarizes exposures to untreated oil and chemically treated oil of several fish and a crustacean, expressed as ppm-hr, and lists the salient laboratory studies that measured dissolved or added hydrocarbon components to determine their effect on fish. This table shows the 96-hr LC_{50} values and the exposures (in ppm-hr) required to cause 50 percent mortality among several Alaskan species. Most of the species survived in the maximum possible concentration of dissolved hydrocarbons from Cook Inlet crude oil; a condition that never occurs in marine spills.

The largest number of studies with various species that measured aromatic or total dissolved hydrocarbons in exposure waters was reviewed by McAuliffe (1986, 1987a). The most extensive studies were conducted by Rice et al. (1979, 1981), using subarctic species in southern Alaska.

Rice et al. (1981) also determined the 96-hr LC_{50} for the larvae of five marine species to total aromatic hydrocarbons in the ballast water treatment effluent at Port Valdez, Alaska. The relative proportions of the dissolved aromatic hydrocarbons in the effluent discharge were very similar to those measured for water equilibrated with Cook Inlet and Prudhoe Bay crude oils. As for the studies reported in Table 3-15, saturate hydrocarbons present in the ballast water effluent were not measured.

McAuliffe et al. (1981) compared exposures with those found in the best chemically dispersed oil plumes of the 1979 California

TABLE 3-14 Laboratory Studies of Toxicity of Dispersed Oils to Mollusks

Organisms	Method of Analysis		Oil Type	Dispersant Type	Response
	Nominal	Measured			
BIVALVES					
Brachidontes variablis, Donax trunculus	x		Arabian crude	Corexit 7664	Mortality, respiratory rates
Mya truncata, Serripes groenlandicus		x	Venezuelan Lago Medio crude	Corexit 9527 (O:D=10:1)	Metabolism scope-for-growth (enzymes), hydrocarbon uptake
Serripes groenlandicus		x	Venezuelan Lago Medio crude	Corexit 9527 (O:D=10:1)	Behavioral effects (wide range of responses)
Arctic clams (Mya truncata, Astarte borealis)		x (UV, fluor.)	Venezuelan Lago Medio crude	Corexit 9527	Scope-for-growth (physiology)
Bay scallop (Argopecten irradians)	x		Kuwait crude	Corexit 9527	Predator-prey relationships, e.g., predator discrimination
Mussel (Mytilus galloprovincialis)	x		Hydrocarbons (toluene, n-hexane)	Three (Atlantic-Pacific, Corexit 9527, Corexit 7664)	Lethality
GASTROPODS					
Patella vulgata	x		Wide range of crudes and refined oils	Corexit 9527, others	Lethality
Patella vulgata	x		North Sea crude (WSF)	BP1100X, BP1100WD	Histopathology of gill epithelium
Littorina littorea	x		Bunker C	Corexit 8666	Behavior (crawling); respiration (oxygen uptake)

KEY: Fluor. = fluorescence; HC = hydrocarbon; MAFF = Ministry of Agriculture, Fisheries, and Food; SFG = scope-for-growth; UV = ultraviolet; WSF = water-soluble fraction.

[a] O+D>O = toxicity of oil plus dispersant is greater than oil alone.
[b] O+D<O = toxicity of oil plus dispersant is less than oil alone.

Toxicity Comparison			
O+D>O[a]	O+D<O[b]	Comments	References
O+D>O		Respiration rates declined in all treatments	Avolizi and Nuwayhid, 1974
O+D=O		Sensitivity varied with species	
O+D=O		Flow-through exposure, followed by recovery period; concentrations: 0.5-500 ppm-18 hr; enzyme function altered, SFG decreased, HC accumulated, then lost	Englehardt et al., 1985
O+D=O		Dose-response relationship clear; many responses reversible	Englehardt et al., 1985
	x	Comparison was WSF vs. O+D mixture (10:1); reduced SFG at 0.4-2.1 mg/liter; Mya did not recover over 14 days	Hutcheson and Harris, 1982
x		Dispersant and mixture had similar lethal toxicity curves; dispersant and mixture affected scallop discrimination similarly; susceptibility lower in winter	Ordzie and Garofalo, 1981
x		Very-high-exposure concentrations (100 mg/liter to 10%)	Ozelsel, 1983
x		Test was MAFF beach toxicity test	Franklin and Lloyd, 1982
		Comparison of O vs. O+D was not made; both WSFs and dispersants caused damage to surface microvilli, cilia, and epithelial cell structure	Nuwayhid et al., 1980
x		Dispersant was considered responsible for decreased behavior and respiration rates	Hargrave and Newcombe, 1973

TABLE 3-15 Comparison of Measured Dissolved Hydrocarbons in a California Sea Trial With Those Observed in a Number of Laboratory Mortality and Behavioral Studies

Crude Oil or Hydrocarbon and Organism Used	Dissolved Hydrocarbon Exposures (ppm-hr)[a]		References
	Untreated Oil	Chemically Dispersed Oil	
Prudhoe Bay Crude Oil			
1979 California field trials	0.002	0.060	McAuliffe, 1987a; McAuliffe et al., 1981
Salmon homing			
Chinook in fresh water	0.27 (1 hr)	1.50 (1 hr)	Brannon et al., 1986
Coho in salt water	0.07 (1 hr)	0.90 (1 hr)	Nakatani et al., 1985
Salmon bioassays			
Coho adults	>5 (4 hr)	12 (4 hr)	Nakatani et al., 1985
Chum fry	19	9	McAuliffe, 1987a
Herring eggs			
Hatching success			
Before fertilization	>19 (2 hr)		Pearson, 1985
During fertilization	>19 (2 hr)		Pearson, 1985
Percent hatching	>227		Pearson, 1985
Time to hatch	>227		Pearson, 1985
Increased abnormal larvae	>227		Pearson, 1985
Coonstripe shrimp			
Adults	15	7-20	Anderson et al., 1987
Larvae	4[b]	19	Anderson et al., 1987
Sand lance	67	68	Anderson et al., 1987
Pacific herring larvae	223	179	Anderson et al., 1987
Calanus hyperboreus	1,750	4,700	Foy, 1982
Lago Medio Crude Oil			
Artemia sp. nauplii	984-1,560	1,970-2,880	Mackay and Wells, 1980, 1981

Crude Oil Isopod (<u>Gnorimosphaeroma oregonensis</u>)	575–4,870	310–1,870	Duval et al., 1980, 1982
Port Valdez Ballast Water			
King crab larvae	38		Rice et al., 1981
Coonstripe shrimp larvae	32		Rice et al., 1981
Kelp shrimp larvae	14		Rice et al., 1981
Dungeness crab larvae	31		Rice et al., 1981
Pacific herring larvae	36		Rice et al., 1981
Pink salmon fry	35		Rice et al., 1981
Kelp shrimp adults	32		Rice et al., 1981
Cook Inlet Crude Oil			
Thirty-nine species of fish, crustaceans, echinoderms, mollusks, annelids, and nemerteans	16–>221		Rice et al., 1979
Juvenile pink salmon	29		Rice et al., 1979
Dungeness crab, first instar zoeae	60		Rice et al., 1979
Georges Bank (see text)	0.005 to 1.4 (24 hr)		Farrington and Boehm, 1987
Other Observations	<0.2 (24 hr)		Howarth, 1987

[a]Exposures are all 24 hours unless other times are shown in parentheses. Exposures in ppm-hr are products of the measured mean concentration (in ppm total hydrocarbons, by spectroscopy or gas chromatography) and time to 50 percent mortality (in hours). It is equivalent to the toxicity index of Anderson et al. (1984) for constant or nondiluting exposures and is based on the hypothesis that time of exposure is a toxicity variable equivalent in importance to concentration of exposure.

[b]Uncertain value, see text.

SOURCE: Modified from McAuliffe (1987a).

tests, and with exposures measured under untreated Prudhoe Bay crude oil slicks. The total of C_1 to C_{10} dissolved hydrocarbons under untreated slicks were reported to be 1 ppm or less (McAuliffe et al., 1981). In later publications McAuliffe (1986, 1987a) quoted an average integrated exposure of 0.002 ppm-hr, as listed in Table 3-15.

In other field observations, including those at oil spills, concentrations were as much as 100 times higher (Howarth, 1987; NRC, 1985).

On the other hand, the relatively unpolluted area of Georges Bank was sampled at 20 stations four times during 1977 (Farrington and Boehm, 1987). The concentration of dissolved hydrocarbons in the water column ranged as follows:

- February—14 to 60 μg/liter (ppb);
- May—1 to 25 μg/liter (ppb);
- August—less than 1 to 5 μg/liter (ppb); and
- November—less than 0.2 to 2 μg/liter (ppb).

The February values were high because of the *Argo Merchant* oil spill in December 1976 and January 1977. Over a 24-hr period, these data give exposures of 0.005 to 1.4 ppm-hr. If only the latter part of the year is considered, the range is 0.005 to 0.12.

Under the best dispersed oil plumes of the 1979 California tests (McAuliffe et al., 1981), the dissolved hydrocarbon concentration was taken to be 150 ppb for the first 30 min (although samples as high as 54 ppm were reported from 1 m depth 15 min after spraying, more typical samples were 1 ppm) and because the concentration decreased as a result of evaporation and dilution, the integrated exposure over 24 hr is listed in Table 3-15 as 0.06 ppm-hr. Nevissi et al. (1987) and McAuliffe (1987a) determined the exposures to dissolved hydrocarbons required to cause 50 percent mortality of chum salmon fry with untreated and chemically dispersed Prudhoe Bay crude oil. The total C_1 to C_{10} hydrocarbon content of the water was measured 8 times during the 24-hr exposures.

University of Washington researchers (Brannon et al., 1986; Nakatani et al., 1983, 1985) measured the effects of untreated and chemically dispersed Prudhoe Bay crude oil on the homing of adult chinook salmon (*Oncorhynchus tshawytscha*) in fresh water, adult coho salmon (*O. kisutch*) in seawater, and the amounts of these oils to cause mortality of adult coho salmon. Waters were monitored from 4 to 6 times during the 1- and 4-hr exposures, and the results expressed as ppm-hr dissolved hydrocarbons.

At Battelle Pacific Northwest Laboratories, Pearson (1985) studied the effect of Prudhoe Bay crude oil and chemically dispersed oil on Pacific herring egg fertilization, hatching, and larval abnormalities using a flow-through system. Anderson et al. (1987) of Battelle measured acute toxicity of the WSF and chemically dispersed Prudhoe Bay crude oil to coonstripe shrimp (*Pandalus danae*), Pacific herring (*Clupea harengus pallasi*), and sand lance (*Ammodytes hexapterus*). Constant concentrations of oil were selected using a flow-through system and determining the time (from 12 to 96 hr) to 50 percent mortality. Dissolved hydrocarbons were monitored. These bioassays showed little difference in toxicity between total C_1 to C_{10} hydrocarbons from Prudhoe Bay crude oil and dispersed oil. The one exception is the 4 ppm-hr exposure from the WSF of untreated oil that caused half kill of coonstripe shrimp larvae, which is not clearly accounted for. Otherwise, the toxicities, based on the C_1 to C_{10} hydrocarbon fraction, were about the same from untreated and chemically dispersed oil. Dissolved hydrocarbons were monitored in both Battelle studies referenced in Table 3-15.

The WSF of physically and chemically dispersed oil to *Artemia* nauplii with Lago Medio crude oil and Corexit 9527 was compared by Mackay and Wells (1980, 1981). The toxicities were similar. Physical dispersion was a little more toxic than chemical dispersion. Greater toxicity to the dissolved hydrocarbon fraction from untreated oil than from chemically dispersed oil was also observed by Foy (1982). The opposite was observed by Duval et al. (1980, 1982) for the isopod *G. oregonensis* using a crude oil (Table 3-15).

Other studies that measured the dissolved hydrocarbons or added pure hydrocarbon components individually include Morrow et al. (1975), who studied young coho salmon and individual hydrocarbons, and Struhsaker et al. (1974), who tested benzene and the eggs and larvae of Pacific herring and northern anchovy in 24-hr exposures. These additional studies generally showed less toxicity than found by Rice and coworkers.

Table 3-15 supports the following observations:

• The field exposures in the water column for both untreated and chemically dispersed oils generally are much lower than exposures required to cause mortality or behavioral effects on a large number of species and life stages.

• The dissolved hydrocarbons from the WSF of untreated oil and chemically dispersed oil produced similar organism mortalities.

Dispersed oil was about twice as toxic to chum salmon fry compared with dissolved hydrocarbons from untreated oil.

• Coonstripe shrimp, sand lance, and Pacific herring larvae were about the same sensitivity. *Artemia* nauplii and an isopod were also affected to about the same degree by dissolved hydrocarbons from dispersed oil.

Summary

The laboratory studies summarized above, comparing lethal and sublethal toxicities of dispersed oil to various organisms, demonstrate the wide range of responses that may occur when dispersants have been used to treat oil, and the many factors influencing the responses.

In general, the results fall into three categories:

1. those employing nominal concentrations (total oil per unit volume), which find that dispersed oil is more toxic; many (nearly 30 percent) of the tests (usually the earlier ones) fall into this category. Test results stemming from use of this technique are in error, and much data are of little use.

2. those analyzing for the water-soluble fraction, which find no difference in toxicity between physically and chemically dispersed oil; and

3. those comparing dispersant to dispersed oil toxicity that find dispersed oil to be more toxic when a relatively nontoxic dispersant is used, and find dispersant alone to be more toxic when a toxic formulation is used.

When the WSF of the oil has been analyzed, there is seldom evidence for synergism (i.e., greater than additive toxicity) between oil and dispersant components, validating the general conclusion that oil is as acutely toxic as dispersed oil.

These laboratory studies also demonstrate some of the difficulties of accurately controlling the exposure of organisms to complex organic mixtures in small tank systems. Such experimental approaches have been used because they are suitable for specific test organisms, and because they offer some control over experimental variables. It is recognized that such approaches do not simulate field conditions. To date, laboratory studies have been most valuable in exploring the types of responses and the duration of effects under "high exposure" conditions, and offering guidance to the design and conduct of field studies on dispersed oils.

MICROBIAL DEGRADATION

A potentially important factor for planning dispersant use is whether it will significantly enhance or retard degradation—particularly microbial degradation—of spilled oil. The ultimate fate of spilled petroleum depends primarily on the ability of microorganisms to use spilled hydrocarbons as sources of carbon and energy (NRC, 1985).

All marine waters appear to contain mixed natural microbial populations with the genetic ability to grow on petroleum hydrocarbons. However, ocean waters that have continuous oil inputs, as from seeps or discharges from populated areas, are likely to have greater numbers and types of oil-degrading microorganisms. Biodegradation begins after evaporative losses have ceased and continues for a week to a year. Evidence suggests that chemical or mechanical dispersion in the water shortens the time period during which microbial degradation assists oil removal.

Biodegradation appears to be limited primarily to paraffinic and aromatic fractions, although studies by Rontani et al. (1986) have shown some degradation of asphaltenes. To date there is no evidence of biodegradation of polar fractions, or nitrogen-, sulfur-, and oxygen-containing compounds (Westlake, 1982).

Dispersants applied effectively increase the rate and possibly the extent of biodegradation by

- creating more oil surface area;
- reducing the tendency of oil to form tar balls or mousse (Gunkel and Gassman, 1980; Daling and Brandvik, 1988); and
- enabling dispersed oil droplets to remain in the water column instead of beaching or sedimenting (Gilfillan et al., 1985).

They may also diminish biodegradation rates by

- adding new bacterial substrate (the dispersant) that microbes might preferentially attack over the oil; or
- increasing concentrations of dispersed oil and dispersant in the water column, which may have temporary toxic or inhibitory effects on the natural microbial populations.

Creation of new surface area is the most important factor relating to biodegradation. Because chemical dispersion of oil increases surface-to-volume ratios of the oil, and because degradation occurs at the oil-water interface, the use of dispersants should enhance the environmental conditions required for suitable microbial growth.

Ideally, key questions relating to possible differences in the rate and extent of degradation of chemically dispersed and nondispersed oil should be addressed by direct field comparisons. However, these comparisons are extremely difficult to accomplish (Green et al., 1982). As a result, knowledge of dispersed oil degradation is limited mainly to laboratory studies, pond and mesocosm studies, and information on physical and chemical changes that are known to occur mainly when dispersants act on spilled oil.

Laboratory Studies

Laboratory studies are useful for observing such important phenomena as mechanisms of degradation; changes over time of type and numbers of oleoclasts—petroleum-degrading bacteria (Atlas, 1985; Lee et al., 1985); relative degradability of various petroleum components; biodegradability of various commercial dispersants; effects of nutrient supplements; and enhancement or retardation of degradation rates with dispersant use. Laboratory studies, including innoculations of field collections, have shown that degradation rates can be enhanced or inhibited when dispersed oil is added to culture vessels. For example:

• Traxler and Bhattacharya (1978) found that chemical dispersants significantly enhanced bacterial degradation of petroleum hydrocarbons.

• Traxler et al. (1983) found that dispersed oil was more effectively metabolized by hydrocarbon-utilizing microorganisms than either untreated oil or dispersant alone.

• Mulkins-Phillips and Stewart (1974) found only slightly enhanced degradation upon dispersion.

• Bunch and Harland (1976) found no difference between untreated oil and dispersed oil.

• Gatellier et al. (1973) found either enhancement or inhibition depending on the dispersant used.

• Zeeck et al. (1984), using 900 ppm (an extremely high concentration) of dispersants, inhibited bacterial growth or decreased glucose uptake rates.

These widely varying and even apparently conflicting results are not conflicting, however, given differences in laboratory techniques, exposure concentrations and durations, nutrient availability in the culture, temperature, and dispersants and oils tested. Generally, the experiments showing inhibition used dispersant concentrations that exceeded the range found in field tests.

The biodegradable nature of some commonly used dispersants has been reported from several laboratory studies (Cretney et al., 1981; Gunkel, 1974; Traxler and Bhattacharya, 1978). Because some dispersants are preferentially utilized over the oil as the carbon source, some experiments have shown initial oil degradation rates in the laboratory to be inhibited by the addition of dispersants (Bunch et al., 1983; Foght and Westlake, 1982; Foght et al., 1983; Mulkins-Phillips and Stewart, 1974). Griffiths et al. (1981) reported decreased uptake of labeled glucose that appeared to be dependent on dispersant concentration. These dispersant-oil concentrations were higher than would usually be observed in situ, although Griffiths et al. (1981) reported one experiment that showed a 10 percent decrease observed at 1 ppm. Generally, inhibition has not been important in pond or mesocosm studies.

The extent to which laboratory studies of biodegradation rate and extent can be extrapolated to the marine environment is severely limited. Major problems include the confining conditions of test vessels and the need to add nutrient supplements. Hydrocarbon degradation rates from laboratory experiments have been several orders of magnitude higher than in situ rates. Conversely, toxic or inhibitory effects are likely to be magnified in the laboratory because the dispersant and dispersed oil mixtures in the test vessels are not able to dilute as they would in nature.

Mesoscale Studies

Pond and mesoscale experiments are seen by many researchers as a way to increase substantially the realism of oil-dispersant experiments. They suffer some of the same shortcomings as laboratory studies (e.g., a limited water volume), but to a lesser extent. Key results of several experiments are summarized below. They consistently show enhanced oil degradation rates of dispersed oil over undispersed oil (see Chapter 4).

In CEPEX bag experiments reported by Cretney et al. (1981) and Green et al. (1982), biodegradation was greatly increased in the dispersed oil bag. Microbial oxidation of the n-alkane component of the oil was completed within 15 days, a rate at least an order of magnitude higher than for undispersed oil (Green et al., 1982). Furthermore, only 0.1 percent of the dispersed oil reached the sediment during these 15 days, and it was in an advanced state of bacterial decomposition. At the surface slick in the CEPEX experiments, microbial degradation had not begun by the end of the 15 days.

In Seafluxes enclosure studies of dispersant and dispersed oil-stimulated bacterial production, Lee et al. (1985) observed increased glucose uptake rates in enclosures with dispersant and dispersed oil. Biodegradation was more important than abiotic processes in the removal of low volatility n-alkanes of dispersed oil in the Seafluxes enclosure.

In freshwater pond experiments, alkane degradation rates of test oils were substantially increased in dispersed oil ponds versus undispersed oil ponds (Dutka and Kwan, 1984; Dutka et al., 1980; Green et al., 1982; Scott et al., 1984). Heterotrophic bacterial counts increased tenfold in oil-dispersant ponds versus oil-only ponds. Also, substantially less oil was found in the sediments of the pond treated with dispersants than in the oil-only ponds after 1 year (Scott et al., 1984).

In seawater pond experiments, Marty et al. (1979) compared dispersed and nondispersed oil in 20-m^2 (24 × 10^3 liters) basins filled with lagoon seawater. Four months after the first treatment, dispersed slicks were no longer visible, while the untreated reference slick did not appear significantly different. Nutrients were not added to the lagoon seawater. Dispersant concentrations were 13 to 130 ppm, significantly higher than manufacturers found in field tests, even immediately after dispersion.

In dispersant-only tests, oil-degrading bacteria increased by 4 to 100 times those in the seawater only (Marty et al., 1979). Although bacterial populations doubled in the oil-only basin after 14 hr of contact, in the dispersant-treated ponds a doubling was not evident until the fifth day of treatment. Despite the delay, microbial populations and extent of degradation were significantly enhanced in the dispersant tanks after 4 months.

Microbial Field Studies

Bunch (1987) studied the effects of chemically dispersed crude oil on bacterial numbers and microheterotrophic activity in the water column and sediments of selected bays at the BIOS experiment site, Cape Hatt, Northwest Territories, from 1980 to 1983. In the release of dispersed oil in 1981, there was a transient decrease in Vmax (maximum velocity) of glutamic acid uptake in water samples. Bacterial numbers were unaffected.

In vitro experiments with water samples demonstrated that a combination of petroleum and dispersant, or dispersant alone, reduced the Vmax of glutamic acid uptake to a greater extent than petroleum alone.

In addition, total organic carbon and bacterial numbers temporarily increased in the sediments impacted by dispersed oils, recovering to normal (control) values by the second year. Effects on the water column were considered inconsequential or marginally deleterious, while effects on the sediment were indirect, long-term, and likely of marginal significance to microheterotrophic activity.

Summary

Some laboratory studies and all mesocosm studies have shown increased oil biodegradation rates when dispersants are used. Temporary inhibition of biodegradation with dispersed oil also has been observed in the laboratory, but appears to occur only at dispersed oil concentrations higher than would occur in the field. Data from pond and mesocosm studies strongly indicate that effective use of dispersants would enhance the biodegradation rate of spilled oil. With limited field data (Bunch et al., 1983, 1985) available, and because biodegradation may be slow or incomplete under some field conditions, this conclusion requires additional verification by field studies.

The primary objectives of dispersant use are to enhance dilution effects, to get oil off the water surface, and to prevent stranding of oil. Hence, any rate enhancement of biodegradation probably should be viewed simply as a secondary benefit to the primary objectives.

Finally, on the question of whether dispersants enhance the extent of biodegradation, available information suggests that refractory compounds would remain undegraded regardless of the addition of dispersants (Lee and Levy, 1986). One aspect of this question that has not been quantified is the extent to which dispersants prevent tar ball formation. Prevention of tar balls and large mousse accumulations possibly could be an important advantage of chemically dispersing oil, because tar balls, especially large ones, trap biodegradable hydrocarbons, and mousse accumulations do not break up before stranding and eventually become buried in intertidal and shallow subtidal sediments (Jordan and Payne, 1980; NRC, 1985).

SEABIRDS AND MARINE MAMMALS

Despite concerns of coastal resource managers and oil spill response teams about the effects of oil spills on seabirds and marine mammals, far more research has been conducted on the effects of oil and dispersed oil on intertidal and subtidal invertebrates, plants, and fish. Because of high susceptibility to damage and high visibility when oiled, however, much recent public policy consideration has been given to seabirds and marine mammals. Unfortunately, many of the critical questions regarding damage to marine mammals and seabirds by oil and possible mitigation by dispersants have not yet been addressed.

There are two primary effects of oil on seabirds and marine mammals (Leighton et al., 1985; NRC, 1985):

1. toxic effects resulting from direct ingestion of oil from the water, or indirectly from grooming or preening; and

2. effects on the water-repellency of feathers or fur needed for thermal insulation.

Research on toxic effects of ingestion is reviewed below.

Seabirds

The few studies of direct toxicity of oil and dispersants to seabirds (Table 3-16; Peakall et al., 1987) show that dispersant and crude oil reduce hatching success and lower resistance to infection to about the same extent, and sometimes less than, oil alone. Studies have been primarily on avian reproduction and physiology. The effects of oil alone on embryos and early development are well known, and the effects of oil-dispersant mixtures have been studied at various stages of the reproductive cycle. Generally, crude oil and Corexit 9527 mixtures and crude oil alone are similarly toxic to bird eggs, based on nominal concentrations.

Work with other species, such as mallard ducks and herring gulls (Table 3-16), also shows a wide range of sensitivities, particularly with duck eggs. For example, Albers (1979) tested Prudhoe Bay crude oil, Corexit 9527, and mixtures (5:1 and 30:1) on the hatchability of mallard eggs (*Anas platyrhynchos*) over 6 to 23 days. All produced diminished hatchability at the 20-μl dose level per egg (external surface). The oil, dispersant, and 5:1 mixture had similar effects, but the 30:1 mixture was significantly less toxic. At reduced dosage, only the Corexit mixture caused significant effects.

TABLE 3-16 Studies of the Effects of Prudhoe Bay Crude Oil, Corexit 9527, and Combinations of Avian Reproduction

Stage of Reproductive Cycle	Species	Protocol	Combination Studied	Finding	References
Hatchability of eggs	Mallard (Anas platyrhynchos)	Applied to surface of egg with syringe	Oil: Corexit, 5:1 and 30:1 mixtures	Toxicity ranking, Corexit= 5:1, oil = 30:1; Corexit alone similar to control	Albers, 1979
		Oil slick on water sprayed with Corexit	Oil: Corexit, 10:1 combination	Oil and combination showed similar decrease in hatchability	Albers and Gay, 1982
Weight gain and survival of nestlings	Leach's petrel (Oceanodroma leucorhoa)	Emulsion or oil painted on plumage or given internally	Oil: Corexit, 10:1 combination	Combination applied externally to adults caused greater decrease of survival and weight gain of chicks than did oil alone	Butler et al., 1987
	Mallard	Given in diet	Oil: Corexit, 10:1 combination	Weight gain and survival were not affected	Eastin and Rattner, 1982
	Herring gull (Larus argentatus)	Single internal dose	Oil: Corexit, 10:1 combination	Corexit alone similar to control; oil and combination decreased weight gain to a similar extent	Peakall et al., 1982
		Single internal dose; birds' food stressed	Oil: Corexit, 10:1 combination	Oil and combination birds both lost weight faster than control	Peakall et al., 1985
		External, painting on feathers; not food stressed	Oil: Corexit, 10:1 combination	Birds exposed to combination lost weight; oil and control birds maintained weight	

SOURCE: Peakall et al., 1987.

The results of coating experiments are important to note, particularly for breeding birds, which may transfer dispersant and oil back to their nests. For example, a field study with Leach's storm petrels (Butler et al., 1988) showed no effects of internal dosing with Prudhoe Bay crude oil or mixtures with Corexit 9527 (10:1), but the highest dose of externally applied dispersant-oil mixture (1.5 ml per bird) significantly increased the percentage of brooding birds deserting the nesting burrow. No significant effects were seen with oil alone. Hatching success was decreased to the same extent with both oil and dispersant-oil treatment.

A mathematical model for the exposure of diving and surface-feeding seabirds to surface oil and dispersed subsurface oil (Peakall et al., 1987) led to the conclusion that the exposure resulted from the surface slick, and that "a highly effective dispersant significantly reduces oil exposure for both types."

The literature review by Peakall et al. (1987) on dispersed oil effects on seabirds concluded that the hazard of chemically dispersed oil to seabirds depended primarily on differing exposures under naturally and chemically dispersing conditions. Their evaluation of the toxicology, based on sublethal responses at the biochemical and physiological level, showed similar responses to oil components, with and without dispersants.

Other studies have examined the toxicity of dispersed oils to seabirds; these include Butler et al. (1979, 1982, 1987), Albers (1980), Lambert and Peakall (1981), Miller et al. (1981, 1982), Peakall and Miller (1981), Butler and Peakall (1982), Trudel (1984), and Ekker and Jenssen (1986). Collectively these studies, including those by Peakall et al., show the range of responses of birds to oil and dispersed oils, the similarity in responses to oil and dispersed oils, and the obvious need to reduce surface oiling for bird protection.

There are also occasional concerns regarding the direct effects of the dispersants themselves on seabirds, both on adults and on eggs and young at the nest. These effects, although perhaps fewer than those produced by oil itself, include direct accidental spraying of birds with dispersants (from aircraft) and the potential increased risk of oiling to seabirds from slicks that have spread after dispersant application.

The seabird-dispersant issue, following from the above summary, seems to be one of exposure to the dispersant and the dispersed oil, rather than one of enhanced toxicity of the oil as perceived until recently.

Marine Mammals

Effects of oil spills on marine mammals include physical fouling, thermal and compensatory imbalance due to oil coatings, uptake, storage and depuration of hydrocarbons, changes in enzymatic activity in the skin, interferences with swimming, occasional mortalities, eye irritation and lesions, and oiling of young (Engelhardt, 1985).

Reviews by Geraci and St. Aubin (1980), Smiley (1982), Engelhardt (1983, 1985), and NRC (1985) describe the effects of oiling on the fur of sensitive marine mammals (sea otters), based on laboratory and mesocosm toxicology experiments and observations of oiled animals in the field. More than a twofold increase in thermal conductance (over baseline), and therefore a 50 percent reduction in insulating capacity, has been reported for polar bears (Hurst and Oritsland, 1982), sea otter pups and fur seals (Kooyman et al., 1977), live adult sea otters (Costa and Kooyman, 1982), and sea otter pelts (Hubbs Marine Research Institute, 1986; Kooyman et al., 1977).

Dispersants have been used experimentally like "shampoos" to remove crude oil from marine mammal fur, but such attempts removed natural skin oils along with the crude oil, thus destroying the fur's water-repellency (Williams, 1978). Surface-active agents, such as those used in dispersants, can increase the wettability of fur or feathers, which in turn allows cold water to penetrate and increase the thermal conductance of the pelt. This is particularly dangerous to animals that are buoyed or insulated by their fur or feathers. In the case of the sea otter, unless grooming can quickly repair the damage, cold water leaks through the fur and against the skin, causing fatal chilling. If the animal grooms excessively, however, it can scratch away large amounts of underfur, further complicating the restoration of body insulation (McEwan et al., 1974).

Direct toxicity is also a potential problem. Polar bears died from toxic effects of oil ingested during grooming (Engelhardt, 1981) as did river otters examined after an oil spill at Sullom Voe, Shetland Islands (Richardson, 1979).

To date, only Hubbs Marine Research Institute (1986) has addressed chemically dispersed oil effects. The critical work by the American Society for Testing and Materials (ASTM, 1987) reviews the literature on oil damage and other human disturbances to marine mammals, but cites none regarding the effects of dispersed oil.

Attempts to remove oil or dispersed oil from sea otter pelts showed that any residue of oil or dispersant left on the fur, even if the fur was dry, permitted water to penetrate into the fur upon immersion (Hubbs Marine Research Institute, 1986). This confirmed the earlier studies of the damaging effects of increased wettability of sea otter fur after contact with crude oils, detergents, and dispersants (Costa and Kooyman, 1982).

Research on thermal responses was conducted by rubbing fresh oil or 5-day weathered Santa Barbara crude oil on adult California sea otter pelts (Hubbs Marine Research Institute, 1986). Fresh oil alone, or with Corexit, easily penetrated the fur, which quickly saturated upon immersion in water. Thermal conductance was more than twice as high as in untreated control pelts. There was no difference in conductance between fresh crude alone or with oil combined with the dispersant. Based on such sparse information, oil dispersant chemicals may not reduce the physical threat of spilled oil to some fur-insulated sea mammals.

Smiley (1982) stated that

> Nonetheless, dispersion of large oil slicks is probably a useful countermeasure tool, assuming that both the floating oil and the applied chemical are effectively diffused into the water column. The risks of direct fouling and of inhalation toxicity when swimming at the sea surface would be reduced, especially in cold icy situations where natural weathering and evaporation of oil slicks is slow.

In addition, the ASTM (1987) concluded that

> Use of chemical dispersants and mechanical methods is recommended to prevent these habitats from being contaminated or to reduce contamination. . . . Because sea otters and polar bears are very sensitive to oil contamination, dispersant use is recommended even if application must occur near or in a habitat used by these animals.

However, available data do not seem to support this recommendation, and Smiley's conclusion assumes complete dispersion and disappearance of surface oil after dispersant application, which may not occur.

In view of the enormous public interest in, and concern for, the fate of seabirds and marine mammals, it is surprising that so little research with dispersants has been done with these animals, and that the conclusions on the use of dispersants for protecting these animals can only be tentative. Clearly, there is a great need for more laboratory and field studies, particularly in order to determine whether the use of dispersants will lessen the adherence and impact of oil on the fur of marine mammals and the feathers of birds. Thus far the data only appear to indicate that there is no difference between the effects of oil with dispersants or alone.

4
Intermediate-Scale Experiments and Field Studies of Dispersants Applied to Oil Spills

Because the magnitude, type, and duration of effects on aquatic organisms and ecosystems caused by oil spills depend directly on exposure to toxic components of the oil, the effects of oil are expected to be less if the spill is rapidly diluted by chemical dispersion. A number of experiments at scales larger than normal laboratory size (mesoscale) as well as field studies at sea have been conducted to determine the physical dispersion and the subsurface concentrations of oil components. They and their associated biological effects are reviewed in this chapter.

PHYSICAL AND CHEMICAL STUDIES

Although laboratory tests can rank various dispersant formulations as to their relative effectiveness and can be used to investigate the effects of parameters, such as temperature, water salinity, and oil viscosity, the real test of dispersant effectiveness is a full-sized spill in a test at sea. However, rigorous sea tests are expensive and difficult to conduct, and results have often been disappointing.

An absolute measure of effectiveness in the field would require that a very large set of water samples be taken, covering the entire water mass within which the oil might become dispersed, as well as an accurate measurement of the amount of oil that evaporates from the slick under the field conditions. Very few experiments have attempted this approach, but it is these that provide the most direct evidence of dispersal of oil at sea (Brown et al., 1987).

Some studies have been set up to obtain typical water samples from beneath a dispersant-treated or untreated slick, so as to assess whether the concentrations of oil components exceeded potentially toxic levels. The emphasis of others is to see if dispersants ranked better or poorer by laboratory test would perform in the same order on a larger scale.

As will be discussed in Chapter 5, much thought has been given to remote monitoring systems, but there is as yet no standard method for determining effectiveness outside the laboratory. As a result, dispersant operations at spills of opportunity have provided only a limited and ambiguous set of effectiveness data. The compilations of Nichols and Parker (1985) or Fingas (1985) if taken uncritically could be rather discouraging. In many cases, dispersal was observed but could have been due to natural processes—adequate control spills without chemical dispersant were unavailable. In other tests, different observers at the same site reached different conclusions about how much of the slick had been dispersed. The reported effectiveness at any but the most carefully planned field trials is extremely dependent on the types of observations or samples, the location of the observers or sampling devices, and the dispersant application technique.

Intermediate-Scale (Mesoscale) Studies

Some studies intermediate in size between laboratory and field (microcosm and mesocosm)—although not without limitations—can provide useful information with greater control and at less expense than a full field study (Adams and Giddings, 1982). An example is the Controlled Ecosystem Pollution Experiment (CEPEX) in British Columbia, Canada. Two 13-m deep CEPEX plastic* enclosures moored at Saanich Inlet were treated with 3 liters of oil (Green et al., 1982). In one system, Corexit 9527 was added, producing a stable emulsion with average droplet size about 1 μm or less (measured by underwater photomicroscopy).

---

*Typical experiments are conducted using plastic (usually polyethylene) enclosures in an open-water area. In one study (Laake et al., 1984), the function of Ekofisk crude oil suffering different fates was measured using tritium tracers. Only 0.0037 percent of the oil was adsorbed in the plastic walls.

In contrast to expectations, evaporation was inhibited in the dispersed system, which required 10 to 15 days to lose volatiles, whereas the undispersed slick lost volatiles in 1 to 2 days. After 7 and 27 days, the amount of oil reaching the sediment increased by about a factor of 10 in the enclosure where dispersant had been used. However, the dispersed oil was more rapidly biodegraded, by a factor of 10, with alkanes essentially oxidized in 15 days, so that in each case only about 0.1 percent of the oil remaining in the water column eventually was found in the sediment (Green et al., 1982).

To test effects of dispersants in a littoral ecosystem simulating the shallow rocky Baltic Archipelago, Linden et al. (1985, 1987) used six pools, 8 m^3 each, with a flow-through seawater system. Two pools were exposed to 20 ppm (average initial nominal concentration) North Sea Forties crude oil, two were exposed to 20 ppm oil with Corexit 9550 added, and two served as controls. The differences in biological effects between treatments were attributed to dispersed oil remaining in the water column longer, without adhering to particles or organisms or settling to the bottom. Because seawater flowed continuously through the systems, the dispersed oil was more rapidly washed out.

In a trial using intertidal enclosures, Farke et al. (1985a,b) oiled sand by contaminating inflowing seawater on 12 successive rising tides with ultrasonically dispersed Arabian light crude oil, Finasol OSR-5 dispersant, and an oil-dispersant mixture (ratio O:D of 10:1). With or without dispersant, average oil concentration in the water (sampled at high tide) was about 10 ppm, and core samples showed that less than 5 percent of the oil penetrated the deeper sediment. Maximum concentration of oil in the top 2 cm of sediment was 560 ppm in both the oil-dispersant experiment and oil-only experiments. After contamination was stopped, the oil concentration returned to baseline values within 4 to 6 weeks.

Although the Farke et al. (1985a,b) study using premixed dispersant and oil does not fully simulate the impact of oil dispersed at sea coming in on the tide, it is closer to that ideal than other intertidal studies in which dispersant was applied after a beach was oiled. In the latter studies, dispersant increased the degree of oil penetration. It is clear from this example that dispersed oil is more easily washed out of an enclosed system than untreated oil, and there appears to be evidence that dispersed oil adheres less to particles and sediment than untreated oil (see Chapter 2).

Recent examples of mesoscale studies of physicochemical characteristics of dispersed oils include those at the Esso Resource wave
basin in Canada (Brown et al., 1987; To et al., 1987). These studies
show that the dispersed oil plume often is highly irregular in shape
and nonuniform in concentration. This nonuniformity can lead to
serious errors when attempting to estimate dispersant efficiency by
analyzing chemical samples from the water column. Further, these
mesoscale experiments reemphasize the need to apply dispersants
preferentially to the thicker portion of the slick in order to achieve
good overall efficiency.

American Petroleum Institute Research Spills

Open-ocean tests sponsored by the American Petroleum Institute in 1978 and 1979 evaluated several factors bearing on dispersant
performance and fate: oil type, sea state, and dispersant type and
dosage. The studies compared the fates of untreated and chemically
treated oil and measured total hydrocarbons in water under slicks
(McAuliffe et al., 1980, 1981).

In four spills conducted off New Jersey, 11.7 bbl of Murban and
LaRosa crude oil were sprayed with dispersant by helicopter. Murban
crude oil changed rapidly when dispersant was immediately applied.
A distinct whitish-brown subsurface plume appeared quickly. Over
several hours, this plume grew in area and diminished in color and
visibility as the dispersed oil diluted. Rough mass-balance calculations, supported by visual and photographic observations, indicated
that Murban crude oil was almost completely dispersed (McAuliffe
et al., 1980).

The highest total oil concentration measured under the low-
viscosity Murban (39° API gravity) crude oil was 18 ppm at 1 m at 23
min after dispersion, decreasing to less than 1 ppm at 6 m after 1 hr.
The highest dissolved hydrocarbon concentrations—40 to 50 ppb—
occurred in the samples with the highest total oil concentrations.
After 110 min the highest hydrocarbon (C_1 to C_{10}) concentrations
were 2 ppb at 1 m.

When dispersant was sprayed on the fresh La Rosa (24° API
gravity) crude oil, no sudden change was apparent. However, in time
this oil became a thin sheen, as contrasted with the thick, black,
asphaltic appearance of the thicker oil in the downwind, leading edge
portion of undispersed oil. About half the slick was estimated to have
been dispersed. The highest total oil concentrations measured in the

dispersed oil plume were 2 to 3 ppm from the surface through 3 m at 23 min after spraying. These concentrations also existed after an hour, and thereafter decreased. The highest dissolved hydrocarbon concentrations were 14 ppb at 47 min, and 9 ppb after 94 min.

The two slicks that were allowed to weather for 2 hr before dispersant spraying showed low concentrations of oil in the water. This was probably due to the greater area of the slick and the fact that most of the oil was in the downwind portion of the slick. Since the overall slick was uniformly sprayed, the thick portion where most of the oil resided may have been undertreated. Weathering also would have increased oil viscosities, thereby decreasing dispersant effectiveness. These effects were clearly demonstrated during the 1979 API studies off southern California.

In the September 1979 API studies off the coast of southern California (McAuliffe et al., 1981), nine separate 10- or 20-bbl releases of 0.90 specific gravity (26.6° API gravity) Prudhoe Bay (Alaskan North Slope) crude oil took place over 2 days. These tests were unusual in the large number of water samples taken (900), which allowed contours of subsurface concentrations to be obtained and a more accurate mass-balance made. The slick areas of the 20-bbl oil discharges were 2 to 3 ha (5 to 7 acres) at the start of aerial spraying, 10 to 30 min after release, but increased during the 30-min multipass spraying. The average slick thickness was 0.1 to 0.2 mm initially, but decreased as the slick area increased.

The most effectively dispersed 20-bbl slick was sprayed with dispersant concentrate from a DC-4 aircraft; the results of this test are discussed in detail here. Figure 4-1 shows that the oil concentrations for the first sampling run under the remaining slick and through the dispersed oil plume (five stations were placed along the length of the slick and two across it). The highest dispersed oil concentrations occurred at Station 2 (center of the downwind thicker part of the slick) with an average of 41 ppm at 1 m and 10 ppm at 3 m. A mass-balance was estimated (by layers) by calculating the water volume (as the slick length multiplied by two-thirds of the width) multiplied by the average oil concentration of the seven stations. The chemical analyses were on a weight basis. Correcting for the specific gravity (0.90) changes the amount of oil discharged from 20 bbl to 18 bbl on a weight basis. It was further assumed that by the time of spraying, 15 percent of the oil had evaporated. Thus the amount of oil in the slick that was sprayed was estimated to be 15.3 bbl. The amount of oil measured in the water column at the various depths (totaled in

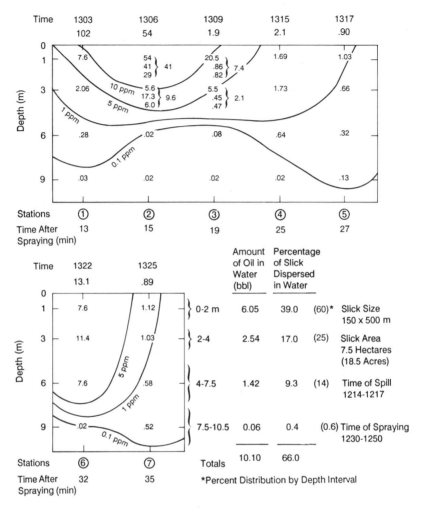

FIGURE 4-1 Concentrations (ppm) of oil in water under a 20-bbl crude oil slick that was sprayed immediately with dispersant by DC-4 aircraft, September 26, 1979; first sample run. Percentage of slick dispersed in water is based on the estimated amount of oil in the slick, not the amount of oil discharged. Source: McAuliffe et al., 1981.

Figure 4-1) was 11.2 bbl, or about 66 percent. Therefore, two-thirds of the oil in the slick (after evaporation) was dispersed, and one-third remained on the surface. Only 0.8 percent of naturally dispersed oil was found under untreated (control) slicks during single sampling runs immediately after the control slicks were released. The highest amounts of naturally dispersed oil generally are found under fresh oil slicks.

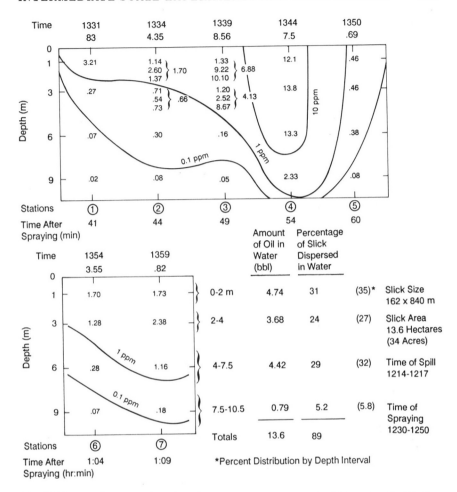

FIGURE 4-2   Concentrations (ppm) of oil in water under a 20-bbl crude oil slick that was sprayed immediately by DC-4 aircraft, September 26, 1979 (day 1); second sample run. Source: McAuliffe et al., 1981.

Figure 4-2 shows the results of a second sampling run through the slick about an hour after spraying. During this time, the slick had elongated as the remaining wind-driven surface slick separated from the dispersed oil plume. The highest concentration, found at 1 m at Station 2 after 15 min (Figure 4-1), occurred from 0 to 6 m at Station 4 after 54 min. The dispersed oil had diluted by mixing downward. The mass-balance calculation at that time provided an estimate of about 89 percent of the unevaporated oil dispersed.

A third sample run was started 3 hr after spraying, as a transect

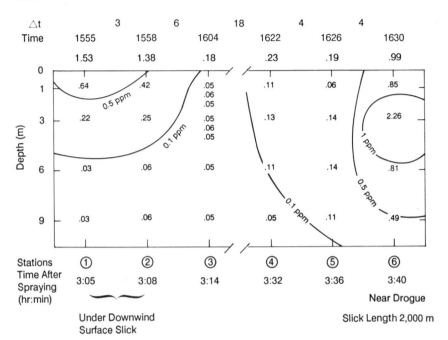

FIGURE 4-3   Concentrations (ppm) of oil in water under a 20-bbl crude oil slick that was sprayed immediately by DC-4 aircraft, September 26, 1979 (day 1); third sample run through small downwind slick and then near drogue where dispersion occurred. Source: McAuliffe et al., 1981.

from under the separated teardrop-shaped slick back to the drogue that was following the dispersed oil plume in the water (Figure 4-3). The distance was about 2 km. Samples taken at the drogue had concentrations of 1 to 2 ppm through 6 m and 0.5 ppm at 9 m. These concentrations represent the further dilution of the dispersed oil with time.

The results of these very elaborate field studies produced some of the few quantitative mass-balances on dispersed oil that have ever been obtained. Although a large number of samples were collected, there can still be errors in the calculated amounts of dispersed oil.*

---

*An independent check of dispersant effectiveness can be made based only on the observed concentrations in the water. An average 0.1-mm-thick slick, if completely dispersed and uniformly mixed in 1 m of water, would produce a concentration of 100 ppm, 33 ppm in 3 m, and 17 ppm in 6 m. The sample stations, with the higher concentrations shown in Figures 4-1 and 4-2, approach these values if the oil measured at greater depths is added back to the shallower depths. The measured concentrations would be even closer to the theoretical concentrations if they were corrected for volume of oil to weight of oil, percentage of evaporation, and an estimate of the amount of slick remaining.

Table 4-1 summarizes the effectiveness of dispersant sprayed on seven slicks and the estimated percentage of oil naturally dispersed under the two control slicks. Table 4-1 shows the effectiveness of aircraft versus boat spraying, a comparison of two different dispersants applied in the same manner, and immediate spraying versus a 2-hr delay. Aerial spraying of the fresh oil slick was more effective than boat spraying (60 and 78 percent versus 62 percent with less dispersant aerially applied). The reduced effectiveness of boat application may be due principally to mixing the dispersant concentrate with seawater (with an induction system to 2 percent concentration) before spraying. Dispersant H (Corexit 9527) was 5 to 6 times more effective than dispersant J (unidentified) (62 versus 11 percent) in boat spraying of just the thick oil part of the fresh slick. (Laboratory testing with both fresh and weathered Prudhoe Bay crude oil showed about this same difference.) Oil that was on the water for 2 hr prior to spraying was not as effectively dispersed, probably due to increased viscosity from greater loss of volatile hydrocarbons by evaporation. Boat spraying the entire slick uniformly was very ineffective compared with spraying just the portion of the slick that contained most of the oil. Too little dispersant was sprayed on the thick areas, too much on the thin ones. This difference in effectiveness is in accord with the basic concept that the dispersant should be sprayed where most of the oil exists. Table 4-1 also shows that generally the higher the rate of dispersant application, the greater the amount of oil dispersed.

Another API-sponsored study of dispersed versus untreated oil was conducted at Long Cove, Searsport, Maine. The results are presented later in this chapter.

Protecmar

The French Protecmar program's studies emphasized middle-scale field tests with two boats because such tests are more realistic than laboratory tests and less expensive than full-scale offshore tests (Bocard et al., 1981, 1984, 1987; Desmarquest et al., 1985). In one test, 45.5 bbl of light fuel oil was treated with undiluted Dispolene 325 sprayed from an airplane. About half the oil was dispersed within 4 hr, the thicker areas slowly disappeared after 7 hours, and only a scattered sheen remained after 20 hr.

In a similar experiment with light fuel oil in the Mediterranean Sea, several dispersants were sprayed from aircraft and boats (Bo-

TABLE 4-1 Effectiveness of Dispersant Treatments During the 1979
Southern California Studies

Treatment	Estimated Percentage of Dispersant Applied[a]	Percentage of Slick Dispersed in Water
Sprayed immediately by plane, day 1, dispersant H	4.9	78 ± 16[b]
Sprayed immediately by plane, day 2, dispersant H	3.6	60 ± 3.5[b]
Sprayed after two hours by plane, day 1, dispersant H	4.0	45
Thick oil part of slick sprayed immediately by boat, dispersant H	8.7	62
Thick oil part of slick sprayed immediately by boat, dispersant J	8.4	11
Entire slick sprayed immediately by boat, dispersant H	1.5	8
Entire slick sprayed after two hours by boat, dispersant H	1.5	5
Untreated oil		0.8 ± 0.4[c]

[a]Estimated percentage of dispersant applied to thicker part of oil slick, estimated to contain 90 percent of the oil.
[b]The mean of the first two sampling runs through the immediately aerial sprayed slicks: day 1 first sample run, 66 percent; second sample run, 89 percent.
[c]The mean of one run through each of the two untreated control slicks.

SOURCE: McAuliffe et al., 1981.

card and Gatellier, 1981). For both chemical and natural dispersion, infrared analysis of water samples detected 0.1 to 0.5 ppm concentrations of oil. Subsurface concentrations obtained in some of the Protecmar tests are summarized in Table 4-2. Several general conclusions were derived from the Protecmar tests:

• The trend of dispersant effectiveness observed was similar to the United Kingdom's standard test (Labofina/Warren Spring Laboratory), but with more viscous oils, different dispersants can hardly be distinguished.

• The French standard test (also a version of Labofina) does not account exactly for the variation in dispersant efficiency with oil viscosity.

TABLE 4-2 Subsurface Oil Concentrations After Dispersant Application in the Protecmar Studies (Maximum Significant Values in ppm)

Trial	1 m			2 m/2.5 m		
	t_1	t_2	t_3	t_1	t_2	t_3
Protecmar 1				2	1	
Protecmar 2	1			1		
Protecmar 3						
Slick A	1	60	5	1	3	3
Slick B	4	3		3	5	
Protecmar 4						
Sprayed by	17	4	4	1	Traces	Traces
Canadair CL215						
All other slicks	7	4	4	3	Traces	Traces
Protecmar 5	3	Traces		3	Traces	
Protecmar 6						
Helicopter	1	2 (1 hr 30 min)		Traces	Traces	
Ship	4			2		

NOTE: The columns t_1, t_2, and t_3 are sampling times corresponding, respectively, to 0 to 30 min, 3 to 3 hr 30 min, and 6 to 7 hr after dispersant application.

SOURCE: Bocard et al., 1987.

- The results of the IFP dilution test (Chapter 2) agree with those obtained at sea when mixing is applied to the treated oil slick. Ranking of dispersants for a given oil viscosity is the same as in the field test (Desmarquest et al., 1985)

North Temperate and Arctic Tests

A number of field trials have been conducted to test effects of dispersants in north temperate coastal and arctic habitats. In 1978, in three experiments at Victoria, British Columbia, oil was spilled in a semiprotected coastal area and restrained with a boom (Green et al., 1982). Ten percent Corexit 9527 was applied by ship using the Warren Spring Laboratory system. Fluorometric monitoring of water samples showed as much as 75 percent dispersal. The highest oil concentration was 1 ppm, which decreased to background levels (less than 0.05 ppm) within 5 hr, in agreement with the field tests cited above. Previous tests in CEPEX enclosures showed that microbial oxidation of dispersed Canadian North Slope oil occurred at least 10 times faster than undispersed oil, and this may have been a factor in the rapid disappearance of the oil (Green et al., 1982). The results of the Royal Roads tests were consistent with an earlier test on Kuwait crude (Cormack and Nichols, 1977), but other tests using the WSL application system produced less objective results because of inadequate observations from aircraft (Smith and Holliday, 1979).

A 1981 field trial held off St. John's, Newfoundland was among the first to employ both remote sensing and a real-time computer simulation model (Gill and Ross, 1982; Intera Environmental Consultants, 1982). Aerial photographs clearly indicated rapid formation of an oil-in-water emulsion immediately following application of Corexit 9527. The waterborne cloud was visible from the air for 3 hr.

Field tests of the effectiveness of Corexit 9527, Corexit 9550, and BPMA700 were conducted near Halifax, Nova Scotia in 1983 (Canadian Offshore Aerial Applications Task Force [COAATF], 1986; Gill et al., 1985; Intera Environmental Technologies, 1984; Swiss and Gill, 1984). An acoustic monitoring system for verifying water column oil concentrations was tested, along with oil spill tracking models to predict the movement of slicks during the trial. Water samples analyzed by fluorescence and radiotracer techniques showed that 2 to 40 percent of the oil was dispersed, but because there were differing sea states for the various tests, the dispersant products were not ranked for effectiveness.

To better understand oil spills in an arctic environment, 16 small-scale tests with Tarsiut crude oil were conducted in Mackenzie Bay, Canada in 1984 (Dickinson et al., 1985). The dispersant BPMA700 was determined qualitatively to be the most effective; Corexit 9527 was also effective, but took longer to disperse the oil; Corexit 9550 caused oil resurfacing and was not recommended for arctic conditions. This study demonstrated that chemical dispersal of Tarsiut crude oil in Beaufort Sea waters is feasible and it suggested improvements in the spray application system.

More extensive arctic tests were held in the Beaufort Sea near Tuktoyaktuk, Northwest Territories in August 1986. Four slicks, each containing 2.5 bbl of aged (by aeration) Alberta sweet mixed blend, were laid down by tugboat as follows (Oil Spill Intelligence Report, 1986c; Swiss et al., 1987a; Jones, private communication):

- treated with a single application of BP Enersperse 700;
- treated with multiple applications of BP Enersperse 700;
- treated with Exxon CRX-8; and
- untreated—served as control.

In the untreated slick, thick and thin portions of the spill were visually easy to distinguish (Jones, private communication). After spraying there was an obvious plume of dispersed oil in the water, and little of the thick oil was left. Wherever dispersant was sprayed a sheen was visible that in some cases covered 10 times the area of the original spill. An observer noted that the thick oil consisted of pea-sized droplets that from time to time rapidly expanded into a larger sheen (Swiss et al., 1987b).

Preliminary conclusions from this study are that Alberta sweet mixed blend crude can be dispersed at 6°C, and that multiple spraying operations using helicopters are feasible. However, a single application at a dispersant-oil ratio of 1:10 appears to be as effective as the "multihit" technique, provided observations are made over several hours. Although the two dispersants appeared initially to have different effects, the amount of oil dispersed after several hours was the same for both (Swiss et al., 1987a).

In addition to the field studies discussed above, a series of experimental spills were conducted on northern Baffin Island, Northwest Territories. These studies, which were primarily conducted for biological observations, comprised the Baffin Island Oil Spill Project, which is described later in this chapter.

In an early test, Cormack and Nichols (1977) measured the following concentrations of chemically dispersed Ekofisk crude oil at a depth of 1 m:

- 16 to 48 ppm within the first 2 min;
- 5 to 18 ppm after 5-10 min; and
- 1 to 2 ppm after 1 hr 40 min.

Lichtenthaler and Daling (1983) reported on seven 13-bbl research spills conducted with topped (initial boiling point 150°C) Statfjord crude oil in May and July 1982. One release at each test time was a control (untreated), and the remainder were sprayed by boat with 200 liters (53 gal) each of three different dispersants diluted to 10 percent concentration by seawater. Forty water samples were collected under each dispersed slick at 1, 2.5, 5, and 9 m. The maximum subsurface oil concentration was 10 ppm at 1 m with dispersant B.

The authors suggested that the concentrations and effectiveness were reduced because the oil was released over 7 to 12 min, resulting in several thicker oil patches spread over wider areas of the slicks. The southern California studies (McAuliffe et al., 1981) released the oil in 3 min, and the slick was quite uniform initially, with the thicker oil located in the downwind portion. In addition, no air surveillance was available in the North Sea tests, and since a spray boat was used, preferred spraying of the thick oil patches was impossible. Lichtenthaler and Daling (1983) indicated that these difficulties resulted in the low dispersant effectiveness observed:

- dispersant A, 6 to 19 percent;
- dispersant B, 17 to 22 percent;
- dispersant C, 2 percent; and
- controls, 0.7 to 2.6 percent.

The dispersant ranking was consistent with laboratory tests conducted by Mackay and Szeto (1981).

In the summer of 1983, Delvigne (1985) studied nine 13-bbl spills of Statfjord crude oil and a light fuel oil in the North Sea. Five of the slicks were controls, one was crude oil premixed with dispersant Finasol OSR-5 (used for all tests), and the remainder were aerially sprayed after being on the water for 1 hr. A single surface measurement indicated that the slicks were hit by dispersant that should have resulted in dispersant-oil ratios of 1:10 to 1:30 in the thick part of the slick. Measurements from a towed fluorometer at depths of 3 and 7 m, as well as 800 water samples analyzed by infrared spectrophotometry, surprisingly indicated that dispersion was not greatly increased by dispersant spraying. Lower than expected concentrations of dispersed oil were measured at 3 and 7 m from the Statfjord oil-dispersant mixture (after 30 min to 1 hr 45 min, 3 m values for three samples shown were 0.14 to 3 ppm; 7 m concentrations were 0.1 to 0.3 ppm). These concentrations were somewhat higher than the aerially sprayed slicks.

Delvigne attributed the ineffectiveness to poor mixing of the sprayed dispersant with the oil layer. He suggested that the dispersant droplets penetrated through the slick or washed away from the oil layer before penetration. However, the low concentrations found from the dispersant-oil premix suggest that the dispersant used was not very effective with this oil, or that the sampling did not find the highest oil slick concentrations. The values contrast markedly with those of Lichtenthaler and Daling (1985) reported below. Thirty minutes after release of the oil-dispersant premix these investigators found concentrations to exceed 100 ppm through 3 m, 12 ppm at 5 m, and 3 ppm at 10 m.

Lichtenthaler and Daling (1985) conducted additional studies in the North Sea in June 1984 using a simulated fuel oil produced from

Statfjord crude oil (8 parts of gas oil, 250° to 350°C; 1 part heavy gas oil, 330° to 400°C; and 1 part residue, greater than 380°C). The already sprayed slicks were sampled 30 to 50 min later. The highest oil concentrations in water were 25 to 40 ppm at a depth less than 2 m. Under two untreated slicks, corresponding concentrations were less than 5 ppm, a mean of 8 samples from 2 m was 0.22 ppm. The dispersed and untreated oil concentrations are in general agreement with the 1979 API southern California field studies (McAuliffe et al., 1981).

The long-term effects on surface oil, 1 to 2 days after treatment with dispersant, could be important in evaluating the effectiveness of dispersants in sea tests. In both the Protecmar tests and in the Norwegian North Sea trials, Lichtenthaler and Daling (1985) reported that dispersants tend to act as de-emulsifiers when weathered (water-oil emulsified) surface oil is treated at low dosages. A small amount of dispersant apparently enhances the "natural" dispersion process by breaking down the water-in-oil emulsions and releasing for further spreading.

The mass-balance calculations made in sea trials (typically 0.5 to 2 hr after treatment) only take into account the short-term effects of dispersants, which often have resulted in uncertain conclusions (see Table 4-3).

Summary of Physical and Chemical Field Test Results

The tests discussed in the preceding four sections are summarized in Tables 4-3 and 4-4. Many pieces of quantitative information, such as oil viscosity and oil-dispersant ratio, were estimated by Nichols and Parker (1985). Effectiveness, defined as the fraction of oil removed from the water surface, is obtained by integrating the water-column distribution as well as allowing for evaporation from the slick. Numerical effectiveness values are given in Table 4-3 only if the authors of the cited papers gave such values. Fingas (1985) and Nichols and Parker (1985) have estimated quantitative effectiveness values from water-column concentrations; neither of these papers gives the methodology.

One particular study (McAuliffe et al., 1981), in which unusually complete distributions of petroleum hydrocarbons in the water column were measured, gives the clearest indication that chemical dispersion is more effective than natural dispersion in relatively calm seas; that dispersant treatment by air is superior in most cases to

TABLE 4-3 Summary of Dispersant Effectiveness Trials

Oil Type	Viscosity (cSt)	Dispersant Amount (m³)	Application Method	Oil to Dispersant Ratio	Sea State	Effectiveness Percentage	References
Ekofisk	Weathered 2 hr	0.5 10% in salt water	WSL ship	20:1	1	(?)	Cormack and Nichols, 1977
Kuwait	14[a]	(?)	WSL ship	20:1	2-3	(?)	
Murban	Weathered 2 hr	1.7 Corexit 9527	Helicopter	> 5:1	1	Visual, little	McAuliffe et al., 1980
La Rosa	Weathered 2 hr	1.7 Corexit 9527	Helicopter	5:1	1	Effective(?)	
Murban	5[a] Treat immediately	1.7 Corexit 9527	Helicopter	> 11:1	1	100	
La Rosa	80[a] Treat immediately	1.7 Corexit 9527	Helicopter	> 11:1	1	50	
Prudhoe Bay	35[a]	1.6 None (control)		NA	2-3	0.5	McAuliffe et al., 1981
Prudhoe Bay	35[a]	1.6 None (control)		NA	2-3	1.0	
Prudhoe Bay	35[a] Treat immediately	1.6 2% H in salt water	Boat	67:1[a]	2-3	8.0	
Prudhoe Bay	Weathered 2 hr	1.6 2% H in salt water	Boat	67:1[a]	2-3	5.0	
Prudhoe Bay	35[a] Treat immediately	1.6 2% J (poor in lab)	Boat	11:1[a]	2-3	11.0	
Prudhoe Bay	35[a] Treat immediately	3.2 H conc. (1978 test)	D-C4 plane	20:1[a]	2-3	18.0	
Prudhoe Bay	Weathered 2 hr	1.6 H conc.	D-C4 plane	25:1[a]	2-3	45.0	
Prudhoe Bay	35[a] Treat immediately	3.2 H conc.	D-C4 plane	25:1[a]	2-3	60.0	
Prudhoe Bay	35[a] Treat immediately	1.6 2% H	Boat	11:1[a]	2-3	62.0	
Prudhoe Bay	Weathered 2 hr	1.6 H conc.	D-C4 plane	20:1[a]	2-3	78.0	
Light fuel	35	6.5 None (control)		NA	1-2	Visible after 20 hr	
Light fuel	35	6.5 Dispolene 32 S	CL215 plane	> 3:1	1-2	50 After 4 hr[a]	
Light fuel	35	6.5 Dispolene 32 S	CL215 plane	> 3:1	1-2	100 After 20 hr[a]	

Light fuel	35	6.5 Shell	CL215 plane	> 3:1	2-3	Wider area[a]	
North Slope	70[a]	0.2 10% Corexit	WSL ship 9527	1:1	2	"Easily and obviously dispersed"	Bocard et al., 1983, quoted by Nichols and Parker, 1985
North Slope	70[a]	0.4 10% Corexit	WSL ship 9527	1:1	1		Green et al., 1982; Nichols and Parker, 1985
North Slope	70[a]	0.2 10% Corexit	WSL ship 9527	1:1	1	75[a]	
Statfjord	Weathered	0.2 None			2:3	0.6[a]	Lichtenthaler and Daling, 1983, quoted by Nichols and Parker 1985
Statfjord	17[a] Fresh	0.2 None			2:3	2.6[a]	
Statfjord	Weathered	0.2 10% Conc. in salt water			2:3	6[a]	
Statfjord	Weathered	0.2 10% Conc. in salt water	Ship	10:1	2:3	17[a]	
Statfjord	17[a] Fresh	0.2 10% Conc. in salt water	Ship	10:1	2:3	19[a]	
Statfjord	17[a] Fresh	0.2 10% Conc. in salt water	Ship	17:1	2:3	22[a]	
Statfjord	17[a] Fresh	0.2 10% Conc. in salt water	Ship	18:1	2:3	2[a]	
				12.5:1	2:3		
Arabian light	10[a] Fresh	20 None	Aircraft	NA	1	Oil removed from surface but resurfaces[a]	Cormack, 1983c, quoted by Nichols and Parker, 1985
Arabian light	170[a] Weathered 4 hr	20 Corexit 9527	Aircraft	2:1	1		
Statfjord	8[a]	2 None	Airplane		1:2	2% per hr	Delvigne, 1983, quoted by Nichols and Parker, 1985
Light fuel	60[a]	2 None	Airplane		1:2	2% per hr	
Statfjord	8[a]	2 None	Premixed		1	2% per hr	
Statfjord	8[a]	2 Finasol OSR-5	Airplane	10-30:1	1	2% per hr	
Light fuel	60[a]	2 Finasol OSR-5	Airplane	10-30:1	1	2% per hr	
Statfjord	8[a]	2 Finasol OSR-5	Premixed	20:1	2:3	100% per hr	
Light fuel	60[a]	2 None	Airplane		2:3	2% per hr	
Statfjord	8[a]	2 Finasol OSR-5	Airplane 10-30:1	10-30:1	1:2	2% per hr	
Statfjord	8[a]	2 Finasol OSR-5		1:2	2% per hr		

TABLE 4-3 (Continued)

Oil Type	Viscosity (cSt)	Dispersant Amount (m³)	Application Method	Oil to Dispersant Ratio	Sea State	Effectiveness Percentage	References
ASMB	> 7 Weathered 15%[a]	2.5 Corexit 9527	Helicopter	20:1	1	1-3.8	Gill et al., 1985 (range of efficiency is fluorescence vs. isotope method)
ASMB	> 7 Weathered 15%[a]	2.5 None	Helicopter		1	0.5-2.4	
ASMB	> 7 Weathered 15%[a]	2.5 Corexit 9527	Helicopter	10:1	1	15.8-20.5	
ASMB	> 7 Weathered 15%[a]	2.5 None	Helicopter		1	<1	
ASMB	> 7 Weathered 15%[a]	2.5 BP MA 709	Helicopter	10:1	2-3	9.5-41	
ASMB	> 7 Weathered 15%[a]	2.5 None			2-3	4.3-7	
Statfjord	129-317, 2-5 hr	10 None	Airplane	75:1	1		Lichtenthaler and Daling, 1985
Statfjord	< 60 2-5 hr	10 Corexit 9527			1		
Statfjord	600 Weathered 5 hr	10 None			1		
Statfjord	82-317, 1-5 hr	10 Corexit 9527	Airplane	80:1	2		
Statfjord	12 Fresh	12 Corexit 9527	Premix	33:1	2		
Statfjord	< 60	10 Corexit 9527	Airplane	50:1	?		
Fuel oil	40 Fresh	5 None			1	< 10%	Fingas, 1985 and unpublished observations of French Protecmar 6 trials
Fuel oil	40 Fresh	28 Dispolene 35 S	Helicopter		1	10-50%	
Fuel oil	40 Fresh	< 28 Dispolene 35 S	Ship spray		1	10-50%	
Fuel oil	40 Fresh	< 28 Dispolene 35 S	Ship aerosol		1	10-50% trials	

KEY: ASMB--Alberta Sweet Mixed Blend; Conc.--Concentrate; H--hydrocarbon; SW--salt water; WSL--Warren Spring Laboratory; (?)--information is uncertain.

[a]Estimated by Nichols and Parker (1985).
SOURCES: Based on literature cited by Fingas (1985) and Nichols and Parker (1985).

dispersant treatment by boat; that weathered oil is not dispersed as effectively as fresh oil; and that a dispersant (J) that performed poorly in laboratory tests also performed poorly in the field. These results are supported by other quantitative studies (Gill et al., 1985; Lichtenthaler and Daling, 1983, 1985).

It must be emphasized that absolute measurement of dispersant effectiveness or quantitative assessment of application technique was not the primary goal of most of these studies—they wanted to find out if laboratory tests comparing a series of dispersant formulations could predict the order of performance in the field, or whether toxic concentrations of petroleum components would be found beneath chemically dispersed slicks. Both questions have been partially answered: laboratory tests can usually predict which dispersant will be more effective; concentrations of toxic petroleum hydrocarbons under oil slicks dispersed at sea are generally low compared to toxicity levels for most organisms (see Chapter 3). The biologically oriented tests, designed to test this last conclusion directly, are described in the next section.

BIOLOGICALLY ORIENTED MESOCOSM AND FIELD STUDIES

This section discusses mesocosm and small-scale field studies of algal, microbial, and planktonic populations and describes arctic studies. Some mesoscale studies in temperate and tropical shallow subtidal areas are also described.

Three significant long-term field studies provide the bulk of this section: the Baffin Island Oil Spill study of an arctic environment; the Long Cove, Maine, study of a shallow north temperate environment; and the Panama study of a tropical mangrove environment. (These and related investigations are reviewed below and results are summarized in Table 4-6 at the end of this chapter.)

Algal, Zooplankton, and Microbial Populations

Algae, zooplankton that graze on them, and bacteria that mediate recycling processes comprise the base of the marine food web and are of concern in the event of an oil spill. A number of field studies have examined the effects of oil and dispersants on these organisms. The effects of Ekofisk crude oil and Corexit 9527 dispersant on flagellate communities were examined using in situ marine mesocosms, 1 m in diameter by 20-m deep (Throndsen, 1982). Changes

TABLE 4-4 Monitoring Methods Used at Dispersant Field Trials

Trial Location	Methodology	Water Sampling Depths (m)	Analysis	Surface Sampling Method	Use
Britain, North Sea	Knudsen bottles	2-15	Fluorescence	Grab	Water content, emulsification
United States, New Jersey	Pumping to bottles	1, 3, 6, 9	FTIR	--	--
Canada, Victoria	Pumping to fluorometer and to bottles, towed fluorometer	1, 3.5	Fluorometer, GC	--	--
United States, Long Beach	Pumping to bottles	1, 3, 6, 9	IR, GC	--	--
United States Long Beach	Pumping to bottles	1, 3, 6, 9	IR	--	--
France, Toulon (Protecmar 1&3)	Pumping to bottles and instrument	0.2, 0.6, 1, 2.5	IR (bottles), turbidimeter (direct)	Sorbent	--
Canada, St. John's	Pumping to fluorometer and bottles	1, 2, 4 done, spoiled	Bottles not	Sorbent	Thickness
Norway, North Sea	Sample bottle	1, 2.5, 5, 9	GC	--	--
Britain, North Sea	Pumping to bottles	1, 3, 6, 9	GC, fluorometer	Sorbent Skimmer	Thickness Weathering
France, Toulon (Protecmar 4&5)	Pumping to bottles and instruments	0.3, 0.7, 1, 2	IR (bottles), fluorometer and turbidimeter (direct)	Sorbent	--
Holland, North Sea	Pumping to bottles	1, 1.5, 2, 3, 7, 10	GC, IR	Sorbent	Weathering
	Towed fluorometer (Q-instrument)	Various	Fluorometers		
	Pumping to instrument	--	Droplet size analyzer (malvern)		

| Dispersant Sampling | Remote Sensing | | References |
	Sensors	Use	
Kromekote cards	--	--	Cormack and Nichols, 1977
--	Color photographs	Documentation	McAuliffe et al., 1980
--	Photo	Documentation	Green et al., 1982
--	Color photographs	Documentation	Smith et al., 1979
Pans	Color photographs and video	Documentation	McAuliffe et al., 1981
--	IR	Slick area	Bocard and Gatellier, 1982
Kromekote cards	IR, UV	Slick area surface tension	Gill and Ross, 1982
	Laser fluorosensor, photo	Documentation	
--	--	--	Lichtenhaler and Daling, 1983
Kromekote cards	IR	Documentation	Cormack, 1983b
--	IR	Documentation	Bocard, 1985
Sorbent paper	IR, UV	Slick area	Delvigne, 1983
GC analysis of recovered oil			

TABLE 4-4 (Continued)

Trial Location	Methodology	Water Sampling Depths (m)	Analysis	Surface Sampling Method	Use
Canada, Halifax	Pumping to bottles and instrument Radioactive tagging (tritiated octadecane) Acoustic spectrometry	1, 2, 5, 10	Fluorometer (direct), GC, IR, fluorometer, counting (bottles)	Sorbent	Thickness
Norway, North Sea	Pumping to bottles and instrument	0.5, 1, 2, 3 (5, 10)	GC (bottles), turbidimeter (direct)	Sorbent	Thickness
Canada, Beaufort Sea	Pumping to bottles	0-4	GC	Sorbent	Oil remaining
Norway, Haltenbanken	In-situ fluorometer and turbidimeter	Most	--	--	--
Canada, Beaufort Sea	In-situ fluorometer	1 --	--	--

KEY: FTIR--Fourier transform infrared spectroscopy; GC--gas chromatography; IR--infrared spectroscopy; SLAR--side-looking airborne radar; and UV--ultraviolet.

in species composition resulted from the addition of 500 ml (32 ppm if uniformly mixed) of oil, which appeared to promote growth of populations of amoebae. No change in flagellate diversity occurred. Adding 100-ml Corexit 9527, resulting in a concentration of 6.4 ppm, altered the community structure substantially, resulting in a shift in dominance to colorless flagellates and coccoid chlorophycae, and reduced abundance overall.

In the system simulating the Baltic Archipelago (Linden et al., 1985, 1987), oil plus dispersant was added to one tank, beginning at 20 ppm oil (nominal concentration), and oil alone (nominally 20 ppm) was added to another beginning at 1 ppm. Both tanks were diluted to background levels by the fourth day, and resulting total exposure in the dispersed oil tank was approximately 80 ppm-hr. Acute effects were observed on numbers of heterotrophic bacteria, on abundance of zooplankton, and on community metabolism, but long-term effects, such as decreased abundance and shifts in diversity, of untreated oil

Dispersant Sampling	Remote Sensing Sensors	Use	References
Filter paper	IR, UV SLAR	Slick area experimental	Swiss and Gill, 1984
--	IR, UV SLAR	Slick area	Lichtenthaler and Daling, 1985
--	--	--	Dickinson et al., 1985
--	IR, UV microwave photography	Slick area and plume movement	Sorstrom, 1986
--	IR, UV photography	Slick area, effectiveness	Swiss et al., 1987a, b

on zooplankton and community metabolism were greater.

Seawater tanks of the University of Rhode Island's Marine Ecosystem Research Laboratory (MERL) were used to examine microbial responses to Kuwait crude oil alone or dispersed by Corexit 9527. Neither led to a larger bacterial population in seawater, but did lead to slightly more hydrocarbon utilizers being present, except at low temperatures (Traxler et al., 1983; Wilson, 1980). Biodegradation potentials were of the magnitude of nanograms per hour. The effect of oil or dispersed oil on hydrocarbon turnover rates was small. Compared to laboratory experiments, hydrocarbon turnover rates were lower in the MERL experiments, and apparently were inhibited because of decreased wall effects and larger water volumes. Despite this result, which was also observed in some laboratory studies (Chapter 3), both Wilson (1980) and Traxler et al. (1983) concluded that using dispersants increases exposure of the microorganisms to the bulk oil and that "in the right circumstances, chemical dispersion of oil does enhance the overall oil biodegradation potential" (Traxler et al., 1983).

Mesocosms consisting of plastic bags 1 to 30 m³ in volume, suspended in seawater and containing a natural population of plankton, were used to investigate the effects of oil and chemically dispersed oil on phytoplankton communities in shallow water (Scholten and Kuiper, 1987). Oil (60 to 1,000 ppb approximate concentrations) diminished primary productivity per unit of chlorophyll, but caused elevated biomass. This effect was attributed to reduced grazing because of mortality of copepods and planktonic bivalves. After an oil spill, a rapid succession from a diatom-dominated to a microflagellate-dominated community occurred, apparently due to exhaustion of silicate. Dispersant generally aggravated the effects. The limited volume of the plastic bag communities retained a higher concentration of dissolved hydrocarbons (compare to studies in Chapter 3, "Laboratory Studies With Dispersed Oil").

The mesocosm studies have shown that effects of dispersants and dispersed oil can generally be attributed to increased immediate exposures of plankton (including microbes) to hydrocarbons after oil is dispersed throughout the water (see Chapter 3, "Laboratory Studies With Dispersed Oil" and "Microbial Degradation"). The observed effects are increased by longer exposure time, but some mesocosm studies showed oil concentrations decreasing only as the result of weathering and evaporation. Such experiments are more representative of isolated water bodies, for example the freshwater ponds examined by Scott and Glooschenko (1984). In open-ocean waters, dispersed oil would normally be diluted before such long exposures to high concentrations could be experienced.

In general, dispersed oil supported increased microbial growth, especially on plant and algal surfaces, and changed the composition of phytoplankton populations. Zooplankton and sometimes phytoplankton were reduced in abundance as well. It is significant that zooplankton were contaminated with oil droplets in the oil-only treatment but not in the dispersed-oil treatment (Linden et al., 1987). This condition probably reflects rapid loss washout of the dispersion or possibly adhesion of untreated oil to the zooplankton, but not the inability of zooplankters to graze the smaller dispersed oil particles.

BIOS Arctic Studies

A major study in the Arctic, the Baffin Island Oil Spill Project, is an excellent example of a large-scale field project with good controls, and a longer time-scale for monitoring effects. Begun in 1980, it consisted of 4 years of multidisciplinary studies of bays off Ragged Channel, in Canada's Northwest Territories (Figure 4-4). The area is pristine. Before the experimental spills it contained low concentrations of biogenic hydrocarbons, and very low amounts (less than 1 ppb) of polycyclic aromatic hydrocarbons, polycyclic aromatic nitrogen compounds, and polycyclic saturated hydrocarbons, presumably from anthropogenic sources (Boehm et al., 1981).[*]

The four bays monitored during the BIOS studies were labeled:

- bay 7, a control;
- bay 9, treated with an underwater release of 94 bbl of oil plus dispersant;
- bay 10, an untreated control adjacent to bays 9 and 11; and
- bay 11, which received an untreated oil slick.

Bay 11 received 94 bbl of a partly weathered (by oxygen bubbling) Lago Medio crude oil released onto the surface waters over 6 hr beginning at high tide. Onshore winds produced an even coating of oil across the intertidal zone. Immediately after the release, samples revealed that more than 44 of the 94 bbl of oil were present in the oiled intertidal zone. The untreated-oil caused no immediate effects on subtidal benthic organisms, but intertidal amphipods and some larval fish were affected by physical coating (Blackall and Sergy, 1983b; Cross et al., 1983). Fluorometer profiles found oil concentrations of 0.01 to 2.8 ppm in the top 1 m of water. Such low subsurface concentrations are consistent with the field studies described earlier.

In bay 9, which received oil plus dispersant, 94 bbl of oil were premixed with 9.4 bbl of Corexit 9527. One part oil-dispersant mixture was diluted with 5 parts seawater and discharged subtidally via a line placed on the seafloor extending 320 m perpendicular to the shore, over 6 hr beginning at high tide. Overall, the floor of the bay was exposed to approximately 50 ppm of dispersed oil; the highest concentrations were 167 ppm representing a maximum exposure of 300 to 500 ppm-hr (see Chapter 3). Pre- and postspill sediment samples were collected daily, weekly, and yearly from transects along

---

[*]References for the BIOS studies include Blackall and Sergy (1981, 1983a,b), Sergy (1985), Boehm et al. (1985), more than 30 reports issued by Environment Canada, and a special 1987 issue of the journal *Arctic*, Vol. 40, Supplement 1, edited by G. Sergy.

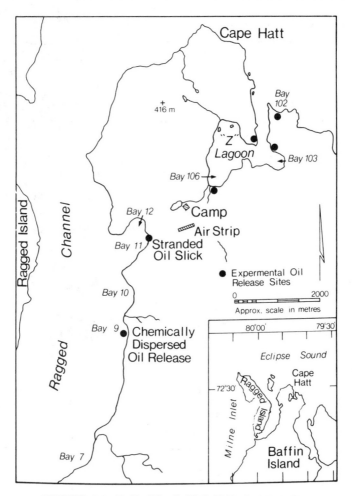

FIGURE 4-4   Baffin Island Oil Spill Project test sites.

the 3- and 7-m depth contours. A few samples were collected from 10-m deep microplots.

The dispersed oil plume moved throughout Ragged Channel (Boehm, 1983). One control area, bay 10, adjacent to the two treated bays, was exposed by currents to dispersed oil from the treated areas. Concentrations at the seafloor were about 5 ppm, one-tenth the exposure in bay 9 (Blackall and Sergy, 1983a,b). Oil was found in bay 9 (dispersed oil) and bay 10 (adjacent control) sediments after 1 to 2 days, and in all four bays after 2 to 3 weeks. Petroleum hydrocarbons in bay 9 sediments reached a peak of 5 ppm, 2 weeks after the

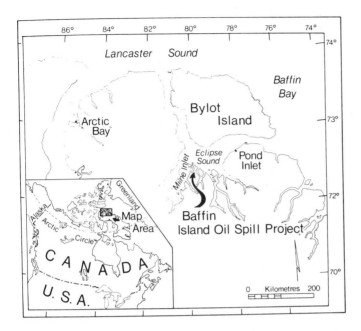

FIGURE 4-4   (Continued).

dispersed oil release. Bay 7, the distant control, had concentrations similar to bay 10 and slightly lower than bay 11. Less than 1 percent of the dispersed oil released in bay 9 ended up in its own sediments.

Macrobenthic organisms were stressed by dispersed oil in bays 9 and 10. Dissolved hydrocarbons—such as hexane, benzene, toluene, and xylene—as high as 9 ppm appear to have caused narcosis (Boehm and Fiest, 1982; Neff and Anderson, 1981). Benthic sediment dwellers, such as clams and polychaetes, surfaced in various conditions of weakness and incapacitation. Elevated concentrations of volatile hydrocarbons in the water column were still measured the day after the spill, but by the second day they were at background levels of 30 to 50 ppb (Boehm et al., 1985, 1987).

Within 1 to 2 weeks, surfaced polychaetes and bivalves reburied themselves and the numbers of sea urchins, previously reduced, were near prespill levels (Cross and Thomson, 1982). Systematic monitoring of benthic populations demonstrated that exposure to dispersed oil did not cause large-scale mortality of benthic animals. After 1 year, there were no statistically significant differences in benthic community composition between bay 9 (dispersed oil) and the control

TABLE 4-5 Hydrocarbons in Sediment Samples (μg/g dry weight[a,b]) Taken at "Tissue Plot" Stations at 7 m Water Depth

Time Period	Distant Control (Bay 7)	Dispersed Oil (Bay 9)	Adjusted Control (Bay 10[c])	Untreated Oil (Bay 11)
Prespill, (1981)	0.47 (0.26, 0.85)	0.18 (0.05, 0.73)	0.33 (0.17, 0.64)	0.51 (0.19, 1.4)
1-3 Days postspill	0.69 (0.49, 0.97)	2.0 (1.4, 2.8)	0.88 (0.52, 1.5)	0.18 (0.07, 0.47)
2-3 Weeks postspill	0.93 (0.58, 1.5)	7.8 (4.3, 14)	1.6 (0.74, 3.7)	1.0 (0.54, 1.9)
1 Year later (1982)	1.2 (0.93, 1.5)	2.2 (1.4, 3.3)	1.7 (1.3, 2.1)	5.3 (2.4, 11.4)
2 Years later (1983)	3.2 (1.1, 9.1)	7.6 (5.7, 10)	--	13 (9.8, 16)

[a] Determined by ultraviolet/fluorescence analysis.
[b] Concentrations presented as geometric means with lower and upper 95 percent confidence limits shown in parentheses.
[c] Bay 10, located between bays 9 and 11, received about 10 percent of the dispersed oil received by bay 9.

SOURCE: Boehm et al., 1987.

bays (Blackall and Sergy, 1983b). Thus, effects from dispersed oil on benthic organisms only occurred over the short term.

Table 4-5 summarizes the combined mean concentrations of petroleum hydrocarbons in sediment samples for the prespill survey, 1- to 3-day postspill samples, 2- to 3-week samples, and yearly averages for 1982 and 1983 (typically 40 to 50 for each bay). The greater persistence in the untreated oil bay is evident and contrasts with the light initial levels in the sediments where oil was chemically dispersed.

After 1 year, petroleum concentrations in almost all water samples were low, less than 0.1 ppb. Exceptions (about 3 ppb) were found in samples from under visible oil sheens emanating from the untreated beached oil in the intertidal region of bay 11 (Boehm, 1983).

In the control (bays 7 and 10) and dispersed oil (bay 9) sediments, hydrocarbon content was only slightly higher than prespill levels, although some bay 9 samples showed relatively undegraded oil, and more aromatics (on an absolute basis) were found in one subtidal sample than were found in the previous year. The increased oil in bays 7 and 9 sediments in 1982 and 1983 may have traveled from bay 11, where the untreated oil was applied.

Subtidal sediments in bay 11 (untreated oil), however, had increased to an average of 10 ppm, with values as high as 66 ppm observed (Boehm et al., 1984). The oil had a patchy distribution, perhaps from subtidal movement of oiled sediment particles. After 1 year more than 31 bbl of the original 94 bbl of oil remained as a source for continued subtidal contamination (Blackall and Sergy, 1983b).

Chemical analysis of the tissue of five species of benthic animals in bays 9 and 10 revealed that they accumulated and depurated significant concentrations of oil the first year after the spill. As an example, concentrations of up to 700 ppb, or μg/liter (dry weight), were attained within 2 days, 2-week postspill body burdens were declining, and after 1 year they were down to 1 to 5 μg/liter.

Increased hydrocarbon content of the subtidal organisms in bay 11 appeared to reflect the increased oil content of the sediments (Cross et al., 1984; Sergy, 1985).

Although the initial impact of dispersed oil was more severe (a subsurface discharge of premixed oil-dispersant mixture certainly introduced more toxic aromatics to the water column than would a surface slick), the persistence of dispersed oil in subtidal sediments was much less (at background level after 1 year) than at the untreated oil site.

Temperate Shallow Subtidal and Intertidal Habitats

Studies of temperate subtidal habitats, like the mesocosm studies of microbes and plankton above, indicate that oil concentration and length of exposure to oil constituents appear to be the controlling factors, not whether the oil is dispersed.

Wells and Keizer (1975) tested the effects of oil and oil plus Oilsperse 43 on sea urchin (*Strongylocentrotus droebachiensis*) populations exposed to samples of water from two shore-based mesocosms (8,000 liters). After 30 days, no mortality was caused by 4-day exposures to water from the mesocosm treated with oil alone, at 40 ppb total extractable organics (by fluorescence). Because of the higher concentration in the dispersed oil mesocosm (250 ppm oil, up to 125 ppm dispersant), greater than 50 percent mortality resulted. Oil plus dispersant was also observed to reduce mobility of the urchins, as measured by the percentage climbing the walls of the test tanks. Such oil and dispersant concentrations could be reached in inshore applications of dispersant to actual oil spills, but would be unlikely to remain so high for a month (Gordon et al., 1976).

In these tests, 50 percent of dispersed oil and 24 percent of undispersed oil was lost by day 22 from the mesocosm. The dispersant itself was well above the acute lethal threshold for urchins (125 ppm), and was considered to be the major cause of mortality.

Mortality, larval settling, spawning, and octopus predation on littleneck clams (*Protothaca staminea*) were examined in small outdoor tanks and in field quadrats (0.5 m²) using Trans-Mountain Western crude oil and Corexit 9527, at 1,000 ppm oil plus 100 ppm dispersant in seawater (Hartwick et al., 1979, 1982). Seven liters of this mixture were poured onto the field plots daily for 5 days. The stock mixture in the tank was allowed to weather naturally outdoors during this period so that each successive application was with a more weathered oil. This experimental design may simulate a usually high and repeated exposure of an intertidal environment to dispersed oil, as might occur in an estuary where a large spill was dispersed but not diluted rapidly by water circulation (Hartwick et al., 1979, 1982).

Reduced siphon activity was apparent on the first day, significant mortality occurred with the dispersed oil after 4 days, but mixing the dispersant with fresh water prior to use reduced mortality. No mortality occurred after 10 days with oil alone. Settlement of larval clams was lowest at highest concentrations of oil plus dispersant (1,000 ppm oil plus 100 ppm dispersants). Octopus predation was substantially reduced when clams were tainted with oil (Hartwick et al., 1979, 1982).

The impact of oil and oil plus Finasol OSR-5 on the benthos in a 13-m² intertidal mesocosm test in West Germany showed that oil-degrading bacterial populations were stimulated by the treatments, particularly by dispersed oil (Farke and Guenther, 1984; Farke et al., 1985a,b). The metabolic activity of benthic diatoms was initially stimulated. Macrofauna populations were undiminished. Feeding activity of clams (*Mya*), cockles (*Cerastoderma*), and polychaetes (*Arenicola*) was initially reduced with dispersed oil, and to a greater degree by oil alone. No penetration of oil into the sediments was observed either with or without dispersant. No toxic effects of dispersed oil were noted.

The effects of oil on blue mussels (*Mytilus edulis*) were studied in a mesoscale experiment simulating the Baltic Archipelago (Linden et al., 1985, 1987). In the dispersed oil tank, oil concentration in the mussels increased more rapidly than in the mussels exposed to oil only. However, by the end of the experiment, the dispersed-oil mussels had added about twice as much shell length as the oil-only mussels, while controls had added three times as much. Mussels exposed to oil plus dispersant exhibited reduced byssal thread production and spawning activity for the first 4 days, but recovered by day 12. With exposure to oil alone, spawning was still abnormal after 12 days (Carr and Linden, 1984).

In 3-month studies in large-scale, flow-through exposure tanks of sublethal responses of invertebrates (i.e., lobsters, scallops, clams, mussels) to nominal 50 ppm light Arabian crude, with and without Corexit 9527, maximum measured concentrations were 1.0 to 2.2 ppm for 6 hr for oil alone, and 12.7 to 19.4 ppm for 6 hr for oil plus dispersant. No mortalities occurred, however, clams (*Mya* sp.) exhibited some reversible changes and reduced shell adductor muscle (Carr et al., 1985, 1986). Other transient sublethal effects were observed in both oil and oil-dispersant treatments. For example, mud snails tried to avoid oil alone and were narcotized in oil-dispersant treatments. Hydrocarbon concentrations were initially elevated in sediments and mussel tissues in the dispersant-treated tanks, with alkylated dibenzothiophenes present in tissue after 21 days. It was not determined from these results whether chemically dispersing the oil was more or less detrimental to the animals than physical dispersion alone (Carr et al., 1986).

When dispersed oil is rapidly removed by water movement, recovery of a habitat can in some cases be more rapid if dispersant is used. In areas with poor circulation, on the other hand, using dispersants can increase the exposure of organisms and habitats to oil, and actually increase damage.

Intertidal Communities

Although the primary focus of this report is the effectiveness of dispersants in open marine areas, a number of valuable intertidal studies have been conducted. They are noted here for the additional understanding they shed on the overall impact of dispersed oil compared with untreated oil.

Eelgrass (*Zostera noltii*) cover decreased when oil alone, dispersant alone, and oil plus dispersant were applied (Baker et al., 1984).

The sand and mud flat experiments showed rapid dispersion of both oil and dispersed oil, with little effect on the meiofauna. However, dispersant led to greater retention of oil in the upper sediment layers, and reduced abundance of the tube worm *Arenicola* sp. (Rowland et al., 1981).

Biological effects of North Sea crude oil and BP1100WD were studied by Crothers (1983). The experiment simulates a situation in which dispersed oil washes ashore. Oil was sprayed onto 2-m² plots during ebb tide, and dispersant was sprayed on the flood tide. Shores covered with seaweed (*Fucus* spp.) were unaffected, and recovery was rapid. As above, limpets and small periwinkles were most affected in the short term, while barnacles showed long-term effects. Of all treatments, oil plus dispersant was most harmful, oil alone had moderate effect, and dispersant alone was not toxic compared with controls.

Dispersants SD LTX and BP1100WD applied to Maui D-sand petroleum condensates* on rocky intertidal plots in New Zealand produced varied results among species for dispersant and weathered versus fresh condensate (Power, 1983). In some cases, dispersant use reduced mortality of barnacles and bivalves exposed to condensate, while in other cases, mortality increased.

A major study in the United Kingdom compared effects of oil and dispersants for a variety of intertidal rocky shore, salt marsh, seagrass, sand, and mud flat habitats. Experimental plots were treated with oil, dispersants (BP1100WD, BP1100X, Corexit 8667, and Corexit 7664), oil plus dispersant, premixed oil plus dispersant, and no treatment (control) (Baker, 1976; Baker et al., 1984; Rowland et al., 1981). Major effects were confined to limpets and periwinkles, which were reduced in population numbers for several months.

Salt Marshes

The dispersant BP1100WD was ineffective in cleaning an oiled salt marsh in the United Kingdom (Baker et al., 1984). Long-term (1 to 2 years) reduction in *Spartina anglica* density and short-term loss of *Salicornia* spp. was noted with both oil and dispersed oil. *Salicornia* recovered after 2 years. In contrast, in Louisiana, *Spartina* salt marsh recovered much more rapidly (Smith et al., 1984). Short-term effects on meiofauna were observed, but by 5 to 10 weeks after

---

*These New Zealand condensates have a high API gravity, low specific gravity.

oiling there were no significant differences between the test plots and the controls.

Although dispersant applied directly to Louisiana salt marsh plants prior to oiling caused reduced biomass by the end of the growing season (Delaune et al., 1984), little evidence of oil or dispersant-caused mortality among the meiofauna was found. The general conclusion was that the Louisiana salt marsh exhibited a low sensitivity to oiling.

Application of weathered Nigerian crude, fuel oil, and mousse to salt marsh, and treatment with a new Type III dispersant (BP Enersperse 1037), showed that total hydrocarbon concentrations in the sediment were less for dispersant-treated oils (Little and Scales, 1987a,b). Dispersant-treated oils were more damaging in the short term, but less destabilizing to the marsh in the long term.

Studies of an Atlantic Coast salt marsh exposed to weathered crude oil and Corexit 9527 provided additional information on the extent of damage and recovery of affected marsh vegetation in three vegetation zones (Lane et al., 1987). Sensitivities of marsh zones ranged from midmarsh (high) to high marsh (low). Both the midmarsh and creek-edge vegetation communities were most sensitive to the oil plus dispersant applications based on a range of morphological, growth, and plant stress parameters; oil alone had the least impact, and dispersant effects were similar to dispersed oil.

Intertidal Areas

Artificial intertidal mudflats were treated with Forties crude oil and Finasol OSR-5 dispersant, and monitored for 10 months (Dekker and van Moorsel, 1987). Dispersant plus oil had more severe short-term effects than oil alone: high mortality in cockles (*Cerastoderma edule*), clams (*Macoma balthica*), and polychaetes (*Arenicola marina*). In both treated areas, *C. edule* was more vulnerable to frost than in the control areas.

Some of these results, which apparently contradict the results of the BIOS and Long Cove studies, are characteristic of dispersant applied directly to oiled shoreline sediments. This is distinct from the situation in which oil dispersed offshore is washed ashore by tides and currents. In the latter case, there is considerably less long-term biological impact than would be observed with untreated oil.

In summary, some experiments reveal more damage to organisms when dispersants are used directly on rocky shores, and some reveal less damage. However, there seems to be no strong, general ecological documentation either for or against dispersant use in these areas.

Dispersant use directly on oiled intertidal and subtidal environments, ranging from mud flats to marshes and seagrass beds, may facilitate the penetration of oil into the sediments and thereby increase ecological damage without decreasing the time necessary for recovery.

Further discussions, including seagrass beds and the recent recommendations of the American Society for Testing and Materials regarding application of dispersants to seagrass habitats, are covered in a later section.

Temperate Shallow Subtidal Studies:
Long Cove and Sequim Bay

A controlled field study involving oil dispersal in a large volume of shallow water is not easy to plan and conduct especially in temperate areas, such as in the United States, where it is difficult to obtain permits to discharge oil for research studies. However, in 1981, such a field study was conducted in Long Cove, Searsport, Maine, comparing the fates and effects of two 6-bbl spills of Murban crude oil, one dispersed and one untreated (Gilfillan et al., 1983, 1984, 1985, 1986). The study was designed to simulate frequent small spills that occur in nearshore Maine waters.

Three shallow (3.5 m) areas were boomed off for the study. In one area, 250 gal of untreated Murban crude oil were released on an ebbing tide. In a second area, 250 gal of crude mixed in an oil-dispersant ratio of 10:1 with 25 gal of Corexit 9527 were released at high-water slack tide and mixed with gates towed by small boats. A third area served as a control, as did samples taken in the two test areas before the oil was released.

The spill of untreated oil, released 1 hr after high tide, coated and adhered to the tidal flat as the tide receded. After two tidal cycles, oil was cleaned from the beach using conventional methods. The spill of crude oil mixed with dispersant was released over the intertidal zone at high tide in a separate section of the cove. The treated oil quickly dispersed as very fine droplets in a light-brown cloud, even under the quiescent (slack tide) conditions. Concentrations of 15 to 20 ppm of dispersed oil (exposure 20 to 30 ppm-hr) were measured 10 cm from the bottom, conforming to the range expected. Water samples taken near the surface and near the bottom showed that chemically dispersed oil lost lower molecular weight hydrocarbons (below n-C_{15}) as the droplets mixed downward (Page et al., 1983, 1984, 1985).

Following the discharge, significant amounts of Murban crude oil were found in sediments exposed to untreated oil, mostly in the upper intertidal zone, but not in sediments exposed to the cloud of dispersed oil (Gilfillan et al., 1983, 1984). Differences between treatments were mostly within one standard deviation (Gilfillan et al., 1986).

Hydrocarbons were found in clams and mussels collected from the untreated-oil site 1 week after the spill, but were absent or near the level of detection in the same species collected from the dispersed oil site (Gilfillan et al., 1984; Page et al., 1983, 1984). In clams and mussels from the untreated oil site, two enzyme systems were markedly elevated after the spill: glucose-6-phosphate dehydrogenase (sugar metabolism) and aspartate amino transferase (protein metabolism). In contrast, the activities of those enzymes at the dispersed oil site were similar to those at the control site (Gilfillan et al., 1984, 1985).

Effects on infaunal communities mirrored the chemical results. At the untreated-oil site some species were reduced in number or eliminated, and there were blooms of opportunistic polychaetes, changes in community structure that are consistent with observations at accidental oil spills. There was no evidence of adverse effects on infaunal community structure from exposure to dispersed oil, but there was clear evidence that exposure to untreated oil adversely affected community structure (Gilfillan et al., 1983, 1984, 1985).

The more severe and long-lasting effects from the untreated oil in the Long Cove study were attributed to greater persistence of oil in intertidal sediments. Dispersed oil, which adheres less to sediments (Harris and Wells, 1979; Little et al., 1980) and is more easily washed from the water column by tidal currents, offered less exposure than untreated oil.

As in the BIOS studies, untreated oil that was stranded on the shore at Long Cove released hydrocarbons slowly, contaminating the subtidal region over a longer time than in the area where the oil had been dispersed. Both the BIOS and Long Cove studies concluded that biological effects attributable to dispersed oil were fewer and more brief. In contrast, significantly reduced diversity and increased dominance by opportunistic species occurred where the spilled oil was not dispersed.

Anderson et al. (1985) studied the fate and effects of Prudhoe

Bay crude, alone and with Corexit 9527, in the intertidal zone of Sequim Bay, Washington. The sediments were placed on trays and oiled in various ways (thorough mixing, layering) and clams were introduced into them. The sediments were placed in the field for 1 to 6 months. The premixing of sand and oil with dispersant exposed the clams to greater concentrations for a longer time than in a typical dispersed oil situation (i.e., Long Cove and BIOS). The dispersant did not affect oil retention time or penetration depth in the sediment. Dispersant presence was correlated with more oil uptake by *Macoma*. Exposures in the field for 1 to 4 months produced equivalent effects in oil and dispersed oil groups (i.e., deaths, contamination, amino acid loss). *Macoma* was more sensitive than *Protothaca* although *Protothaca's* growth was reduced by dispersed oil in the surface sediments. Fate of the oil components was the same in both treatments. This study (Anderson et al., 1985) was the last of several examining the fate and ecophysiological effects of dispersed oils on intertidal mollusks.

Fish

Although fish are one of the primary reasons for deciding whether to employ dispersants to treat oil spills, there are only a few reliable studies that compare the effects of dispersed oil with untreated oil. Laboratory studies on fish eggs and larvae (ichthyoplankton) and on adult fish were surveyed in Chapter 3. In this section, a combined laboratory-field study of salmon homing comes as close to a field study of effects on fish as is currently available.

Salmon have been studied because of their highly developed chemical sense, which might be disturbed by low concentrations of petroleum hydrocarbons (see Chapter 3). The effects of untreated and chemically dispersed Prudhoe Bay crude oil on the homing of salmon were measured in two experiments (Brannon et al., 1986; Nakatani et al., 1983, 1985; Nevissi et al., 1987). In the first study, 24 adult chinook salmon (*Oncorhynchus tshawytscha*) were caught in a freshwater pond, anesthetized, tagged, and divided among four tanks (Nakatani et al., 1983):

- an untreated control group;
- a tank with untreated Prudhoe Bay crude oil as a 0.5 mm thick slick;
- a tank containing 105 ppm of chemically dispersed crude oil (10:1 oil to dispersant); and

- a tank containing 10.5 ppm of freshwater chemical dispersant.

After a 1-hr exposure, the salmon were removed from the exposure tanks, held overnight in a raceway, trucked 9 km downstream, and released. Of the 215 fish released, 154 (72 percent) returned to the pond where they were caught. There was no significant difference in the percentage of return among the four groups, nor in the time it took the fish to return.

Tests with coho salmon (*O. kisutch*) produced similar results. The methods used were the same, except that seawater was used instead of fresh water in holding and exposure tanks, and the tanks were outdoors, not indoors. Again, there was no statistically significant difference in the percentage of returns or time to return.

Nakatani et al. (1985) concluded that, as for fresh water, there was no reason to believe short-term exposure to Prudhoe Bay crude oil had any deleterious effect on homing success. A lower mean percentage return for coho salmon was attributed to an extensive gillnet fishery near their release point (Nakatani et al., 1985).

However, oil avoidance was observed in these studies. Before the untreated oil was added to the exposure tank, the salmon swam throughout the tank. After oil was added, the fish swam to the bottom of the tank and remained there during the 1-hr exposure.

Chemical senses are considered essential to homing in salmon, but these studies concluded that the olfactory systems were not impaired enough from a 1-hr exposure to interfere with homing. Histopathological analysis of the olfactory organs of the salmon showed no anomalies. Thus, while adult salmon might avoid oil (Weber et al., 1981), forced brief exposure to whole Prudhoe Bay crude oil or chemically dispersed oil at high concentrations did not prevent or delay homing.

Tropical Shallow Intertidal and Subtidal Habitats

Seagrasses

The potential impact of dispersing an oil spill over a shallow area with a seagrass-based benthic community is of considerable concern in tropical areas. Such areas are highly diverse and productive, hence inherently valuable. Because dispersal of an oil slick introduces higher concentrations of toxic hydrocarbons into the water column in the short term, but tends to decrease the residence time of oil in the long run, it is important to compare chemically dispersed oil with physically dispersed (untreated) oil on a common basis. As with laboratory studies, when effect threshold was based on total oil per unit volume (nominal concentration), chemically dispersed oil apparently had a greater impact; when effect threshold was based on actual water-accommodated hydrocarbons, there was generally little difference in the toxicity of chemically and physically dispersed oils. In this section, both laboratory and field studies with seagrasses are reviewed.

Baca and Getter (1984) found similar toxicity of Prudhoe Bay crude and chemically dispersed oil to the seagrass *Thalassia testudinum*, using hydrocarbon concentrations measured by gas chromatography. The dispersant (Corexit 9527) had similar acute toxicity to the dispersed oil, but caused more leaf discoloration at sublethal and LC_{50} doses.

Thorhaug and Marcus (1985, 1987a,b) and Thorhaug et al. (1986) studied the effects of a range of concentrations of dispersed oils (Murban and Louisiana crudes, and seven dispersants) on three major Caribbean and Gulf of Mexico species of subtropical and tropical seagrasses grown in 100-liter, aerated seawater aquaria (Table 4-6). The mortality percentage of the three grasses, exposed for 100 hr to 75 ml of Louisiana crude oil in the aquarium (approximately 750 ppm nominal concentration) and 75 ml of dispersant (as instructed for that dispersant), was measured. In the aquaria containing chemically dispersed oil, the measured concentration of total water-accommodated hydrocarbons was 429 to 634 ppm; in the untreated oil aquaria the corresponding concentration was 5.6 to 10.4 ppm (Thorhaug et al., 1986). Therefore it is not surprising that chemically dispersed oil showed a greater negative effect on species survival and growth than oil alone.

In other studies, exposures were measured by gas chromatography and toxicity was measured as health, that is, browning, yellowing, and spotting of young blades (Thorhaug and Marcus, 1985; Thorhaug et al., 1986). Growth rates were reduced only at extremely high exposures: when the three grasses were exposed for 100 hr to 750 ppm (nominal concentration), *Thalassia* was more resistant than *Syringodium filiforme* and *Halodule wrightii*. Effects were observed after 100 hr at 1,250 ppm. Response varied by species, but effects were generally greater when dispersants were present. Mortality followed the same order of sensitivities as growth.

At a dosage level of 1 part (150 ppm) dispersant, 10 parts oil

TABLE 4-6 Mortality Percentage of Seagrasses Caused by
Louisiana Crude Oil and Dispersants

Dispersant[a]	Thalassia	Halodule	Syringodium
Finasol OSR-7	0	7	7
Jansolv-60	10	41	12
Cold Clean 500	16	21	18
Corexit 9550	22	58	63
Corexit 9527	26	76	84
OFC D-609	35	73	86
Conoco K(K)	65	97	98

[a]All treatments contained oil and dispersant.

SOURCE: Thorhaug and Marcus, 1987a,b.

(1,500 ppm), and 10,000 parts water, no significant mortality occurred (Thorhaug and Marcus, 1987a,b).

The American Society for Testing and Materials (1987) has reviewed the literature to 1985 on the effects of oil and dispersants on seagrasses, and recommends the following use guidelines:

- If there is a possibility that an oil slick will strand on intertidal portions of seagrass beds, dispersant use would be most effective while an oil slick is still offshore to prevent the oil from impacting the grass bed.
- Dispersant-use decisions to treat oil already over a seagrass bed should ultimately take into account the depth of the seagrass bed and the potential for dilution of the dispersed oil.
- The use of dispersants over shallow submerged seagrass beds is generally not recommended, but should weigh the potential impact to the seagrass beds against impacts that might occur from allowing the oil to come ashore.
- Dispersant use should be considered to treat oil over seagreass beds in water deeper than 10 m if the alternative is to allow the oil to impact other sensitive habitats onshore.
- Dispersant use is not recommended in shallow lagoons or areas of restricted flushing rates.

Related field and ecological studies of the effects of oil and dispersed oil on salt marshes were discussed earlier.

Coral Reefs

Admired throughout the world for their beauty and biological productivity and diversity, coral reefs are generally considered a habitat worth protecting from oil spill damage. As with seagrass communities, there is concern that dispersal of an oil slick in shallow water near or above a coral reef might cause greater damage than the untreated oil, because of the higher concentrations of hydrocarbons introduced into the water column.

Legore et al. (1983) studied the effects of untreated light Arabian crude oil and chemically dispersed oil (20:1 Corexit 9527) on corals that were submerged 1 m at low tide. The individual plots (2 by 2 m) were boomed with the skirts (3.5 m) extending to the bottom, even at high tide. The initial nominal oil concentrations for the 24-hr exposures were that of a slick 0.25-mm thick. Thus, the dispersed oil concentration (including dissolved hydrocarbons) would have been 250 ppm if mixed uniformly. The slick remained on the surface except for oil that may have adhered to the walls of the boom. In 5-day experiments, oil equivalent to a 0.1-mm thick slick was added initially and the same amount added on days two, four, and five.

One-day treated corals showed normal appearance and tolerated the short exposures. Following 5-day exposure, coral recovered more slowly from seasonal bleaching. There was also some long-term reduction in growth for both oil treatments. The exposures used are high and were produced by restricting water flow by the boom skirts.

A study to simulate the effects of an oil spill moving over a coral reef with and without the use of dispersants was done at Bermuda (Cook and Knap, 1983; Dodge et al., 1984, 1985a,b; Knap, 1987; Knap et al., 1983, 1985; Wyers, 1985; Wyers et al., 1986). Corals were exposed both in the laboratory and on the reef to Arabian light crude dispersed with 1:20 Corexit 9527 or with 1:10 BP1100WD. Some corals were taken to the laboratory and then returned to their home on the reef for recovery studies; some were actually oiled on the reef using an enclosure. Exposures were at levels of 11 to 23 ppm for periods of 6 to 24 hr and were similar for oil alone and dispersed oil. There were no significant differences between the oil and dispersed oil treatments in tissue rupture, contraction or swelling, mesenterial filament extrusion, or pigmentation loss. Most of the observed stress occurred during dosing. Recovery began 24 hr after oiling and was complete in less than a week.

One of the few apparently synergistic effects was reduced photosynthesis of the zooxanthellae (symbiotic algae) within the coral resulting from an 8-hr exposure to 19 ppm dispersed oil and inhibited synthesis of lipids, particularly wax esters and triglycerides (Cook and Knap, 1983). Oil or dispersant alone had no such effect. Carbon fixation was restored within 5 hr after exposure ceased, and lipid synthesis returned to normal within 5 to 24 hr. The smaller droplets in the chemically dispersed oil did not adhere to the corals in contrast to the physically dispersed oil droplets, some of which were found on coral a few weeks after exposure to 20 to 50 ppm oil alone.

In summary, when corals were exposed to oil and dispersed oil under field conditions, exposure to hydrocarbons was greater in dispersant-treated plots than in untreated oiled plots. However, no effects were observed on growth of corals exposed either to oil or to dispersed oil for 24 hr and measured 1 year after the exposure. Corals exposed for 5 days to the oil or dispersed oil showed reduced growth in comparison to the controls. "Exposure to dispersed oil also appeared to delay recovery of corals under stress by cold temperatures" (ASTM, 1987, based on Birkeland et al., 1976 and Fucik et al., 1984).

As with other organisms and habitats, the primary factor is exposure to the water-soluble fraction of the oil. When toxicities are based on such analytical measurements, there are no major differences between physically and chemically dispersed oil. Knap et al. (1985) concluded that "in the long term, *Diploria strigosa* appears relatively tolerant to brief exposures to crude oil chemically dispersed in the water column." However, these experiments were limited to only one coral species, and other reef organisms, such as crustaceans, echinoderms, and other invertebrates, are known for their sensitivity to both chemically and physically dispersed oil. Therefore protection of the entire reef community, perhaps by dispersal of the oil offshore, is a priority.

The American Society for Testing and Materials has published a standard guide for the use of chemical dispersants in the vicinity of coral reefs (ASTM, 1987). It recommends the following:

• Whenever an oil spill occurs in the general vicinity of a coral reef, the use of dispersants should be considered to prevent floating oil from reaching the reef.

• Dispersant-use decisions to treat oil already over a reef should take into account the type of oil and location on the reef.

• Coral reefs with emergent portions are high-priority habitats for protection during oil spills.

• The use of dispersants over shallow submergent reefs is generally not recommended, but the potential impacts to the reef against impacts that might occur from allowing the oil to come ashore should be weighed.

• Dispersant use should be considered to treat oil over reefs in water depths greater than 10 m if the alternative is to allow the oil to impact other sensitive habitats on shore.

• Dispersant use is not recommended to treat oil already in reef habitats having low-water exchange rates (e.g., lagoons and atolls) if mechanical cleanup methods are possible.

Mangroves

Mangroves grow along tropical and subtropical shorelines. As intertidal forests, they are important protectors of shorelines from storms and currents, they provide habitat and nursery areas for many organisms, some of which are commercially important, such as lobsters, prawns, shellfish, and finfish (Teas, 1979), and they contribute to the overall productivity of tropical marine environments. Their importance as nursery areas and nutrient sources cannot be overstated.

Experiments on mangroves are relevant to how the organic fraction of sediment and suspended particulate matter interacts with oil and dispersed oil (Getter and Ballou, 1985; Getter et al., 1985). A mangrove study begun in 1978 in Panama, indicated that exposure of a mature red mangrove forest to oil and dispersant resulted in many of the effects observed in the laboratory and at other spill sites: changes in growth, respiration, transpiration, and uptake of petroleum hydrocarbons (API, 1987). These effects were reduced at the site treated with oil and dispersant compared to the site treated with oil alone (Getter and Ballou, 1985; Getter et al., 1985).

In the Panama study two similar mangrove sites were selected and boomed, and mangrove trees, seagrasses, and associated biota were surveyed. One site received 380 liters (2.4 bbl) of Prudhoe Bay crude oil, and the other site received the same quantity of oil mixed with 19 liters (5 gal) of Corexit 9527. This oil-dispersant mixture was formulated to simulate oil dispersing on the water within or immediately adjacent to the mangrove forest. Thus, the dispersed oil concentrations in the water were presumably higher than if the oil had been dispersed before reaching the mangroves.

Oil premixed with dispersant did not readily adhere to sediments or algal mats; it lightly coated prop roots and was evenly distributed on the forest floor within two tidal cycles. The dispersed oil was mostly removed from surface sediments and algal mats within 1 week, after which only a noticeable sheen remained. Similar results were obtained in the BIOS and Long Cove studies. The three studies should be distinguished from those that treated oiled intertidal sediment directly with dispersant.

Although less oil was retained in the mangrove peat or organic sediments when dispersant was used, the proportion of aromatic compounds retained was greater. Dispersed oil was taken up much more rapidly by mangrove seedlings, but it did less long-term harm because tidal flushing rapidly removed the dispersed oil from the surfaces of the sediments and algal mats. However, some undispersed oil accumulated on the wrack at the high-tide swash line.

Untreated oil formed a slick over the enclosed area and eventually was pushed well into the mangrove forest by waves and rising tides, and was retained there, contaminating substrate, prop roots, beach wrack, intertidal algal mats, crab burrows, and depressions in the forest floor. Sediments and prop roots at the outer edge of the forest were cleaned of heavy oil contamination after several tidal cycles, but interior areas remained heavily oiled after 1 week. Heavily oiled wrack was evident throughout the site, especially at the high-tide swash line.

Inspection of the sites 120 days later confirmed these observations. Where untreated oil contacted the mangrove site, 60 to 70 percent of the plants were dead or defoliated. Where the equal volume of chemically dispersed oil came ashore, there was no evidence of oil and very little damage (5 percent defoliation). This is one of the more dramatic examples of how chemical dispersal of oil can diminish the impact compared to untreated oil (Getter and Ballou, 1985).

In Malaysia, another study involving mangrove trees, using oil and Corexit 9527, gave variable results between oil versus oil plus dispersant (Hoi-Chaw, 1986; Hoi-Chaw and Meow-Chan, 1985; Lai and Feng, 1985). Treatments were most toxic to seedlings. Most trials showed either no difference or that oil alone was more toxic than dispersed oil. Mortality was found to be related to surface deposition and uptake. The largest accumulation of oil components occurred in leaf tissue (Lai and Feng, 1984, 1985).

In experiments to find how to save mangroves that were already oiled, Teas et al. (1987) treated mangrove trees with oil and after 1 day washed them with high-pressure seawater or with a nonionic water-based dispersant. Oil killed many of the trees (within 30 months) whether or not they were spray-washed the next day.

To simulate the case in which the mangrove forest was impacted by oil dispersed offshore, some plots received oil plus glycol ether-based dispersant. These plots did not show significantly more deaths than untreated control plots. The results of this study imply that it is not possible to save trees already oiled by washing them with dispersant, but that if the oil is dispersed offshore, the trees can be protected.

In summary, mangroves can be protected from oil damage by dispersing the oil offshore, but if the untreated oil comes into the forest it can cause serious harm that may require 20 years or more for mature tree regrowth.

SUMMARY

Boehm (1985) in summarizing the BIOS studies concluded, in agreement with Mackay and Hossain (1982), that chemical dispersion reduced the affinity of oil for solid particles as long as the dispersant-oil micellar association persisted. Further, dispersal reduces the probability of an oil mass coming ashore, and often reduces the long-term impact of that which does reach the shore or intertidal zone. This is borne out by most of the field studies reviewed above (Table 4-7).

Dispersants never increased sorption of oil to sediments and in a few experiments decreased sorption of oil to organisms. The distinction must be made between dispersed oil coming ashore, which generally did not penetrate sediment, and dispersant applied to an oiled intertidal sediment, where treated oil usually penetrated more deeply than the untreated oil.

In subtidal areas, use of dispersants may increase toxicity to benthic fauna and plants, at least in the short term, but may reduce long-term effects of oil. In some intertidal areas, such as mud flats, dispersed oil coming in with the tide had little or no effect, while untreated oil caused longer-term effects (e.g., Long Cove, Searsport, Maine study). Once oil has penetrated salt marshes, the best approach is to leave it alone. Dispersant applications in a marsh show little benefit, as is the case with oil stranding on a beach.

Benthic organisms respond differently to oil over the short versus

TABLE 4-7 Summary of Major Microcosm and Field Studies of Chemically Dispersed Oils, 1970-1987

Treatments	Results	References
	Mesocosm Studies--Water Column	
Norway 500 ml Ekofisk crude, 1,200 ml Corexit 9527, in 10 m^3 in situ enclosures; D, O, O+D	O caused changes in flagellate community, increased amoeba. O+D led to significant community alteration, reduced flagellate numbers	Throndsen, 1982
Rhode Island (MERL) United States Kuwait crude (15 ppm); Corexit 9527 or BP1100WD in mesocosms; D, O, D+O	Small response by hydrocarbon utilizers to O, O+D. Dispersion may enhance oil biodegradation	Traxler et al., 1983; Wilson, 1980
CEPEX, British Columbia Prudhoe crude (200 g); Corexit 9527 or BP1100WD in mesososcm; C, O, O+D	O+D depressed primary production over short term. O+D stimulated heterotrophic bacteria, reduced predators; more toxic than O to ecosystem overall, reducing centric diatoms and all zooplankton. D should not be used during spring bloom	Acreman et al., 1984; Lee et al., 1985; Parsons et al., 1984; Whitney, 1984
Bedford Institute of Oceanography, Nova Scotia, Canada Guanipa crude, Oilsperse 43 (250 ppb) in mesocosms; O+D	O+D reduced mobility and increased mortality of sea urchins as compared with O	Gordon et al., 1976; Wells and Keizer, 1975
	Mesocosm Studies--Fish (Salmon)	
Washington State, United States 105 ppm Prudhoe crude, 10.5 ppm dispersant; C, O, O+D, D lab exposures	No effects on homing behavior of O, O+D. Lethal concentrations much greater than those found in field spill situations	Brannon et al., 1986; McAuliffe, 1986; Nakatani et al., 1983, 1985; Weber et al., 1981

TABLE 4-7 (Continued)

Treatments	Results	References
	Mesocosm Studies--Littoral	
Sweden 20 ppm Ekofisk crude, Corexit 9550 (1:10, D:O) in mesocosm pools, C, O, O+D	Bacteria increased by D. Phytoplankton, periphyton productivity increased by O+D. Net community productivity initially decreased by O, O+D; later increased by both. Zooplankton decreased by O, O+D. Mussel bioaccumulation greater with O+D. Recovery from effects in bivalves and crustaceans more rapid with O+D	Linden et al., 1985, 1987; Carr and Linden, 1984
Duxbury, Massachusetts, United States 20 ppm North Sea crude, 2 ppm Corexit 9550; nominal concs., 50 ppm Arabian crude, 2.2 ppm Corexit 9527; C, O, O+D in mesocosms	Sublethal effects to macrobenthos of O, O+D. Use of dispersant not more nor less harmful	Carr et al., 1985, 1986
West Germany 2-4 ppm Arabian crude, Finasol OSR-5 (1:10 D:O) in 13 m² intertidal enclosure, O, O+D	O+D stimulated bacteria, benthic diatoms. Macrofauna feeding reduced with O+D and especially O	Farke et al., 1985a, b; Farke and Guentner, 1984
	Field Studies--Littoral/Sublittoral	
Arctic/Baffin Island (BIOS), Canada Two nearshore spills of 15 m³ Venezuelan crude, one dispersed with premixed Corexit 9527 (1:10 D:O);	O+D caused greater bioaccumulation in filter feeders while O led to long-term bioaccumulation in deposit feeders,	Blackall and Sergy, 1981, 1983a,b; Boehm, 1983, 1984; Boehm et al., 1984, 1985;

dispersed oil measured at between 5-50 ppm in subtidal zone after spill	Bacterial in sediment increased with O, O+D. Use of dispersant recommended to reduce benthic exposure or clean vulnerable shore-line habitat, not recommended without clear need	Bunch et al, 1985; Cross et al., 1983; Cross and Thomson, 1982; Eimhjellen et al., 1983; Engelhardt et al., 1984, 1985; Landrum et al., 1987; Mageau et al., 1987; Neff et al., 1987; Sergy 1985, 1987

Temperate Studies--All Shore Types

Rocky Shore, Wales Forties crude, BP1100WD or Corexit (1:22 D:O), C, O, O+D, D; salt marsh: Forties crude, BP1100WD or Corexit (1:25 or 1:14, D:O), C, O, O+D, D	Rocky shores: limpets and periwinkles reduced by O+D. Salt marsh: dispersant ineffective at cleaning oil. Vegetation reduced by O, O+D. Seagrass: O, O+D reduce vegetation. O+D leads to greater sediment retention. Mud and sand flats: greater retention in sediment of O+D	Baker, 1976; Baker et al., 1984; Levell, 1972, 1973, 1976; Little and Scales, 1987a,b; Rowland et al, 1981
Rocky Shores, Somerset, England Forties crude, BP1100 sprayed onto C, O, O+D, D	O+D most harmful to limpets and periwinkles. Recovery from treatments rapid	Crothers, 1983
Rocky Shores, England Laboratory: light Arabian crude (1-10,000 ppm), Corexit 9527 or BP1100WD (0.1-1,000 ppm)	Toxicity to red algae of Corexit 9527 plus oil greater than BP1100WD plus oil	Grandy, 1984
Rocky Shores, Anawhata, New Zealand Dispersants (Shell SD LTX, BP1100WD, BP1002) applied to field plots at level of 0.6 liters/m^{-2} C, D (no oil)	Newer, less-toxic dispersants showed no effects on barnacles or bivalves	Power, 1983
Sandy Shores, Battelle, Sequim Bay, Washington Clams exposed to 2,000 ppm Prudhoe crude, Corexit 9527 applied directly to mud (1:10, D:O); C, O, O+D	More uptake of O+D by some mollusks. Reduced growth of some mollusks with O+D	Anderson et al., 1985

TABLE 4-7 (Continued)

Treatments	Results	References
Sandy Shores, British Columbia, Canada 100 to 1,000 ppm BC crude; 100 and 1,000 ppm Corexit 9527; C, O, O+D, D in lab tanks and field plots	Mortality to clams with O+D, but not O. Siphon activity reduced by O+D. Larval settling reduced by O+D. Predation on tainted clams lower	Hartwick et al., 1979, 1982
Sandy Shores, Long Cove Searsport, Maine, United States Field release of 250 gal Murban crude with Corexit 9527 (1:10 D:O), C, O, O+D	Greater sediment absorption of O+D. Effects more serious, longer lasting with O. Less bioaccumulation of O+D in bivalves	Gilfillan et al., 1983, 1984, 1985; Page et al., 1983, 1984, 1985
Seagrass, Florida, United States Louisiana crude Corexit 9527 (1:10 D:O) 0.75, 7.5, 12.5, and 550 ppm in lab; C, O, O+D	Effects on mortality and growth generally greater with O+D	Thornhaug and Marcus, 1985, 1987a,b; Thorhaug et al., 1986
Salt marsh, Louisiana, United States Louisiana crude (2 liters/m^{-2}); unspecified dispersant (0.3 liter m^{-2}) in 6 ml field plots; oil and dispersant in lab; C, O, O+D	Salt marsh had low sensitivity to oiling. O+D reduced macrofauna. O reduced macrophyte biomass	Delanune et al., 1984; Smith et al., 1984
Mangrove, Malaysia Arabian crude (0.005-1.2 ml/cm^2) applied in lab tanks and field with Corexit 9527 (1:20, D:O); C,O, O+D	O alone generally more toxic to mangrove than O+D	Hoi-Chaw, Meow-Chan, 1984, 1985; Lai-Feng, 1985

Mangrove, Panama Two 100-gal spills of Prudhoe crude in situ, one treated with Corexit (1:20 D:O); O, O+D	O+D increased oil mobility, reduced amount stranded. O+D produced short-term effects in crabs, fish, and sponges (lethal and sublethal): O affected crabs most. Mangrove mortality higher, sprouting lower with O. Normal with O+D. Defoliation with O, O+D but recovery better with O+D. Dispersant caused more short-term toxicity, but better community recovery	Baca and Getter, 1984; Ballou et al., 1987; Getter and Ballou, 1985; Getter et al., 1985;
Coral Reef, Jurayd Island Arabian crude in 0.25 mm slick (1-day exposure) or 0.1 mm slick (5-day exposure), with Corexit 9527 (1:20 D:O) in 2 x 2 m enclosed plots; C, O, O+D, D	Long-term effects of both O,O+D noted, although coral relatively tolerant of both	Legore et al., 1983
Coral Reef, Berumuda Arabian crude (1-40 mg/liter) with Corexit 9527 or BP1100WD (1:10 D:O) applied in lab tanks; C, O, O+D, D	Direct contact with O alone most damaging. O+D results in short-term behavioral changes, reduced photosynthesis, and possibly reduced growth. Coral generally tolerant of O+D	Knap et al., 1983, 1985; Dodge et al., 1984, 1985a,b; Cook and Knap, 1983; Knap, 1987; Wyers, 1985; Wyers et al., 1986

KEY: C--control; O--oil; O+D--oil plus dispersant; D--dispersant.

the long term. Dispersed oil may occur in sufficient concentrations to cause immediate mortality for various groups, including commercially important shellfish. Reproduction and feeding behavior may also be negatively affected. Bioaccumulation and tainting occur to a greater extent in the short term in filter feeders when dispersants are used, since concentrations in the water are elevated. Over the long term, more bioaccumulation of oil occurs in deposit feeders when dispersants are not used.

In general, it appears that intertidal and subtidal macroalgae (seaweeds) can be damaged by heavy oiling as measured in laboratory studies but are often not damaged in field oiling situations. Dispersant use would not increase damage and, based on one study in the United Kingdom (Crothers, 1983), might decrease it. Experimental observations and results are scarce and often contradictory; in some field experiments, dispersed oil had no effects on fucoids, whereas in other exposures, red algae suffered tissue damage and inhibited growth when chronically exposed to dispersed crude oil. Only first-generation dispersants have been shown to cause major effects on seaweeds on shorelines.

Effects on vascular plants vary depending on species and habitat type. For subtidal seagrasses and vegetation in salt marshes, dispersant use directly on vegetation or in shallow waters with low circulation may increase the harmful effects of oil because of increased hydrocarbon concentration in subsurface waters and subsequent absorption by plants. Dispersants are not especially effective in preventing damage to oil-covered plants in low-energy salt marsh environments.

Tropical mangroves may be protected if the oil is dispersed before an untreated slick strands within the mangrove forest. This is especially true for mature mangroves. Short-term toxicity to individual organisms within the mangrove ecosystem may be higher, but community recovery is enhanced by the oil being dispersed prior to entry.

On coral reefs the use of dispersants, if it is able to reduce exposure to oil, will benefit the reef in the long run even though there may be short-term deleterious effects on photosynthesis of symbiotic algae within the coral and on other reef organisms.

Use of dispersants does not appear to affect homing of salmon.

5

How Dispersants Are Used: Techniques, Logistics, Monitoring, and Application Strategies

DESIGN OF DISPERSANT APPLICATION SYSTEMS

Application systems are designed to meet, within practical limits, the following basic criteria that were discussed in Chapter 2:

- The dispersant must be sprayed on the oil.
- The dispersant must mix with the oil and move to the oil-water interface.
- The dispersant must attain the proper concentration at the oil-water interface, ideally causing maximum reduction of interfacial tension.
- Sufficient energy must be supplied from natural or artificial sources to disperse the slick into droplets.

How these four goals are accomplished in practice is the subject of this chapter.

Another factor that influences the application of dispersants is the surface area of the spill. Most oils spilled on water attain an average thickness of 0.1 mm or less within a few hours. Therefore, the surface area that must be sprayed to control the spill is approximately proportional to the volume of the oil spill.

Spray Systems

Systems must be able to apply dispersant uniformly to the slick; ideally the droplets will be small enough to descend to the oil surface at a relatively low velocity without penetrating through the oil into the water below, yet large enough that they will not be carried away by the wind (Smedley, 1981). Thus, the primary objective in designing a spray apparatus is to achieve relatively uniform application without undue wind drift losses. Oil thickness or other factors do not influence the design of the apparatus (Exxon Chemical Company, 1985). Application rate per unit area is usually varied by changing pump rate, boat or aircraft speed, or number of spray nozzles, or by repeat application (Belore, 1985; Chau et al., 1986). A new approach uses an air jet—called a "dispersant spraying gun"—which is capable of treating a 0.1- to 0.2-mm-thick oil slick from a distance of 30 m (Barbouteau et al., 1987).

Spray systems can be mounted on a boat, a fixed-wing aircraft, or a helicopter. They usually consist of spray nozzles mounted on a manifold (the spray boom), a reservoir for the dispersant, pumps, meters, valves, and other controls (API Task Force, 1986; Exxon Chemical Company, 1985; Lindblom, 1979).

Small-scale applications on offshore structures, along shorelines, and near piers and bulkheads can rely on hand spraying. Typical equipment ranges from a 5-gal (19-liter) portable tank with spray wand (ITOPF, 1982) to gasoline or electric pumps with reel-mounted hoses and sprayers.

Spray Droplet Size

Droplet size strongly influences dispersant effectiveness, but it is not easily controlled. Several studies have shown that if droplet size is larger than the oil film thickness, the dispersant tends to penetrate the film without mixing and is diluted by the water. If larger droplets predominate, effectiveness is diminished (Gill and Ross, 1980; Smedley, 1981).

Gill and Ross (1980) argue that the advantage of spraying a larger-sized droplet, which improves the probability of hitting the target, outweighs other considerations in successfully applying dispersant via aerial spraying. Little quantitative data exist on how droplet size distribution affects the dispersant fraction that actually reaches and interacts with oil on the water surface. Although exact droplet size can be measured in the laboratory by electromechanical methods, spray patterns from aircraft are normally studied over land, using Kromekote cards to preserve the droplet pattern (used in the API 1979 southern California field trials; McAuliffe et al., 1981).

Smedley (1981) found reduced effectiveness with larger droplet sizes, although not as much as would be expected based on film penetration. Penetration is dependent on oil viscosity: higher viscosity requires greater kinetic energy for the falling droplet to penetrate the slick. Smedley also pointed out that given the same droplet size distribution, dispersants tend to be less effective on thinner slicks, possibly because there is greater penetration. He found, however, that all droplets smaller than 400 μm in diameter have about the same effectiveness. This is similar to the typical range of slick thicknesses, 20 to 200 μm (Chapter 2). There is some evidence that hydrophobic dispersant droplets that penetrate a film are not lost; they resurface on the underside of the film and hence are particularly effective (McAuliffe, private communication).

Airborne test data show that 100-μm droplets are not effectively deposited from aircraft because of windage loss, whereas 500-μm droplets are effectively deposited (Figures 5-1 and 5-2; Smedley, 1981). Therefore 350 to 500 μm appears to be a reasonable size range. This range is produced naturally with more viscous formulations under typical air shear and is practical, since the droplets will hit the target and not drift away on the wind (Lindblom and Cashion, 1983). In any event, only limited control over droplet size is possible. Fine droplets can easily be achieved in the air, but windage loss defines the lower usable droplet size. The upper size is controlled within limits by aircraft speed, pump capacity, and nozzle size, but large drops are broken up by air turbulence (Hornstein, 1973).

Distance between drops also plays an important role in determining efficiency. To achieve good coverage (i.e., acceptably small drop-to-drop distances) requires a droplet diameter smaller than 700 μm (Smedley, 1981). For an oil film 0.02 to 0.2 mm thick, generally acceptable for dispersant application (Exxon Chemical Company, 1985), a dosage between 20 and 70 liters/ha (2 and 7 gal/acre) is recommended.

Mechanical system design factors that strongly affect droplet size distribution are nozzle orifice diameter, number of nozzles, fluid pumping rate, and aircraft air speed (Lindblom and Cashion, 1983). These factors affect the shear acting on the fluid as it passes through the nozzle and enters the air stream behind the aircraft. Mechanical shear at the nozzle outlet, under laminar flow, varies inversely with the cube of the orifice diameter:

$$\text{Shear rate (sec}^{-1}) = 8V/d = 1.7 \times 10^5 Q/d^3,$$

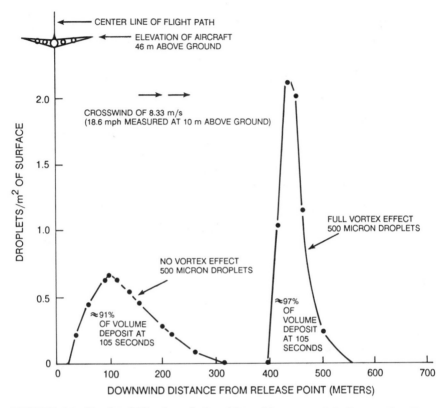

FIGURE 5-1   Droplet (500 microns) deposition with a crosswind. Source: Smedley, 1981.

where V = linear velocity of fluid at the orifice in mm/sec, Q = flow rate in liters/min (or 3.785 × gal/min), and d = orifice diameter in mm.

Orifice, flow rate, and the number of nozzles used are chosen to keep the mechanical shear below 10,000 sec^{-1} (Exxon Chemical Company, 1985). Fortunately, the range of droplet sizes naturally produced with more viscous formulations under typical air shear is in the range recommended above, 350 to 500 μm (Lindblom and Cashion, 1983).

An even stronger effect results from air shear, which increases with aircraft speed or decreasing fluid exit velocity from each nozzle:

$$\text{Fluid exit velocity (m/sec)} = 21 \ L/nd^2,$$

where L = total system flow in liters/min, n = number of nozzles,

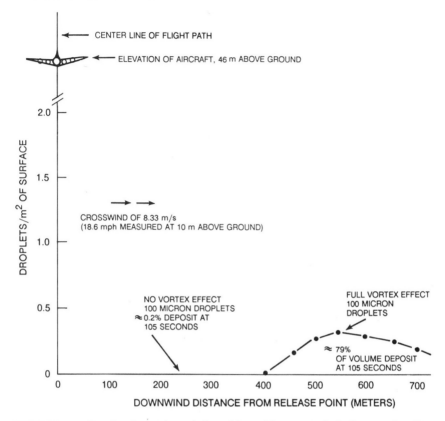

FIGURE 5-2   Droplet (100 microns) deposition with a crosswind. Source: Smedley, 1981.

and d = inside diameter of nozzle (mm). Subtracting the fluid exit velocity from the air speed of the aircraft (also in m/sec) yields the differential velocity. Differential velocity is the single most important factor affecting droplet-size distribution and therefore depositional efficiency. Extensive testing by API and the Exxon Chemical Company has shown that depositional efficiency drops off markedly if differential velocity is more than about 64 m/sec (Lindblom, 1987; Smedley, 1981). Anything that causes a decrease in median droplet size, such as a decrease in dispersant viscosity, increase in differential velocity, or increase in nozzle shear, will decrease depositional efficiency and accuracy.

For a given pump rate, a decrease in the number of nozzles increases mechanical shear but decreases differential velocity. However, since the air shear determines ultimate (deposited) droplet size distribution, it is only necessary that the nozzle orifice be large enough that it will not produce very small droplets. Extremely high pressure drop through the nozzle, which increases flow rate and thus shear rate at the nozzle, should be avoided. Spray-boom pressures on aircraft-mounted systems rarely exceed 2.7 atm (Lindblom and Cashion, 1983).

The preceding effects have been assumed to be independent of aircraft type (Meyers, private communication). Vortex fields, however, may vary from aircraft to aircraft and are known to have a substantial effect (Smedley, 1981).

Dispersant Type

Four physical properties of a dispersant formulation affect droplet size during aerial dispersant spraying (Lindblom and Cashion, 1983):

- *Viscosity* is by far the most important.
- *Volatility* is not likely to be important for any concentrate dispersant formulations, but can be significant for hydrocarbon-based formulations.
- *Density* has an essentially negligible effect, except when droplets penetrate the slick.
- *Surface tension* has a small effect on droplet size, but is of minimal concern for aerial spraying of dispersant.

As the viscosity of sprayed fluid decreases, the median diameter of the droplet distribution also decreases and can become very small, producing a mist or fog that drifts far from the target, which can lower depositional efficiency to less than 50 percent (Figures 5-1, 5-2, and 5-3).

On the basis of viscosity at 60°F, dispersants can be grouped into three classes (Lindblom and Cashion, 1983):

- Five to twenty-five centistokes, typical of most hydrocarbon-based products, generally is not recommended for aerial application.
- Thirty to sixty centistokes including hydrocarbon-based products with high surfactant concentration as well as less-active formulations in other solvents, can be effectively sprayed from helicopters or small airplanes.
- Higher than 60 cSt, including concentrates with high surfactant content, can be applied using all types and speed of aircraft.

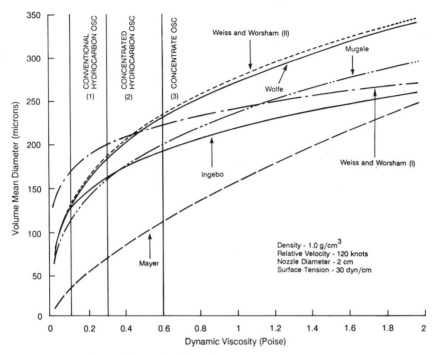

FIGURE 5-3 Comparison of various droplet-size models, volume mean diameter versus dynamic viscosity. Weiss and Worsham presented two possible equations to correlate available data, as noted by (I) and (II). Source: Lindblom and Cashion, 1983.

In most cases, glycol ether- or water-based dispersants are used at sea, but hydrocarbon-based dispersants appear (from laboratory tests) to be more effective on mousse and highly viscous oil (Chapter 2). Undiluted chemicals require a lower flow rate, and nozzles must be properly chosen and sized accordingly (Becker and Lindblom, 1983).

Dosage Control

Dosages recommended by manufacturers are determined by experience and limited in range by viscosity and other properties of the formulation as well as the application system itself. Dosages can also be based on laboratory determinations of optimum dispersant-oil ratios; 1:20 is typical (Cormack, 1983c, and private communication). This calculation requires an assumption of slick thickness to convert observed area to oil volume. Typically a thickness of 0.1 mm is chosen, equivalent to 940 liters/ha (approximately 100 gal/acre) of oil. A dispersant-oil ratio of 1:20 would then require application of 50 liters/ha (5 gal/acre) of dispersant, in agreement with the 20 to 70 liters/ha (2 to 7 gal/acre) guideline quoted above. Since oil slicks are far from uniform in thickness, a constant dose per unit area will produce a correspondingly nonuniform dispersant-oil ratio. Unfortunately, there is no practical way to vary application rate to achieve a constant dispersant-oil ratio.

Another reason dosage is regulated by area sprayed is that the amount of oil being treated is usually not accurately known (Hornstein, 1973). At the time of spraying, the amount of oil the dispersant contacts depends on slick thickness, which is only approximately known, and on the accuracy with which the slick is hit by the spray. Deliberately aiming for the thickest part of the spill, for example, near a wrecked tanker or a blowout, has been recommended as an efficient strategy (Mackay, private communication). Injecting dispersant directly into a blowing well through a previously installed port below the wellhead has been proposed, and some field tests have been encouraging (Audunson et al., 1987; Kolnes, 1986).

Recent tests of depositional accuracy were made using Kromekote cards to obtain total amount deposited per unit area and the percentage of total fluid pumped that reached the target (Lindblom, 1987). Data from these tests indicate that 45 to 90 percent depositional efficiency can be expected (most of the data were in the 65 to 85 percent range), with spray altitude on the order of 15 m and differential velocity (air-speed minus exit velocity of fluid from spray nozzle) of 64 m/sec (210 ft/sec) or less.

Boats

Workboats, barges, fireboats, and tugboats have been adapted to spray dispersants, but the use of surface vessels is complicated by slow speed and vessel motion caused by sea conditions. Only relatively small spills (less than 1,000 bbl) can be treated by boats, even with a nearby chemical reloading point. For example, a boat operating at 5 kn, spraying a 12-m (40-ft) swath can only treat about 1.3 km² (0.5 sq mi) in 12 hr. If slick thickness averages 0.1 mm, the boat could treat about 830 bbl per day assuming that the boat has dispersant storage and fuel capacity to operate the entire day without having to interrupt spraying operations (API Task Force, 1986; Belore, 1985; Chau et al., 1986; McAuliffe, 1986). Use of multiple craft or vessels

TABLE 5-1 Amount of Oil Spray Boats Could Disperse in 10-hr

System	Spray Width		Amount Dispersed[a]	
	Meters	Feet	Meters[3]	Barrels
One boat	5	16	56	350
One boat	10	23	111	8700
One boat	80	100	333	2,100

[a] Calculation based on boat traveling at 6 kn, an average slick thickness of 0.1 mm, and 100 percent efficiency.

with greater swath widths increases treatment area proportionately (Table 5-1).

Many improvements have been made since the early Warren Spring Laboratory work (Chapter 1) in the design of dispersant spraying equipment for boats. Examples include spray booms that do not require overhead rigging, low-drift nozzles, electric-start pumps, and flow meters (Frank Ayles & Associates, 1983, 1984).

Most systems now use water-compatible concentrate dispersants diluted with seawater during application. A seawater pump allows the dispersant to be added to the water stream where it is mixed with the water and applied through the nozzles of a spray boom mounted as far forward as possible. This is done to avoid the bow wave, which may push oil out of reach of the spray at typical boat speeds of 2 to 10 kn (Exxon Chemical Company, 1985). High-pressure water jets have been tested in the laboratory as a means of improving dispersant effectiveness by applying mixing energy to the dispersant and water system. In the laboratory, at a dispersant-oil ratio of 1:100, 80 to 100 percent of the slick was dispersed at the equivalent of low boat speeds (e.g., 2 kn), compared to 10 percent using the Warren Spring breaker-board system (Belore, 1987).

If the dispersant is to be added from a positive displacement metering pump, only the minimum flow rate of water necessary to fill and maintain pressure in the spray system is used, thus keeping water dilution to a minimum.

Booms should be rigged with multiple nozzles arranged to produce a fan-shaped spray pattern, with sufficient pressure to maintain this pattern until it reaches the water level. The required pressure depends on the distance between the nozzles and the water. The spray pattern should be flat and perpendicular to the direction the boat is traveling, and the spray should strike the water in droplets like fine rain. The fan-shaped sprays from adjacent nozzles should overlap just above the oil-water surface and the inboard spray strike the hull just above the water line (API Task Force, 1986; Exxon Chemical Company, 1985; Lindblom, 1979; Lindblom and Barker, 1978; Spraying Systems Co., 1984, 1985). Cone-shaped spray patterns are less efficient because they concentrate the dispersant unevenly along the edges rather than in the middle (Spraying Systems Co., 1984, 1985).

In a properly designed system, the effect of wind on the spray pattern, spray-boom location relative to the bow wave and distance above water, vessel speed, nozzle design, pump rate, and pressure are considered (Exxon Chemical Company, 1985). However, even with a properly designed system, there are disadvantages to using water as the carrier for the dispersant. Field tests with boats have indicated much lower effectiveness rates with diluted dispersants when compared to neat applications (McAuliffe et al., 1981). In heavy seas, application becomes progressively less effective due to pitch and roll, which alters the spray pattern on the water (Cormack, private communication). Direction of spraying operations from the boat deck is limited by poor visibility.

In an emergency, water-dilutable (glycol ether-based but not hydrocarbon-based) dispersant concentrate can be educted into a boat's firefighting system, but dosage control and distribution is difficult (Exxon Chemical Company, 1985). Eduction of dispersant depends on the flow rate of the water stream. The fire system normally provides much more water than is required so its flow must be reduced by valve control or by providing a "bleed" in the stream. For either of these procedures, flow calibration is required.

Aircraft

Aircraft can be equipped with chemical tanks, pumps capable of spraying 20 to 100 liters/ha (2 to 10 gal/acre), digital readout flow meters, spraybooms (generally airfoil in shape attached to the aircraft), and nozzles. Unless the dispersant viscosity is higher than approximately 60 cSt, aircraft speed should not exceed 100 kn (185 km/hr or 115 mph). At faster than about 100 kn, spray nozzle tips are not necessary because the wind shear at the nozzles is sufficient to break the stream into small droplets, and the droplet size tends to be less than 350 μm if the dispersant viscosity is less than 60 cSt. The spray-boom altitude should not be over 9 m (30 ft) for helicopters

TABLE 5-2 Dispersant Spray Capabilities of Various Aircraft

Aircraft	Tank Volume		Amount Dispersed per Flight[a]	
	Meters^3	Barrels	Meters^3	Barrels
Various helicopters	0.4-2.3	2.5-15	8-46	50-290
Agriculture spray planes				
Piper Pawnee, Cessna				
Agtruck, and Ayres Thrush	0.4-1.1	2.5-7.3	8-22	50-140
Turbo Thrush	1.5-2.6	9.5-16	30-52	190-330
DC-3, Fokker F-27, and				
Canadair CL-215	3.0-4.5	19-28	60-90	380-570
Four-engine aircraft				
(DC-4,				
DC-6)	5.7-11	36-72	115-230	730-1,400
Hercules C-130, ADDS				
System	21	13	420	2,600

[a]Assumes dispersant-oil ratio is 1:20 and 100 percent efficiency of dispersion.

SOURCE: Adapted from McAuliffe (1987b).

and small aircraft, although the pilots of larger aircraft prefer to fly at 15 m (50 ft) or higher (Lindblom and Cashion, 1983). Table 5-2 shows the dispersant spray capabilities of various aircraft.

Small airplanes used in agricultural spraying generally have very small load capacity, often less than 380 liters (100 gal), although some newer planes have 1,500- to 2,300-liter (400- to 600-gal) tanks, as shown in Table 5-2. For dispersant spraying, agricultural spray aircraft need to be fitted with adequate pumps, meters, and aft-facing nozzles. Most pumps and nozzles used for agricultural spraying produce too fine a spray, although the airplanes may be quickly converted if the correct equipment is included in a contingency plan.

Small aircraft are best used for rapid response to small spills and test applications while larger equipment is being readied and brought to the spill site. Large two- and four-engine airplanes (i.e., DC-3, DC-4, DC-6, CL-215, and C-130) are most useful for large spills because of their greater range, capacity, speed, and potential for great areal coverage (Lindblom and Barker, 1978).

The Airborne Dispersant Delivery System (ADDS, also known as ADDSPACK) unit developed for the Lockheed C-130 is the only spray system currently available that does not require permanent installation of wing-mounted booms or other integral parts. It can be used with almost any available C-130 that has a rear cargo door that can be opened to horizontal in flight. Liquid capacity is 20,800 liters (5,500 gal), and the boom width is 12.5 m. Spray rates from 0.4 to 90 liters/ha (0.04 to 9.6 gal/acre) can be achieved; typical coverage is 30 ha/min (74 acres/min) at 130 to 150 kn. This system has been tested extensively over land and water and was used effectively during a recent major spill incident (Oil Spill Intelligence Report, 1986a). Like all other systems, it must be well maintained and carefully operated (Fingas, 1986; Lindblom, 1987; Lindblom and Cashion, 1983).

Logistics are not trivial for aircraft deployment. Even though ADDS systems are on standby in Florida (agreement between Biegert Aviation and the Clean Caribbean Cooperative), Singapore (Oil Spill Intelligence Report, 1987a), and in England (Oil Spill Intelligence Report, 1986b), an aircraft must be found to carry the system at the time of a spill. In general, civilian aircraft must be relied upon.

Helicopters

Helicopters are limited by the volume of dispersant they can carry and their relatively short range over water. Integral (airframe-attached) spray systems require a specially equipped helicopter. Suspended bucket units are the most versatile (see Figure 5-4). Most of them have a capacity of under 1,100 liters (300 gal) (Cormack, 1983c), although the French "Sokaf" bucket has a capacity of 3,000 liters (794 gal) (Bocard et al., 1987).

The most important advantages of helicopters are maneuverability and convenience—they do not need a nearby airport. They can also be cost-effective for small spills, especially where offshore platforms provide staging areas and landing sites, although local ordinances that prohibit helicopters from carrying sling loads across highways may pose potential problems in some areas. These situations necessitate transporting spray equipment by truck to a place where it can be picked up at the shore (Clean Seas, Inc., 1981; Onstad, private communication).

Hydrofoils and Hovercraft

Hydrofoils and hovercraft have been suggested as potential vessels for dispersant application because of their speed and intense

FIGURE 5-4 Rotortech Spray Bucket and booms; capacity approximately 100 gal used during Beaufort Sea Trials by Environment Canada. Photograph by Colin Jones (committee).

agitation of nearby water (Flaherty, private communication). Tests using hydrofoils suggest the following advantages (Vacca-Torelli et al., 1987):

- High speed permits the hydrofoil to be on the scene in less than half the time of conventional boats. Furthermore, at higher speeds, hydrofoils do not have a bow wave, which breaks up the oil slick ahead of standard vessels traveling faster than 10 kn.

- The hydrofoil has greater stability for its speed than other waterborne craft.

- Very long spray booms can be used since the hydrofoil is less susceptible to wave motion. Because of the elevation of the craft, however, it may be difficult to position nozzles close enough to the water and the spray would be susceptible to wind-induced drift.

At one time, dispersant spraying gear was carried aboard the Canadian Coast Guard's cargo-carrying hovercraft *Voyageur* as a possible response to oil well blowouts, but it was never used (Gill,

1978). The Warren Spring Laboratory has employed sidewall hovercraft on an experimental basis, but they are more vulnerable to rough seas than most workboats (Nichols, private communication). Another problem specific to hovercraft is that spray is blown away from the craft by the propulsion system's strong air currents (Jones, private communication). It also appears that turbulence caused by the hovercraft may reduce the dispersant contact time enough to lessen effectiveness.

Boats Versus Aircraft

Boats, aircraft, and helicopters all have a role to play in the proper application of dispersants. For applying dispersant to large spills in the open ocean, large aircraft appear to be the only practical means. This is indicated by Table 5-2, which shows the area typically covered by an oil spill as a function of the size of the spill. For some small spills, for example, less than 38,000 liters (10,000 gal), the most practical and cost-effective method of dispersant application is a helicopter with a spray bucket. Boats can spray dispersants on small spills, but their effectiveness may be diminished by limited range and coverage area, weather and sea state (spray-boom length is limited by vessel roll), and the difficulty of determining where to spray. Spotter aircraft generally should be used to direct aircraft and boat spraying operations.

The necessity (in most boat-mounted systems) to spray with diluted dispersant generally diminishes the overall effectiveness of the dispersant. In some situations, however, such as a spill caused by a shipwreck where there is an ongoing salvage operation, boats will not only be available but may be best suited to spray close to the wreck. This may also be true for oil well platforms where a minor spill occurs.

Calibration

Both boat and aircraft application systems require calibration to provide proper dosage control, allow optimum dosage, vary dosage as conditions change, and provide accurate documentation. Dosage from boat-mounted systems is determined by swath width, boat speed, and water or chemical pump rates (Exxon Chemical Company, 1985). Published pump curves, tables of nozzle output versus pressure, eductor percentage settings, and even calculated swaths provide only general guidance, not true calibration. Manufacturers of spray units generally do not provide the necessary calibration (Lindblom, private communication). Water can be used to calibrate units designed for diluted dispersants and for aerial application systems (Exxon Chemical Company, 1985).

For calibration of eductors and chemical metering pumps, the dispersant to be used should be employed if possible. It is especially important that eductors be properly cleaned and maintained and frequently recalibrated (Onstad and Lindblom, 1987). Calibration of commercial equipment at U.S. EPA's Oil and Hazardous Materials Simulated Environmental Test Tank (OHMSETT) facility showed that flow rates among nozzles varied significantly, some as a result of plugged or defective internal parts. The flow meter integral to the system was also inaccurate. Calibration procedures and development of operating charts are described by Onstad and Lindblom (1987).

The results of calibration tests should be made available to equipment operators and response coordinators in easy-to-use form, such as tables, charts, and curves. This format allows for rapid actions in dosage control that do not depend on numerical calculations or interpolations (Onstad and Lindblom, 1987).

System evaluation should be conducted under realistic operating conditions and data recorded during the tests so that a basis for calibrating and adjusting the system can be formulated. Field trials can help quantify the effects of pumping rates, swath width, and speed (Shum and Nash, 1987). Recommended eductor settings and pump curves, or calculated swath widths, provide a starting point for calibration, but do not substitute for field tests.

Aerial swath width can be estimated as the width of the spray boom at 9-m (30-ft) altitude, or 1.2 to 1.5 times the distance between terminal nozzles for an aircraft operating into the wind at 15 m (50 ft) (Exxon Chemical Company, 1985; Lindblom, 1987; Lindblom and Barker, 1978). Crosswind swaths can only be estimated roughly by guidance aircraft.

MONITORING EFFECTIVENESS OF DISPERSANTS

Directly monitoring the fraction of oil removed from the surface and dispersed in the water is the best method of determining effectiveness if it can be done reliably. This concept of effectiveness, equivalent to obtaining a mass balance on the amount of oil dispersed in the water column, has given good estimates in a few elaborate field tests; but it is complicated, requires set-up time, and is not practical in real spills. However, the concentrations of dispersed oil in the water are also relevant to assessing potential toxicity.

Water sampling programs have been useful for scientific purposes and have been proposed as a regulatory requirement by local agencies; for example, in California (API Task Force, 1986; Boehm and Fiest, 1982; McAuliffe et al., 1980, 1981). But water sampling programs are expensive and of limited value unless carefully planned and executed. Even when the protocols, mechanisms, and equipment for such a program are set up beforehand, as in research field tests, the logistics of getting the appropriate people and equipment to the scene and carrying out the program satisfactorily are difficult (Meyers and Onstad, 1986).

Distribution of oil in the water column during oil spill experiments off the coast of Norway in June 1985 was determined by use of instruments that detected light scattering by oil droplets in water and the ultraviolet fluorescence of dissolved oil components (Genders, 1986). Fluorescence in water is the sum of natural fluorescent materials as well as oil. Backgrounds in the North and Baltic seas varied from 0.5 to 10 ppb and occasionally up to 50 ppb. Attempts have been made to separate backgrounds by using different excitation and reception wavelengths. Background is usually a factor of at least 10 less than the signal in the case of significant oil pollution (Hundahl and Jojerslev, 1986).

Remote Sensing

Other real-time monitoring methods that might lead to practical measures of effectiveness have been proposed. Ideally a spill-detection and monitoring system should be able to

• provide continuous, day and night, all weather, wide area, real-time surveillance;

• detect any oil spill that occurs in the marine environment both on and below the surface;

• confirm that the detected substance is in fact oil;

• map the areal extent of the spill;

• obtain the thickness distribution and quantify the amount of oil spilled;

• identify the source and type of pollutant being discharged;

• provide precise navigation information for spill source location and for positioning vessels;

- provide objective data on dispersant effectiveness and on the presence and behavior of threatened organisms (especially birds and mammals) in the area; and
- document all collected data.

Since no one sensor can do all these things, a multisensor system is required. A good sensor will detect oil, identify it as oil, and determine the oil type (Garrett et al., 1986; Intera Environmental Consultants, 1984; Moniteq, 1985). Multispectral devices and active laser fluorosensors are now used for more sophisticated detection systems. The laser fluorosensor is an active device that gives basic oil type and operates in any weather, day or night (Hoge and Swift, 1980).

Aircraft provide adequate spatial coverage and can be dispatched to a spill with all the necessary systems operative. They are often the platform of choice for monitoring (Fast, 1985).

Aerial or remote sensing has a distinct advantage over shipboard observations because of the rapid and extensive coverage possible. Oil can be detected from surface vessels by a variety of methods (e.g., visible color, fluorescence, and radar), but the limited area visible from a boat makes these techniques less than ideal.

Visual and Near-Visible Observation

Reflection, absorption, and scattering from water with and without spilled oil, and the ability of oil to affect the interference and polarization of reflected light allow visual and near-visible observation of oil. During experiments in the North Sea, for example, observers aboard aircraft noted that 1 day after treatment a chemically dispersed slick had disappeared, but an untreated control slick was still clearly visible (Lichtenthaler and Daling, 1985). Videotaping is one simple, inexpensive, and available technique that can be used from airplanes or boats to document the extent and configuration of a spill. It can also record how oil responds to dispersants, the effects of wind, waves, and currents, and the presence and behavior of birds, marine mammals, and other organisms. Using a polarizing filter to enhance the reflectance differences, even thin oil films can sometimes be observed (Burns and Herz, 1976).

The primary drawback of videotape, like visual or photographic observation, is dependence on natural lighting. Sunlight reflecting off water can reveal the shape of very thin slicks, but lack of contrast between oil and water on overcast days greatly reduces the effectiveness of visual observations.

Although there are many examples of qualitative observations, obtaining ground (or sea) truth to calibrate a quantitative measurement system is a difficult challenge. One day after the 1986 Beaufort Sea tests, various observers could not agree about how much surface area was covered by the thick portions of the slick. The two observers who were in aircraft estimated 20 percent of the slick was thick oil, but the observer on the water estimated 80 percent. In addition, observers on the water saw numerous oil particles or flakes, or small tar balls, detail that could not be discerned in the remote-sensing imagery (Fingas and Jones, private communications).

Under relatively calm conditions, experimental slicks tend to spread more extensively when treated with dispersant than do untreated controls (Lichtenthaler and Daling 1985; Sergy, 1985). Because of spreading, herding, and the irregular shape and thickness of a large slick, it is difficult to tell visually how much of the oil has been removed from the surface, and, therefore, it is not always possible to determine effectiveness accurately by this method (Cormack et al., 1986/87; Fingas, 1985).

Infrared Sensing

Because oil absorbs and retains more heat (from solar infrared radiation) than does water, it is possible with the use of infrared-sensing devices to detect oil on the water surface by means of temperature differences. Sensitivity is greatest when the oil is warmer (in sunlight, early afternoon); but when the oil cools to water temperature (in the evening or morning), sensitivity may drop to zero.

High-resolution devices, including television-type cameras, are now available that detect the thicker parts of oil slicks. The successful use of infrared image intensifiers for night vision in combat suggests that night flights might track a slick with infrared monitors in darkness or low visibility, which would allow response forces to position themselves for the next day's activities. Such a procedure was tested at night and detected oil at the natural oil seeps of the Santa Barbara Channel (Onstad, private communication). However, the 1986 Beaufort Sea tests, using a different camera, cast doubt on the value of night flights in tracking slicks (Jones, private communication).

The most commonly used sensors are infrared plus ultraviolet. Remote-sensing airplanes are used in Canada, France, the Netherlands, Norway, Sweden, the United Kingdom, and the United States (Cormack et al., 1987; Fast, 1987; McColl et al., 1987; Lavache, private communication). The ideal system is portable from plane to plane, fitting into window openings, screens, or camera hatches (Fingas, private communication).

Microwave Sensors

Both microwave emissivity and reflectance vary with oil spill presence. Dual-channel-imaging microwave radiometers have mapped oil spills thicker than 100 μm (Hollinger and Menella, 1973). Active microwave sensors have been used successfully, especially in Europe (Croswell et al., 1983; Loostrom, 1986/87).

Radar

Side-looking airborne radar (SLAR) is used as a means of detecting oil slicks at sea. SLAR can operate through a cloud cover and outline the slick, but cannot distinguish a thick slick from a thin sheen. It is a key element in remote-sensing packages used to detect oil spills in the Netherlands, Norway, Sweden, the United Kingdom, and the United States (Cormack et al., 1987; Fast, 1987; Schriel, 1987; White et al., 1979; McAuliffe, private communication).

Detection of slicks by SLAR depends on damping of capillary waves by the floating oil. Reliable detection depends on a variety of factors such as the size of the slick and environmental conditions, but slicks have been detected at distances up to 20 km (Cormack et al., 1987). In the context of monitoring dispersant effectiveness, it may be that the primary use of SLAR would be the important job of locating a slick rather than measuring the subsequent disappearance of the oil from the water surface when treated by dispersants.

Summary of Monitoring Techniques

The best method of surveying appears to be from an aircraft using SLAR for initial mapping, followed by infrared line scans to determine relative slick thickness and ultraviolet line scans to produce a picture that can be correlated with visual observation (Fast, 1985, 1987). With such data, aircraft can direct spraying units to the correct portion of the slicks (Cormack et al., 1986/87; Fingas, private communication).

Remote-sensing information on very thin slicks has produced evidence for the enhanced spread of dispersant-treated oils (Cormack, et al., 1986/87; Lichtenthaler and Daling, 1985; Cormack, private communication). Remote sensing is partially successful in experimental situations, and appears to be at the stage of commercialization of the needed measurement devices. Equipment has been developed for the U.S. Coast Guard (Kim and Hickman, 1975), the Swedish Coast Guard (Fast, 1985), the Canadian Environmental Protection Service (Intera Environmental Consultants, 1984), and the British government (Nichols and Cormack, private communications). This equipment can measure from aircraft some of the parameters necessary to determine slick thickness and configuration, but it is very expensive compared to other response equipment and considerable development is required before remote sensing becomes routine. One or more backup systems have been suggested because there are many opportunities for sensitive instruments to malfunction (Fingas, private communication).

Regulatory Requirement

The U.S. National Contingency Plan does not require or regulate documentation of cleanup, and subjective visual evaluation is normally sufficient. The Region IX contingency plan (Pavia and Smith, 1984) requires documentation, as do some local policies, but does not specify detailed methodology. Canadian regulations imply that effectiveness of cleanup should be monitored, but no techniques are specified (Fingas, private communication).

STRATEGY OF DISPERSANT APPLICATIONS

Development of contingency plans for spill response requires a great deal of thought in advance of any incident. The necessary equipment, personnel, and chemical dispersant must be available to go to work immediately, whenever the need arises. Contingency planning strategy should consider different spill sizes, weather conditions, and all available control measures (see Chapter 6).

Since timely application is essential for success, it is necessary that all personnel involved in decisions be educated in the scientific and technical information and be briefed on their role and the consequences of failure to act promptly or decisively. It would also be desirable for them to have participated in field trials, spraying real dispersants on real oil.

Dispersibility of Oil

Dispersibility of the oil is the logical place to start in the planning process. Many important characteristics of an oil from a known source can be obtained from existing data. Laboratory or field testing of dispersant on a sample of the spilled oil can give even more information, but it is difficult and time consuming to obtain an adequate oil sample from an actual slick. An exception would be if the source were known ahead of time, as with a production rig or a port where a well-defined type of oil is handled. Spraying a portion of the slick is probably the best method of determining effectiveness (Smith and Pavia, 1983; Stacey, 1983).

Spill Size and Configuration

Spill size and configuration must be known for effective dispersant application. Generally, the goal of dispersant use is to protect sensitive marine areas or coastline by controlling the most threatening portions of the slick. This can be done by treating portions of the slick, such as its leading edge or selected windrows of oil. The portions of the slick most effectively treated with dispersant are about 0.02- to 0.2-mm thick (Exxon Chemical Company, 1985).

The state of California Contingency Plans require clear evidence of a leading edge (an unusual configuration for a large spill) and threat to marine mammals or shoreline before permission for dispersant application can be granted. By the time a clear threat can be demonstrated, however, it may be too late to apply dispersant.

The Exxon Chemical Company (1985) suggests that boat spraying operations generally proceed from the spill's outside edge and work gradually toward the center, so that the slick is not disrupted by the boat's wake. However, most of the oil volume is in a relatively small thick portion, while the outlying portions consist of thinner sheen layers. Circling from the outside in results in much of the dispersant being sprayed on the sheen, rather than on the bulk of the oil.

Belore (1985) and Chau and Mackay (1985) have suggested repeated application of dispersant to thicker portions of the slick. This has been verified in field tests (Lichtenthaler and Daling, 1985; McAuliffe et al., 1981).

Lindblom, Meyers, and Onstad (private communications) recommend that small spills generally be treated through their thickest
part and large ones treated across a leading edge threatening shorelines or environmentally sensitive areas. They recommend that sheen
be allowed to disperse naturally. Often the thick areas consist of a
number of separate patches or windrows, and must be identified and
located by aerial reconnaissance. Visual observations must be interpreted cautiously (as discussed earlier)—after treatment, oil can
spread more thinly over a larger area and the slick may appear to be
about the same size as before, even though much of the oil has been
dispersed.

Large spills cannot be treated in their entirety, but dispersants
can be used tactically under favorable conditions to protect sensitive
shoreline areas. Even without dispersant application, the concentrations of oil in a relatively self-contained body of water can remain
high for months after a spill. For example, 2 months after the *Argo
Merchant* spill on Georges Bank in December 1976 the dissolved
hydrocarbon levels were as high as 14 to 60 μg/liter; and 1 to 25
μg/liter 5 months afterward. The "normal" unpolluted level for that
area appears to be 5 μg/liter (Farrington and Bochum, 1987). Would
the use of dispersants have increased or decreased the amount of hydrocarbons retained in the water column? That question cannot be
answered without a great deal of additional knowledge about water
circulation, exchange with the open ocean, and sedimentation and
resuspension rates in that area.

Aerial Spraying Strategy

When flying directly into the wind, the effective swath width
is roughly 1.2 to 1.5 times the overall length of the spray boom.
Crosswind application, which is common in agricultural spraying,
may be useful when very large spills are treated by large aircraft
with high-volume pumps (Smedley, 1981). This technique gives a
much greater swath width than upwind spraying but has a potential
drawback for smaller spills—unacceptable and off-target dispersant
drift. For crosswind application, increased pump rate and nozzle
adjustment are necessary to allow for a wider swath.

Other Strategies

It is logical to be prepared, especially for small operational spills,
that is, on offshore exploration or production platforms, or loading

platforms. Here oil and dispersant can be pretested. This is a condition for "maximum effectiveness."

During the Chevron Main Pass Block 41 *Platform C* spill, a water spray system consisting of a large number of fire monitors mounted on a barge (used for fire control), applied dispersant on the blowing wells and on the oil slick in the immediate vicinity of the platform (McAuliffe et al., 1975). The same type of system was also used at other platform blowouts. The barge was kept upwind to prevent the spray from covering the barge and to produce the most effective downwind coverage of the platform and surrounding water.

During the early stages of an offshore continuous discharge, the oil can be treated initially, at least during daylight hours, by aircraft and boat spraying. Oil that is released during the night and moves as a slick away from the platform can be sprayed the next day. This technique may be sufficient to control smaller oil releases, but it may not control large spills.

An offshore platform blowout, either at the surface or subsea, provides a special opportunity for developing techniques to apply chemical dispersants (Audunson et al., 1987). There is time to assemble a spray system at the source of the oil release, and the platform or surrounding support vessels allow for 24-hr operation of spraying.

Applying dispersant near the oil discharge point has the added advantage of spraying fresh oil and generally thicker oil, thereby more efficiently using the chemical dispersant and better controlling the slick. As discussed in other sections of this report, chemical dispersion becomes more difficult when the oil has weathered and as the slick thins and breaks into patches and particles.

One proposal is to add dispersant or emulsion inhibitor to the cargo of a tanker before the oil is released to the water (Gordon and Milgram, 1986; Ross et al., 1985). This approach does not seem practical—it would require enormous amounts of dispersants, could affect the chemistry of oil refining and combustion, and the dispersant would have to be mixed with the oil before it was loaded in tankers.

Command and Control

An aircraft, given a cloud base of more than 60 m (200 ft) and an indication of where to look, will usually be able to find oil floating on the water. A boat or ship with dispersant spray equipment may find some oil, but has little hope of locating major patches in the area without spotter aircraft. Radio contact between aerial spotters and spraying boats is therefore essential (API Task Force, 1986; ITOPF, 1982; McAuliffe et al., 1981).

Spotter aircraft are also necessary when aerial spraying is conducted. The spraying aircraft flies adjacent tracks and turns the spray on and off in order to hit the slick without wasting chemical on open water. This requires coordination between spotter and spray units (ITOPF, 1982). The potential use of remote sensing by spotter aircraft as discussed earlier can possibly assist in locating the thicker portions of the slick and provide an indication of effectiveness of the dispersant.

Weather

Wind and waves not only affect the physicochemical processes of dispersion, they also affect spraying operations. High winds may blow dispersant spray off target. Heavy seas, however, can be advantageous since breaking waves more rapidly disperse the oil (Bouwmeester and Wallace, 1986a; Raj and Griffith, 1979). Fog and low clouds cause the most difficulty because they obscure visibility and can stop aircraft operations. Application techniques in adverse weather conditions are included in some contingency plans. (EPA Regions IX and X plans allow consideration of dispersant application when mechanical means are impossible.)

6
Technical Basis of Decision Making

This chapter considers the scientific and technical information reviewed in the previous chapters and uses that information to recommend what to do when an oil spill occurs.

FINDINGS FROM PREVIOUS CHAPTERS

The preceding chapters have shown the following:

- Recent chemical formulations can effectively disperse an oil that spreads on water if the oil viscosity is lower than approximately 2,000 cSt. Dispersion becomes progressively more difficult with increasing viscosity until, at viscosities higher than around 10,000 cSt, little oil is dispersed.
- For small, medium, and most large spills, dispersed oil concentrations in open waters tend to decrease rapidly owing to tidal currents and other transport processes.
- Very large spills, such as *Ixtoc I,* may introduce such a large, continuous flow of oil that normal, open-sea current cannot provide rapid dispersal. However, for most spills, unless water circulation is limited, organism exposure to dispersed oil is likely to be low compared with the exposures required to cause behavioral changes or mortalities.
- The principal benefit of oil spill control by chemical dispersion or mechanical recovery is the prevention of oil from stranding

on shore, entering sensitive shoreline habitats, or entering sensitive areas such as seabird colonies or sea otter locations. Serious adverse biological effects from untreated oil have been documented on seabirds (if present) at many spills, and by oil that concentrates on shores.

- Dispersants are most effective when applied early. Oil becomes progressively less dispersible with time as its viscosity increases by loss of volatile hydrocarbons and by formation of water-in-oil emulsions (for a number of oils). Thus, the decision to use dispersants should be made as rapidly as possible after a spill occurs, preferably within the first few hours.

- Spilled oils generally attain an average slick thickness of 0.1 mm or less in an hour or two, and this thickness appears to be relatively independent of spill size for those oils that spread on water. However, it should be noted that the distribution of oil on water is usually not uniform, and there may be some areas within the slick that are significantly thinner or thicker than 0.1 mm.

- As water temperature decreases, oil viscosities increase. Thus, oils that spread in tropical or temperate climates are less able to spread at arctic water temperatures. Lower temperatures may also cause additional oils to be solid or semisolid because the temperature is below their pour point. Some oils have pour points in excess of the highest likely ambient temperatures; little spreading occurs when they spill.

- The dispersant spray must hit the thicker part of the slick. Aerial or boat spraying usually requires direction by spotter aircraft.

- Dissolved hydrocarbons in the water column after dispersion of an oil slick are largely limited to areas close to the spill source, because most of the volatile and soluble hydrocarbons in the oil evaporate rapidly from the slick before dispersion. Hydrocarbons dissolved in the water also evaporate into the atmosphere and are diluted rapidly in the water column. These dissolved hydrocarbons (many of which are aromatic) appear to produce the most immediate biological toxicity.

TECHNICAL QUESTIONS

A number of technical questions must be answered when considering dispersant use as an oil spill countermeasure. These questions are discussed in this section.

Response Options

Whether a countermeasure is needed or whether the spill will be dissipated by natural forces before it can impact a sensitive resource must be determined. Natural dissipation can be expected if the seas are rough, the oil is thinly spread on the water surface, the spill is not threatening a shore or sensitive area, or the volume of oil spilled is small.

Alternative countermeasures, their availability, and determination of their ability to remove more or less oil than dispersants are further considerations. It should be noted that mechanical containment and recovery are generally ineffective if the oil layer is relatively thin (less than about 0.05 mm), or if the sea is moderately rough, typically sea state 4 or greater.

Environmental Considerations

The use of a chemical dispersant may not be appropriate on all portions of a spill. While laboratory and mesoscale tests have shown that the acute biological effect of dispersed oil is no worse than of untreated oil per unit of oil, there are species and habitats, such as benthic organisms and mollusks, that may suffer greater damage than that caused by untreated oil. However, several nearshore studies (Chapter 4) have shown that dispersal of oil offshore reduces its impact on intertidal and benthic communities.

The problem of anticipating environmental damage is tied to an assessment of natural populations and habitats that could be threatened by an oil spill. This environmental assessment should be done and the results incorporated into scenarios for areas of concern as a component of the prespill information base supporting the decision-making process. Since inaction in undertaking spill treatment may cause the greatest environmental harm, the environmental assessment data and information base should be sufficient, and operational scenarios that include this information should be understood and accepted as part of prespill planning. The desirable objective in the decision-making process is to be able to focus on operational details, such as the location of aircraft and boats relative to the spill, at the time of an accidental spill.

Other Factors That Affect Decision Making

Spill size is important because the area covered by the slick may be so great that it overwhelms mechanical response capabilities and possibly even dispersant spray capabilities. Thus, making logical decisions concerning oil spill control requires evaluation of the capabilities of available methods. For purposes of this discussion, an average slick thickness of 0.1 mm is used.

Method capabilities are limited also by operating conditions, which imply that operations should be carefully monitored during a spill. Monitoring, control, and evaluation usually can best be done from the air by spotter aircraft. Thus, operations, whether by skimmers, spray boats, or spray aircraft, are limited to daylight and adequate flying conditions. Night operations are seldom possible, except possibly for spray barges (and boats) and skimmers operating at the source of a continuous spill.

Skimmers with 100 percent efficiency encountering a 0.1-mm-thick slick at 1 kn, with sweep widths of 10 m (3.3 ft) and 100 m (33 ft), would collect, respectively, 116 bbl, and 1,160 bbl of oil in a 10-hr day. Thus, it would take all day for one skimmer with a 10-m sweep width to collect about 100 bbl of oil unless it can operate in areas where the oil thickness is greater than 0.1 mm. A large oceangoing skimmer system with a 100-m encounter width (heavy seaboom, three ships, and collection barge) might handle a 1,000-bbl spill in a day under ideal conditions. If the oil has a high viscosity, and has not been spread by wind and waves, skimmers may have greater collection potential. Skimming systems are also limited by wind, currents, and sea state. It should be noted that the percentage of oil recovered at accidental spills has been low, particularly with large spills.

Spray boats, moving through a slick at 6 kn with spray widths of 5 to 10 m (16 to 33 ft) and operating with 100 percent efficiency (although this is unlikely), might disperse, respectively, 350 to 700 bbl of oil over 10 hr. Although a spray boat can operate in sea states where skimming systems are ineffective, larger waves reduce its efficiency. The boat may have to decrease speed, and the outboard nozzles may dip into the water. Larger boats roll less and can carry large amounts of dispersant. Spray planes have the advantage of spraying dispersant rapidly, but may have the disadvantage of not carrying large amounts of dispersant. They also are capable of rapid response, and of response to more remote areas (perhaps the only response). Small planes and helicopters have limited range from a support base. A large plane flying at 140 kn with a spray swath width of 100 m could cover 28.5 km^2 in 1 hr. Thus, the capacity of the spray tanks, not the slick area, is the controlling factor. Ideally, a C-130 (Hercules) aircraft with ADDS, which has a 130-bbl tank that can spray 2,600 bbl of oil with a dispersant-oil ratio of 1:20 on each flight, could make six to eight flights per day depending on the distance from base to slick.

The above analysis for spray boats and aircraft has assumed 100 percent dispersion of the slick. Generally, that is not the case. Higher dispersant application rates might be required, and correspondingly larger spray capabilities required for oils that are not so readily dispersible. Because water-in-oil emulsion formation hinders or prevents effective chemical dispersion, to be effective, oil slicks should be sprayed before the oil incorporates water. In practice if control of the entire slick is not possible, spraying should be directed to the slick closest to shore or a sensitive resource.

Weather Conditions

In general, oil is dispersed more readily when the sea is rough than when it is calm. Mackay (1986) suggests that chemical dispersion may be less effective at wind speeds under about 7 m/sec, although this is not a precise threshold nor is its value firmly established. This does not mean that dispersants should not be applied, but they are likely to be less effective. Conversely, if the seas are very rough (sea state 5 or higher), treatment may not be necessary because wind and wave action might be adequate to remove the spilled oil from the water surface quickly and application may not be practical under rough conditions. However, two other factors should be considered in rough seas:

1. The spill will move relatively quickly (rapid advection) at high wind speeds, so time available for response may be less.

2. Some of the naturally dispersed oil may resurface as the weather moderates and the seas subside.

ADVANCE PLANNING

Although some of the information needed for decision making will only be available at the time of the spill, much can be obtained well in advance and incorporated into an advance plan for oil spill control. The following information would be desirable for dispersant use, but much of it applies to other control methods as well:

- potential sources of crude oils and products that may be spilled—type of oils produced in or transported through the area of interest, volumes involved, routes traveled (tankers and pipelines), and locations of oil production platforms;
- environmentally sensitive resources that might be impacted by spilled oil—relative sensitivities, local priorities for protection, and relative importance, that is, to the resource management agencies;
- available dispersants and storage locations—dispersant properties and performance with oils of concern, and appropriate application rates;
- available equipment—type and location, with proper calibration for dispersants to be used, and availability of adequately trained operators; and
- monitoring—available means to monitor dispersant application and their effectiveness, other appropriate measurements or observations, needed instruments, and trained operators.

Additional site-specific data are also needed, such as spill location, volume and type of oil, and local meteorological and hydrographic information. Finally, one more component is needed in order to prepare for dispersant use: a well-conceived system for making the dispersant-use decision, and acceptance of this system by the regulatory agencies that are involved.

DECISION SCHEMES

The use of the technical information discussed above may be illustrated by decision-making diagrams, accompanied by extensive footnotes and text. Examples are shown in Figures 6-1 to 6-4. They are similar in some ways, but each was developed for a different purpose and each emphasizes different aspects of spill response. (It should be noted that these decision diagrams are used for illustrative purposes and do not by themselves comprise complete decision-making tools.)

The decision-making diagrams shown have been selected from those that are in use primarily in the United States. However, they are similar to diagrams that have been published elsewhere, e.g., by the International Maritime Organization (1982) and the International Petroleum Industry Environmental Conservation Association (1980).

These diagrams have been proposed for use by spill response coordinators at the time of a spill, but it appears likely that such use will only be effective if the spill response coordinator has experience with their use, for example, through training sessions in advance of a spill. This is because dispersant-use decisions should be made promptly; any delays can result in serious loss of dispersant effectiveness. Thus, those who provide and assemble the background data should be trained in its use, and regulatory decision makers should also be trained so that they will understand the decisions made and the need for speedy action. Ideally, the decision to use dispersants should be made prior to a spill.

U.S. EPA Oil Spill Response Decision Tree

The U.S. EPA procedure, programmed for use on personal computers, is one of the more detailed and complete decision-making procedures available (Flaherty et al., 1987). At each node in the decision diagram the user may request an explanation of the factors involved in each option. Help menus include information on mechanical containment and recovery, observation techniques and needs, and conditions that would lead to a decision to let natural processes clean up the spill. Consideration is given to the effectiveness of different countermeasures, weather conditions, spill site, oil type, and other factors.

Although it is not shown in Figure 6-1, the text of the program explains that simultaneous use of more than one countermeasure may be appropriate. Little or no guidance is given on evaluating the environmental trade-offs that usually must be made between untreated versus dispersed oil.

The most time-consuming component of a dispersant-use decision is the question of environmental damage: Will dispersant use result in more or less damage than nonuse? This question should preferably be addressed prior to any spill, when decisions should be made about the locations and the conditions under which dispersant use should be considered or when their use would be inappropriate.

API Decision Diagram

The API decision diagram is one of the less complex. It is based on the concept that spraying the oil slick will have little or no adverse biological effects based on a comparison of field hydrocarbon exposures with laboratory bioassays and behavioral studies. It also brings in spill size as it relates to the spill control capabilities of skimmers, spray boats, and spray aircraft.

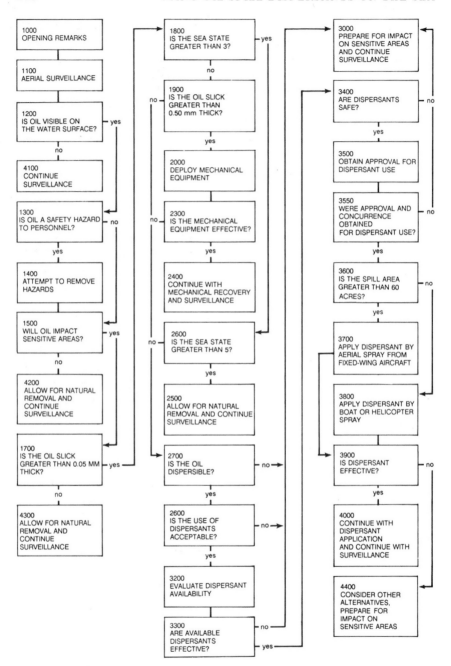

FIGURE 6-1 U.S. Environmental Protection Agency Oil Spill Response Decision Tree. Source: Flaherty et al., 1987.

Figure 6-2 is an oil spill control diagram that outlines the realistically available options. If the estimated spill volume is less than 1,000 bbl, a choice can be made between mechanical recovery and dispersant spraying. This choice depends on availability of mechanical equipment and suitability of winds, waves, currents, and response time; or availability of spray planes and dispersibility of the oil (Figure 6-2, lower left). If neither option is available, the shoreline or sensitive habitats can be cleaned using appropriate methods, such as those suggested by API (1985), or the oil can be left to weather naturally.

Spills much over 1,000 bbl per day have little possibility of being controlled by mechanical means unless conditions are ideal (waves less than 1.3 m and surface currents less than 1 kn) and a large amount of equipment is available. Dispersant application by large aircraft spraying systems would appear to be the only serious control possibility for large oil spills (Figure 6-2, lower right). Because it is unlikely that there will be sufficient mechanical equipment available to control larger oil spills, equipment that is available should be used to collect or divert spilled oil as it approaches critical locations.

Mechanical equipment can be used effectively on spills of oils that have pour points above the ambient temperature, are highly viscous, do not spread, or have formed a viscous mousse. If the oils have not spread, mechanical recovery devices have less area to cover.

Health hazards must be considered. Mechanical cleanup and spray boat personnel must be protected from volatile hydrocarbons when operating in an oil slick downwind near, for example, a well blowout. Special precautions must be taken if the oil and associated gas contain hydrogen sulfide (H_2S). Operations also must be outside the zone in which gas and air forms an explosive mixture.

SLR Dispersant Decision-Making Workbook

The objective of this decision-making method is solely to indicate whether or not dispersant use is environmentally appropriate. The S. L. Ross (SLR) workbook (Figure 6-3) gives methods for characterizing, on a numerical basis, the environmental impacts on populations that may be at risk from either dispersed or untreated oil (Trudel and Ross, 1987). Using these computed values, methodical and objective decisions can be made regarding the advisability of dispersant use or nonuse from an environmental perspective. Other

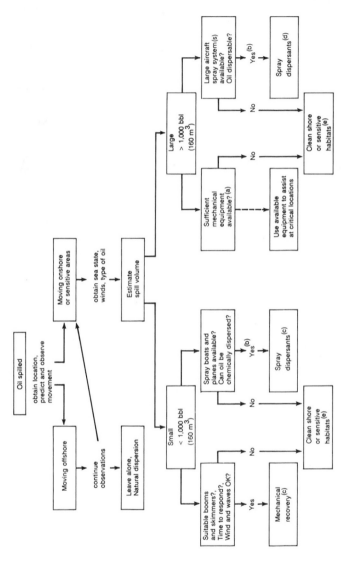

(a) It is unlikely that sufficient mechanical equipment will be available to clean up a large spill.
(b) With the approval of the federal on-scene coordinator and the concurrence of the EPA and the state(s).
(c) Small spills normally should be completely controlled, particularly if both mechanical and chemical methods are used. However, under some conditions some oil may need to be removed from the shore.
(d) Large spills, particularly 10,000 to 30,000 bbls per day, will be difficult to control. Only large aircraft spray systems are suitable and some oil may still strand. However, oil that is kept off the shore will lessen adverse effects.
(e) Appropriate methods should be used to clean shorelines and sensitive habitats. See, for example, API (1985).

FIGURE 6-2   API dispersant-use decision diagram. Source: API Task Force, 1986.

aspects of spill response (i.e., mechanical recovery and natural removal) are deliberately not considered, because they are considered to be separate parts of the oil spill countermeasures problem. No guidance is given on dispersant application rates, effects of weather conditions, spill size, or oil condition.

State of Alaska Dispersant-Use Guidelines

The State of Alaska's guidelines are illustrated in Figure 6-4. The user must assemble a significant amount of information prior to making a dispersant-use decision, including a comparison of the effects of dispersed oil and untreated oil on populations at risk (Regional Response Team Working Group, 1986). However, this system gives no guidance as to how to make the comparison and appears to assume a fairly high level of expertise by the user. Accompanying the decision tree are maps and text showing zones in which dispersants

- may be used with approval by the federal on-scene coordinator (OSC);
- may be used only with concurrence of the EPA and the state plus consultation with the Regional Response Team; or
- may not be used.

Federal Region IX (California) has dispersant-use guidelines that are similar in many ways to those of the State of Alaska, except that maps have not been prepared in California showing areas where the OSC may approve dispersant use unilaterally. It may be noted that the Region IX guidelines have been used on two occasions to reach decisions favorable to dispersant use—in 1984 at the Puerto Rican spill (Zawadzki et al., 1987) and at the 1987 *M/V PacBaroness* spill (Oil Spill Intelligence Report, 1987b,c). However, it should also be noted that on both occasions it took more than 24 hr to come to this decision (Onstad, private communication).

The objective of this method is solely to indicate, from a regulatory perspective, whether dispersant use is or is not appropriate to consider. Note that the OSC must notify the U.S. EPA and the State of Alaska as soon as possible if he or she authorizes dispersant use. The zones are defined by bathymetry and currents, biological parameters, nearshore human activities, and time required to respond. The zones were defined by a subcommittee of the Alaska Regional Response Team. The zones were not evaluated by procedures such as those in the SLR workbook. In the event that dispersant use may

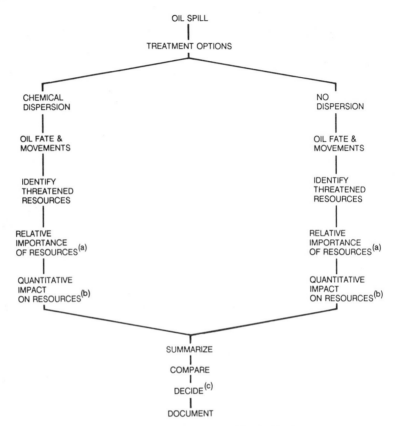

(a) Relative importance of sensitive resources is determined by the
    affected regulatory agencies in terms of "High," "Medium," and Low";
    the determination is based on local priorities.
(b) Quantitative impact on resources is calculated using environmentally
    based algorithms; these algorithms yield a quantitative estimate of
    the degree of impact on each resource in terms of "Major," "Moderate,"
    "Slight," or "Negligible."
(c) The dispersant use decision is based on a comparison of the impacts on
    affected resources by the spilled oil if chemically dispersed versus the
    impacts (usually on a different set of affected resources) by the
    untreated oil.

FIGURE 6-3   SLR decision-making method. Source: Trudel et al., 1983.

be authorized, no guidance is given as to application rates or effects
of conditions such as weather, spill size, and oil condition.

Comparison of Decision-Making Diagrams

The four decision-making diagrams shown in Figures 6-1 through
6-4 are compared in Table 6-1. From the comparison, it appears that

Dispersant Use Decision Matrix. (The Following Questions
Must be Answered Before Deciding to Use Dispersants.)

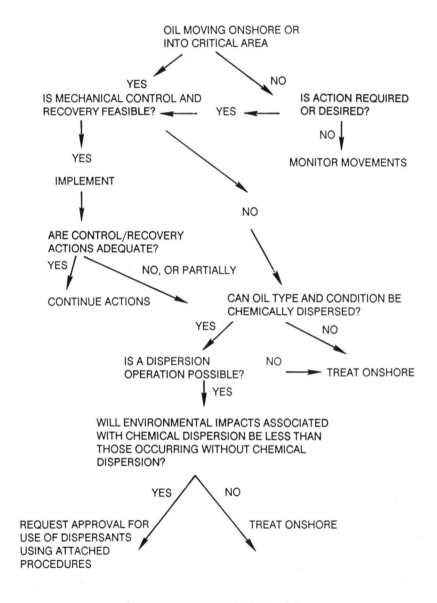

NOTE: Immediate threat to life PREEMPTS the
necessity to use this matrix.

FIGURE 6-4 State of Alaska dispersant use decision matrix. Source: Regional Response Team Working Group, 1986.

TABLE 6-1 Comparison of Decision-Making Diagrams

Factor	EPA, Figure 6-1	API, Figure 6-2	SLR, Figure 6-3	Alaska, Figure 6-4
Surveillance	1	2		2
Personnel hazards	1			
Danger to sensitive areas	1	1	1	1
Is natural removal appropriate?	1	2		2
Is oil thick enough to be a concern?	1		2	1
Spill size	2	1	1	1
Is mechanical recovery feasible?	1	2		1
Is mechanical recovery effective?	1	2		1
Is the oil dispersible?	1	1		1
Are dispersant resources available and effective?	1	1		2
Need to obtain approval	1	2		1
Are environmental impacts of dispersed oil less than those of untreated oil?	1		1	1
Is dispersant use effective?		1	2	2
Application rates	1			

KEY: 1 = Primary consideration or guidance is given; 2 = Included only indirectly or by inference.

the U.S. EPA Oil Spill Response Decision Tree (Figure 6-1) is more complete and detailed than the others. It was developed as an overall tool to guide response to an oil spill. As reported by Flaherty et al. (1987), a user can reach a decision within a few minutes, providing the data are available. The speed of use of this process results in part from its having been programmed for a personal computer, which makes it particularly suited for training purposes.

The API decision diagram (Figure 6-2) emphasizes the need for dispersant use as the only really feasible means of responding to spills that exceed the capabilities of available booms and skimmers. In many cases mechanical cleanup capabilities may be only on the order of 1,000 bbl per day. Figure 6-2 points out the serious limitations to mechanical containment and recovery for extremely large spills (over 1,000 bbl per day). The concepts embodied in the API decision diagram could be effectively incorporated into the EPA computer program, which would be especially useful for training response personnel.

The SLR decision-making method (Figure 6-3) addresses almost exclusively the question of biological trade-offs. It is relatively unique in its approach to comparing the environmental (biological) effects of dispersed oil with the those of untreated oil. This methodology appears to be needed in order to make the judgments called for both in the U.S. EPA computerized Oil Spill Response Decision Tree and the Alaska decision matrix, which is designed as a means of regulating and controlling dispersant use.

7
Conclusions and Recommendations

The major conclusions and recommendations provided below stem from the findings discussed in Chapters 2 through 6. General conclusions and recommendations are followed by those for physics and chemistry, biological effects, ecological effects, birds and mammals, and techniques, logistics, and contingency planning. Supporting text is cross-referenced in parentheses following each conclusion.

GENERAL CONCLUSIONS

Use of chemical dispersants can be an effective spill response and control method, especially to minimize environmental damage, but the application technique is critical to the success of dispersant use.

Dispersants have been shown, in laboratory tests, to be effective; that is, dispersants can remove oil from the water surface (pp. 70–79). In a few carefully planned and monitored field tests, high effectiveness has been documented (p. 174). At other field tests (pp. 179–187) and at accidental spills (p. 317), reported effectiveness has been low—perhaps because application technique was inadequate, the oil was not dispersible, the methods of measuring effectiveness were inadequate, or untested remote-sensing methods were used. Inconclusive visual results have occurred when different observers looking at the same dispersant treatment have provided widely differing estimates of effectiveness (p. 232).

> RECOMMENDATION: Use of dispersants as a first response option to oil spills should be considered along with mechanical cleanup. Implementation of this recommendation must consider spill size, logistical and contingency planning, equipment and dispersant performance and availability, appropriate regulations, and personnel training.

Prompt response with dispersants is essential because the dispersibility of oil decreases rapidly with weathering of oil components by evaporation (pp. 47–49, 54), which may occur in the first few hours, and photooxidation, which may occur over several days (p. 49).

> RECOMMENDATIONS: Regulations and contingency planning must make provision for the associated decision-making process and the need for rapid response.
> Prior approval to field-test a dispersant immediately after a spill, to establish dispersability when it is in doubt, should be included in contingency plans.

The principal biological benefits or objectives of chemically dispersing oil effectively are to:

- prevent stranding of oil in intertidal zones (pp. 193–198);
- reduce hazards of discharged oil to marine birds (p. 162) and mammals (p. 164);
- enhance degradation of oil components (p. 155); and
- reduce chronic impact on some habitats, such as mangroves (p. 206), because of the shorter persistence of oil.

Biological concerns focus on the possible expansion of the surface area of slicks (p. 57) treated by dispersants and the effect of this expanded slick on marine mammals and seabirds (pp. 162, 164), the effect of dispersed oil on marine life in the water column near the sea surface (p. 188), and the effect of oil dispersed offshore that may reach coastal marine habitats and communities.

Laboratory bioassays at measured concentrations show that the toxic effects per unit of dispersed oil are usually the same for chemically dispersed oil as those for physically dispersed oil (pp. 130–155). Acute biological effects are expected to be slight in most open-sea applications because the dispersed oil mixes into a relatively large volume of water, resulting in concentrations and times of exposures that are low compared to those showing effects in laboratory studies.

Hence the overall ecological impact of oil will likely be reduced by dispersion.

In shallow water with poor circulation, and in protected bays and inlets, the acute biological effects on some organisms and habitats from high concentrations of dispersed oil may be greater than the effects of untreated oils (pp. 208–214).

> *RECOMMENDATION: Additional ecological assessments at sites where water is shallow and circulation is well defined are needed to clarify the differences in effects between dispersed and untreated oil.*

PHYSICS AND CHEMISTRY

Oil dispersibility depends on:

• physical properties of the spilled oil, such as viscosity which increases at lower temperatures and with time of weathering (pp. 54–57);

• the structure of the slick, such as nonuniform oil thickness (pp. 36–41) and water-in-oil emulsion (mousse) formation (p. 49);

• oil composition, including the relative concentrations of hydrocarbons and the concentrations of natural surfactants and asphaltenes (pp. 51-54);

• dispersant formulation (pp. 31–35); and

• actual dosage rate, which in turn depends on the equipment and technique used to apply the dispersant, wind, dispersant droplet size distribution, and rate of application.

The spreading and distribution of oil on the water surface depends on complex surface circulation patterns as well as properties of the oil, wind and sea conditions, and distribution and thickness of the original spill (pp. 38–47). These circulation-related phenomena have been observed qualitatively, but predictive theories are not yet dependable (pp. 63–69).

The size distribution of oil droplets dispersed in the water column affects the stability of a dispersion. The smaller the droplets, the more stable the dispersion (p. 58).

Partial resurfacing of dispersed oil, after agitation ceases, occurs under laboratory conditions. There is disagreement about its occurrence in practical situations because little quantitative field evidence exists. Resurfacing of dispersed oil may be less likely than that of untreated oil because of the smaller droplet size resulting from the use of dispersants (pp. 59, 65).

Chemically dispersed oil appears to adhere less to suspended particulate matter (pp. 62, 167), biological materials (pp. 188, 199, 207), and shorelines (pp. 208–214) than does untreated oil.

Application of what is well known about the physical properties and molecular action of surfactants to the problem of dispersing oil requires further experimentation and technical experience to guide development in formulating and applying dispersants.

RECOMMENDATIONS: The following topics for experimental research should be supported:

- *interaction of oil and dispersed oil with suspended particulate matter, sediments, plankton, and benthic organisms;*
- *resurfacing, spreading, and photooxidation of dispersed oil;*
- *oil-water interaction phenomena, such as the formation of mousse, and the influences that surfactants have on this process;*
- *analysis of how turbulent diffusion, surface circulation, and wave motion affect distribution of dispersed oil as functions of depth, time, and volume of spilled and treated oil. These analyses will advance understanding about the concentration of oils in the water column under various dispersion conditions; and*
- *scientific research concerning the mechanisms by which droplets of dispersant contact an oil film, mix and penetrate into it, how their surfactants interact with the oil and migrate to the oil-water interface, and the microscopic processes by which emulsions actually form.*

BIOLOGICAL EFFECTS

Toxicity of Dispersants

The acute lethal toxicity of most dispersants currently considered for use in the United States and Canada (pp. 100–123) is low compared to the constituents and fractions of crude oils and refined products (pp. 123–130). This conclusion is based primarily on laboratory tests. Although the most effective dispersants tend to be more toxic, lower concentrations of these are required (p. 85).

Water temperature has a profound influence on the toxicity of dispersants; there are significantly higher sensitivities of organisms in warmer waters or in summer as compared to winter (pp. 116–117). Screening tests for a dispersant should account for the expected seawater temperature range.

A wide range of sublethal responses, usually at high exposures, has been observed (pp. 118–122). The sublethal effects of dispersants at realistic concentrations are only partially understood. It is unlikely, based on concentrations of dispersants that would result from spraying in marine waters at common rates, that dispersants would contribute significantly to lethal or sublethal toxicities (pp. 122–123). Direct application of dispersants to seabirds or marine mammals is not recommended because dispersants may diminish their water-repellancy and would thus be an exception to this conclusion (pp. 160–164).

> *RECOMMENDATIONS: Biological research and toxicity screening in the laboratory should use exposure conditions more closely reflecting dispersant use and probable dilution in the water column.*
>
> *New products should be screened for short-term toxicity using standard methods that would consider the physicochemical characteristics of the dispersant solutions and lethal and sublethal responses of test organisms. In addition, such standards, if they were international, would permit reliable comparisons of data among nations.*

Knowledge of the chemical composition of formulations is necessary for making responsible decisions about the use of oil spill dispersants (pp. 97–100). The chemical compositions are indicative of toxicity and surfactant properties.

> *RECOMMENDATION: Information on dispersant chemical structures and formulations should be made readily available to researchers. This information is particularly important to studies concerning toxicological and ecological effects of dispersants and dispersed oils.* *

---

*Some committee members expressed the view that much of the information about dispersant composition is proprietary, but a knowledge of the general type of compound, as would be obtained from the patent literature, would be sufficient for toxicological assessments. Other members felt that the detailed structure of the major and minor components is essential to achieve a biochemical understanding of toxic effects.

Toxicity of Dispersed Oils

Many earlier laboratory studies of the joint toxicity of oil and dispersants erroneously concluded that dispersed oils were more toxic than oil alone. These tests used nominal concentration (weight of oil per unit volume of water in the experimental system) that incorrectly included the untreated oil floating on the water surface as well as the oil fraction in the water column to which the organisms were exposed. The amount of oil in the water includes dispersed and dissolved oil, as well as oil that adheres to particulates. The fraction of the oil in the water was higher when dispersants were used. A proper comparison of the effects of chemically treated and untreated oil must be made using only the measured oil fraction in the water, exposing test organisms to the same actual concentration of oil in both cases (pp. 126–128).

Based on laboratory studies, acute lethal toxicity of chemically dispersed oils resides not in the dispersant but primarily in the oil droplets (for some species) and the low molecular weight and dissolved, aromatic, and aliphatic fractions of the oil (for most species).

Acute toxicity of chemically dispersed oils is generally similar to that of the portion of oil in the water column alone (p. 154).

Different species and life stages show sensitivity to chemically dispersed oils at exposures varying by 3 to 4 orders of magnitude (pp. 130–154).

Laboratory tests of the toxicology of dispersed oil should cover the ranges of exposures that would be expected in the field, but there is no commonly accepted technique for comparing laboratory bioassay data with field exposure data. One approach is to express exposure as an integral of concentration over time based on a "toxicity index" concept that concentration and time of exposure are of equal importance in short hydrocarbon exposures.

Several studies have shown that chemically dispersed oil does not adhere as much as untreated oil to some organisms or habitats (e.g., mangrove trees, pp. 206–208).

Greater adverse effects have sometimes been observed from a few untreated oil droplets than from many chemically dispersed oil droplets (e.g., abnormal larvae from exposed herring eggs, pp. 136–137, 153).

RECOMMENDATIONS: Methods for comparing laboratory and field exposures should be developed based on sound physiological, toxicological, and physicochemical principles.

Additional research should be conducted, or observations made, to compare the effects of untreated oil adhesion to organisms as compared with effects from chemically dispersed oil. Particular emphasis should be placed on organisms that reside at or near the sea surface.

Biodegradation

The biodegradation of dispersant components has been demonstrated in the laboratory and in mesocosm experiments (pp. 155–158).

Some laboratory studies and all mesocosm studies have shown that the rate of biodegradation of dispersed oil is equal to or greater than that of undispersed oil (p. 159).

RECOMMENDATION: Further field studies of biodegradation rates of dispersed oils and hydrocarbon components should be undertaken, together with studies of critical factors controlling biodegradation in the field.

ECOLOGICAL EFFECTS

Exposure of communities of organisms and their habitats to oil, dispersant, and dispersed oil depends on many factors, including (pp. 186–214):

- amount of oil dispersed;
- amount of dispersant used and effectiveness of dispersion;
- volume of water available for dispersal;
- exchange of water through turbulence and vertical and horizontal transport, such as local currents, tides, convergences, and upwellings; and
- distribution and abundance of affected organisms; for example, exposure is higher in restricted bays and estuaries than in open-ocean situations.

The best strategy for protecting sensitive inshore habitats (i.e., littoral and shallow subtidal, polar to tropical) is to prevent undispersed oil from contacting them. Dispersion of oil before it reaches these habitats may keep them from becoming oiled, or may reduce the persistence of oil that contacts them. Thus offshore chemical dispersal may be the best technique for reducing overall, particularly chronic, impact of the oil in those habitats.

Acute effects of chemically dispersed oils on organisms and habitats vary.

• Organisms near the water surface are exposed to higher concentrations of undispersed oil than are organisms in the water column, but organisms in the water column, particularly in the upper layers, will experience greater short-term exposure to oil components if the oil is dispersed (pp. 60, 168–172). Benthic organisms, such as mollusks, may also be exposed to higher concentrations of dispersed oils, resulting in short-term bioaccumulation. However, long-term harmful effects may be reduced by chemically dispersing oil rather than not treating the spill (pp. 192, 199).

• In a habitat with restricted water exchange, the acute effects of dispersed oil on some organisms or marine plants may be greater than that of oil alone. However, mangroves (pp. 206–208) and some other intertidal habitats, such as north-temperate mudflats, are less damaged by dispersed oil and can recover faster. Therefore, if a slick is treated before it enters these areas, the community recovers faster.

• Intertidal areas on rocky coasts, sand and mud flats, and salt marshes can be protected if the oil is dispersed offshore, but do not benefit from dispersant application after the oil reaches shore. Once oil has penetrated a salt marsh, it is best left alone (pp. 193–198).

• Studies of the persistence of hydrocarbon fractions in a range of habitats, with and without dispersants, show that reduction of chronic exposure is a key to reducing biological damage (pp. 209–213).

• Toxic effects of untreated oil on fish have been reported in some shallow-water environments, but the effects of dispersed oil have not been studied (pp. 134–137, 150–154, 200–201).

• In offshore open water, concentrations of dispersed oil will be much lower than in shallower waters, or in waters with poor circulation, and the resulting impacts will be correspondingly less (pp. 168–172).

• No measurable effects of dispersed or untreated oil on commercial fisheries and their supporting food webs have yet been found. However, such effects, if they occur, would be difficult to detect and measure effectively because of the mobility of most fish and many invertebrates, the natural variability of their populations, and the effects of overfishing on stocks.

RECOMMENDATIONS: Additional ecological studies under controlled or established water circulation in shallow en-

vironments should be conducted to define the conditions under which dispersant use can be environmentally safe.

Long-term studies of the recovery of selected ecosystems exposed to oil are desirable, including continued studies of those sites where the impact of oil and dispersed oil has been compared already.

BIRDS AND MAMMALS

Adverse effects of oil, dispersants, and dispersed oil have been shown in laboratory tests on seabirds and on mammals (otter, polar bear, and fur seal pelts) and from limited field tests and observations, such as (pp. 160–164):

• reduced water-repellancy of fur and feathers (critical for thermal insulation);
• reduced hatchability of eggs; and
• physiological and biochemical effects.

These laboratory results are consistent with field observations for untreated oil, but may not appropriately represent exposures to dispersants and dispersed oil in the field. However, over a short period, residual sheen from dispersed oil slicks may cover a greater area than untreated oil, and more birds rather than fewer may be oiled; this effect may possibly offset some biological benefits resulting from dispersant use.

Theoretical considerations indicate that the exposure of birds and mammals to dispersed oil in the water column may be less damaging than exposure to untreated floating oil (p. 162). No definitive field studies, in which birds and mammals were exposed to dispersed oil, have been carried out.

Concern about the effects of dispersants on birds and marine mammals centers on a question of the extent of exposure rather than on enhanced toxicity of the oil.

RECOMMENDATIONS: Laboratory studies should be undertaken using realistic exposure conditions (e.g., initial concentration of dispersed oil of 10 to 20 ppm, decreasing with time) to assess the ability of fur and feathers to maintain the water-repellancy critical for thermal insulation under dispersed oil exposure conditions comparable to those expected in the field.

Laboratory tests, under exposure conditions similar to those expected in the field, should be conducted of the hatchability of seabird eggs that have come into contact with dispersed oil and untreated oil.

The possible effects of ingested oil from exposure of birds and sea mammals to dispersed and untreated oil should be observed when possible (e.g., after an accidental spill).

TECHNIQUES, LOGISTICS, AND CONTINGENCY PLANNING

Logistical constraints in some situations may dictate that dispersant use (if a dispersant can be accurately applied) will be the preferred spill response, compared to the use of mechanical methods. Distance to the spill, time required to mobilize equipment, the size of the spill, and the roughness of the seas are major factors that influence the use of dispersants (pp. 241–243).

Oil spills of all sizes can be treated by the use of undiluted dispersant, which is usually applied by various techniques:

• Major spills (greater than 2,500 bbl) encompass large areas and require rapid treatment response; the only technique that may be capable of countering these spills is to spray from helicopters or large aircraft (pp. 224–226).

• Moderate or small spills can be countered by spraying from boats, but this technique is less efficient on spills greater than 1,000 bbl because of logistical limitations (pp. 222–224).

Dispersant droplet size produced by aerial spray systems can be controlled only within limits determined by nozzle design and the effects of air currents produced by aircraft motion. For boat systems, nozzle selection and flow rate determine droplet size (pp. 216–220).

There is, as yet, no practical means of varying the application rate of dispersants to achieve a constant dispersant-oil ratio in treating a spill (pp. 218–219).

Operators of aircraft or boats spraying dispersant need guidance to locate the oil and apply dispersant accurately. This guidance is normally provided by spotter aircraft (pp. 237–238). Remote-sensing methods are being developed, but they are still difficult to interpret (pp. 230–233).

For documentation and evaluation of effectiveness, photography, including videotape, is useful but must be interpreted cautiously because of the influence of available sunlight and sea-state conditions on the detection of oil or dispersed oil (pp. 231–232).

In contingency planning, as well as in its implementation, the primary questions to be answered in deciding whether to use dispersants are:

- Is the oil dispersable (pp. 50–57)?
- Can dispersant be effectively applied to the oil slick (pp. 215–228)?
- Will the use of dispersants reduce environmental damage (pp. 208–214)?

Contingency planning is fundamental to the effective use of dispersants for the following reasons:

- Equipment is specialized and availability must be assured (pp. 241–243).
- Calibration of the spraying system is essential (pp. 228–229).
- Training is needed on all aspects of dispersant use.
- Response must be fast because weathering decreases effectiveness (pp. 54–57, 244–253).

> *RECOMMENDATIONS: More reliable remote-sensing equipment and easily interpreted processing systems need to be developed.*
>
> *More accurate monitoring and documentation of dispersant effectiveness, on spills of various sizes and under environmental conditions covered by contingency plans, are encouraged.*
>
> *Detailed contingency planning for the use of dispersants should include:*
>
> - *past experience in applying dispersants and evaluating their effectiveness;*
> - *hazard evaluation for sensitive species, populations, communities, and habitats;*
> - *familiarity with available dispersant chemicals;*
> - *training and practice in decision making; and*
> - *training of observers to monitor dispersant application and results.*

References

Abbott, F. S. 1983. A simple field effectiveness test for dispersants and recent dispersant effectiveness test results. Spill Technol. Newsletter 8:197-108.

_____. 1984. Guidelines on the use and acceptability of oil spill dispersants—2nd edition. Spill Technol. Newsletter 9(1):6-10.

Abbott, R. A. 1972. Acceptability criteria and testing procedures for oil spill treating agents. Toronto: Environment Ontario, Ministry of the Environment, Industrial Wastes Branch. 15 pp.

Abel, P. D. 1974. Toxicity of synthetic detergents to fish and aquatic invertebrates. J. Fish. Biol. 6:279-298.

Abernethy, S. A., M. Bobra, W. Y. Shiu, P. G. Wells, and D. Mackay. 1986. Acute lethal toxicity of hydrocarbons and chlorinated hydrocarbons to two planktonic crustaceans: The key role of organism-water partitioning. Aquat. Toxicol. 8:163-174.

Acreman, J. C., H. M. Dovey, L. Guanguo, P. J. Harrison, C. M. Lalli, K. Lee, T. R. Parsons, C. Xiaolin, and P. A. Thompson. 1984. An experimental marine ecosystem response to crude oil and Corexit 9527: Part 2—biological effects. Mar. Environ. Res. 13(4):265-275.

Adams, S. M., and J. M. Giddings. 1982. Review and evaluation of microcosms for assessing the effects of stress in marine ecosystems. Environ. Internat. 7:409-418.

Ahsanullah, M., R. R. C. Edwards, D. G. Kay, and D. S. Negilski. 1982. Acute toxicity to the crab *Paragrapsus quadridentatus* (H. Milne Edwards) of Kuwait light crude oil, BP/AB dispersant, and an oil-dispersant mixture. Aust. J. Mar. Freshwat. Res. 33(3):459-464.

Åkesson, B. 1975. Bioassay studies with polychaetes of the genus *Ophryotrocha* as test animals. *In* Sublethal Effects of Toxic Chemicals on Aquatic Animals, J. H. Koeman and J. J. Strik, eds. New York: Elsevier. pp. 121-135.

Albers, P. H. 1980. Transfer of crude oil from contaminated water to bird eggs. Environ. Res. 22(2):307-314.

265

_____. 1979. Effects of Corexit 9527 on the hatchability of mallard eggs. Bull. Environ. Contam. Toxicol. 23(4/5):661-668.

Albers, P. H., and M. L. Gay. 1982. Effects of a chemical dispersant and crude oil on breeding ducks. Bull. Environ. Contam. Toxicol. 29(4):404-411.

Allen, T. D., ed. 1984. Oil Spill Chemical Dispersants: Research, Experience, and Recommendations. STP 840. Philadelphia: ASTM. 465 pp.

American Petroleum Institute (API). 1985. Oil Spill Response: Options for Minimizing Adverse Ecological Impacts. Publication No. 4398. Washington, D.C.: API. 98 pp.

_____. 1987. Effects of a Dispersed and Undispersed Crude Oil on Mangroves, Seagrasses, and Corals. Publication 4460. Washington, D.C.: API.

American Petroleum Institute (API) Task Force. 1986. The Role of Chemical Dispersants in Oil Spill Control. Publication 4425. Washington, D.C.: API. 39 pp.

American Society for Testing and Materials (ASTM). 1987. Ecological Considerations for the Use of Chemical Dispersants in Oil Spill Response—Coral Reefs. *In* Annual Book of Standards, v. 11.04. pp. 909-923, 932-938, 941-961. Standard F929085, Marine Mammals; F930-85, Rocky Shores; F931-85, Seagrasses; F932-85, Coral Reefs; F971-86, Mangroves; F972-86, Nearshore subtidal; F 973-86, Tidal Flats; F990-86, Sandy Beaches; F999-86, Gravel or Cobble Beaches; F1008-86, Salt Marshes; F1009-86, Offshore; F1010-86, Bird Habitats; F1012-86, The Arctic. Philadelphia: ASTM.

Anderson, J. W. 1986. Predicting the effects of complex mixtures on marine invertebrates by use of a toxicity index. *In* Workshop on Hazardous Assessment for Complex Effluents. New York: Pergamon Press. pp. 115-122.

Anderson, J. W., J. M. Neff, B. A. Cox, H. E. Tatem, and G. M. Hightower. 1974. Characteristics of dispersions and water-soluble extracts of crude and refined oils and their toxicity to estuarine crustaceans and fish. Mar. Biol. 27:75-88.

Anderson, J. W., S. L. Kiesser, and J. W. Blaylock. 1980. The cumulative effect of petroleum hydrocarbons on marine crustaceans during constant exposure. Rapp. P.-v. Reun. Cons. int. Explor. Mer 179:62-70.

Anderson, J. W., S. L. Kiesser, R. M. Bean, R. G. Riley, and B. L. Thomas. 1981. Toxicity of chemically dispersed oil to shrimp exposed to constant and decreasing concentrations in a flowing system. Proc. 1981 Oil Spill Conference. Washington, D.C.: API. pp. 69-75.

Anderson, J. W., S. L. Kiesser, D. L. McQuerry, R. G. Riley, and M. L. Fleischmann. 1984. Toxicity testing with constant or diluting concentrations of chemically dispersed oil. *In* Oil Spill Chemical Dispersants: Research, Experience, and Recommendations, T. E. Allen, ed. STP 840. Philadelphia: ASTM. pp. 14-22.

Anderson, J. W., S. L. Kiesser, D. L. McQuerry, and G. W. Fellingham. 1985. Effects of oil and chemically dispersed oil in sediments on clams. Proc. 1985 Oil Spill Conference. Washington, D.C.: API. pp. 349-353.

Anderson, J. W., D. L. McQuerry, and S. L. Kiesser. 1985. Laboratory evaluation of chemical dispersants for use on oil spills at sea. Environmental Science and Technology. Sequim, Wash.: Battelle Northwest Laboratories. pp. 454-457.

Anderson, J. W., R. Riley, S. Kiesser, and J. Gurtisen. 1987. Toxicity of dispersed and undispersed Prudhoe Bay crude oil fractions to shrimp and fish. Proc. 1987 Oil Spill Conference. Washington, D.C.: API. pp. 235-240.

Araujo, R. P de A., K. Momo, E. Gherardi-Goldstein, M. G. Nipper, and P. G. Wells. 1987. Marine dispersant program for licensing and research in São Paulo state, Brazil. Proc. 1987 Oil Spill Conference. Washington, D.C.: API. pp. 289-292.

Aravamudan, K. S., P. K. Raj, and G. Marsh. 1981. Simplified models to predict the breakup of oil on rough seas. Proc. 1981 Oil Spill Conference. Washington, D.C.: API. pp. 153-160.

Atlas, R. M. 1985. Effects of hydrocarbons on microorganisms and petroleum biodegradation in arctic ecosystems. *In* Petroleum Effects in the Arctic Environment, F. R. Engelhardt, ed. London: Elsevier Applied Science Publishers. pp. 63-99.

Atwood, D. K., ed. 1980. Preliminary Results from the September 1979 *Researcher/Pierce* Ixtoc-I Cruise. Proceedings of a symposium at Key Biscayne, FL, June 9-10, 1980. Boulder, Colo.: NOAA Office of Marine Pollution Assessment. pp. 10-11.

Audunson, T., V. Dalen, J. P. Mathisen, J. Holdorsen, and F. Krogh. 1980. SLIKFORCAST—a simulation program for oil spill emergency tracking and long term contingency planning. Proc. Petromar 1980, Monaco. London: The Oil Industry International E & P Forum.

Audunson, T., H. K. Celius, O. Johansen, P. Steinbakke, and S. E. Sorstrom. 1984. The Experimental Oil Spill on Haltenbanken 1982. Publication No. 112. Trondheim, Norway: Continental Shelf Institute. 109 pp.

Audunson, T., O. Johansen, J. Kolnes, and S. E. Sorstrom. 1987. Injection of Oil Spill Chemicals into a Blowing Well. Proc. 1987 Oil Spill Conference. Washington, D.C.: API. pp. 335-340.

Auger, C., and J. Croquette. 1980. L'utilisation des dispersants. Bull. Cedre. 1:7-12.

Avolizi, R. J., and M. Nuwayhid. 1974. Effects of crude oil and dispersants on bivalves. Mar. Pollut. Bull. 5(10):149-153.

Ayers, R. R., and A. V. Barnett. 1977. SOCK—an oil skimming kit for vessels of convenience. Proc. 1977 Oil Spill Conference. Washington, D.C.: API. pp. 361-365.

Ayers, R. R., J. P. Fraser, and L. J. Kazmierczak. 1975. Developing an open-seas skimmer. Proc. 1975 Oil Spill Conference. Washington, D.C.: API. pp. 401-408.

Baca, B. J., and C. D. Getter. 1984. The toxicity of oil and chemically dispersed oil to the seagrass *Thalassia testudinum*. *In* Oil Spill Chemical Dispersants: Research, Experience, and Recommendations, T. E. Allen, ed. STP 840. Philadelphia: ASTM. pp. 314-323.

Backus, R. H., G. R. Flierl, D. R. Kester, D. B. Olson, P. L. Richardson, A. C. Vastano, P. H. Wiebe, and J. H. Wormuth. 1981. Gulf Stream cold core rings: Their physics, chemistry, and biology. Science 212:1091-1100.

Baker, J. M. 1976. Environmental effect of oil and the chemicals used to control it on the shore and splash zones. *In* The Control of Oil Pollution on the Sea and Inland Waters, J. Wardley-Smith, ed. London: Graham and Trotman Ltd. pp. 57-72.

Baker, J. M., and G. B. Crapp. 1974. Toxicity tests for predicting the ecological effects of oil and emulsifier pollution on littoral communities. *In* Ecological Aspects of Toxicity Testing of Oils and Dispersants, L. R. Beynon and E. B. Cowell, eds. Barking, Essex, U.K.: Applied Science Publishers Ltd. pp. 23-40.

Baker, J. M., J. H. Cruthers, D. I. Little, J. H. Oldham, and C. M. Wilson. 1984. Comparison of the fate and ecological effects of dispersed and non-dispersed oil in a variety of marine habitats. *In* Oil Spill Chemical Dispersants: Research, Experience, and Recommendations, T. E. Allen, ed. STP 840. Philadelphia: ASTM. pp. 239-279.

Ballou, T. G., S. C. Hess, C. D. Getter, A. Knap, R. Dodge, and T. Sleeter. 1987. Final results of the API Tropics oil spill and dispersant use experiments in Panama. Proc. 1987 Oil Spill Conference. Washington, D.C.: API. 634 pp.

Bancroft, W. D. 1913. Theory of emulsification. J. Phys. Chem. 17:501-519.

Barbouteau, G., M. Angles, Y. Le, and G. La Salle. 1987. Dispersant spraying gun. Proc. 1987 Oil Spill Conference. Washington, D.C.: API. pp. 313-316.

Bardach, J. E., M. Jujiya, and A. Hull. 1975. Detergents: Effects on the chemical senses of the fish, *Ictalurus ratalis* (le sueur). Science 4:1605-1607.

Bardot, C., and G. Castaing. 1987. Toxicity of chemically dispersed oil in a flow-through system. *In* Fate and Effects of Oil in Marine Ecosystems, J. Kuiper and W. J. van den Brink, eds. Dordrecht, The Netherlands: Martinus Nijhoff. pp. 207-209.

Bardot, C., C. Bocard, G. Castaing, and C. Gatellier. 1984. The importance of a dilution process to evaluation effectiveness and toxicity of chemical dispersant. 7th Arctic Marine Oilspill Program Technical Seminar. Ottawa: Environment Canada, Environmental Protection Service. pp. 179-201.

Barger, W. R. 1973. Laboratory and field testing of surface-film forming chemicals for use as oil collecting agents. Proc. Joint Conf. on Prevention and Control of Oil Spills, 1973. Washington, D.C.: API. pp. 241-246.

Battelle Memorial Institute. 1970. Oil spill treating agents. Test procedures: Status and recommendations. Project OS-7. Report to American Petroleum Institute Committee for Oil and Water Conservation by Battelle Pacific Northwest Laboratory, Richland, Washington.

Battershill, C. N., and P. R. Bergquist. 1982. Responses of an intertidal gastropod to field exposure of an oil and a dispersant. Mar. Pollut. Bull. 13(5):159-162.

Becker, C. D., J. A. Lichatowich, M. J. Schneider, and J. A. Strand. 1973. Regional Survey of Marine Biota for Bioassay Standardization of Oil and Oil Dispersant Chemicals. Publication 4167. Washington, D.C.: API.

Becker, K. W., and G. P. Lindblom. 1983. Performance evaluation of a new versatile oil spill dispersant. Proc. 1983 Oil Spill Conference. Washington, D.C.: API. pp. 61-64.

Bellan, G. L., F. Caruelle, P. Foret-Montardo, R. A. Kaim-Malka, and K. Leung Tack. 1969. Contribution to the study of different physico-chemical polluting factors on marine organisms. I. Action of detergents on the polychaete *Scolelepis fuliginosa*. Tethys. 1(2):367-374.

Bellan, G. L., D. J. Reish, and J.-P. Foret. 1972. The sublethal effects of a detergent on the reproduction, development and settlement in the polychaetous annelid *Capitella capitata*. Mar. Biol. 14(3):183-188.

Belore, R. 1985. Effectiveness of the repeat application of chemical dispersants on oil. Report No. 006. Ottawa: Environmental Studies Revolving Fund and S. L. Ross Environmental Research Ltd.

_____. 1987. Use of high-pressure water mixing for ship-based oil spill dispersing. Proc. 1987 Oil Spill Conference. Washington, D.C.: API. pp. 197-302.

Beynon, L. R., and E. B. Cowell, eds. 1974. Ecological Aspects of Toxicity Testing of Oils and Dispersants. Barking, Essex, U.K.: Applied Science Publishers, Ltd. 149 pp.

Birkeland, C., A. A. Reimer, and A. A. Young. 1976. Survey of marine communities in Panama and experiments with oil. U.S. Environmental Protection Agency Ecological Research Service. EPA-66/3-76-028. 177 pp.

Blackall, P. J., and G. A. Sergy. 1981. The BIOS project—frontier oil spill counter measures research. Proc. 1981 Oil Spill Conference. Washington, D.C.: API. pp. 167-172.

_____. 1983a. BIOS Project—preliminary results. Proceedings of the Arctic Marine Oilspill Program Technical Seminar. Ottawa: Environmental Protection Service. pp. 292-295.

_____. 1983b. The BIOS project—an update. Proc. 1983 Oil Spill Conference. Washington, D.C.: API. pp. 451-455.

Blacklaw, J. R., J. A. Strand, and P. C. Walkup. 1971. Assessment of oil spill treating agent test methods. Proc. Joint Conference on Prevention and Control of Oil Spills. Washington, D.C.: API. pp. 253-261.

Blackman, R. A. A., F. L. Franklin, M. G. Norton, and K. W. Wilson. 1977. New procedures for the toxicity testing of oil slick dispersants. Fisheries Research Technical Report No. 39. Lowestoft, U.K.: Ministry of Agriculture, Fisheries, and Food, Directorate of Fisheries Research. 7 pp.

_____. 1978. New procedures for the toxicity testing of oil slick dispersants in the United Kingdom. Mar. Pollut. Bull. 9:234-238.

Blaikley, D. R., G. F. L. Dietzel, A. W. Glass, and P. J. van Kleef. 1977. SLIKTRAK— a computer simulation of offshore oil spills, cleanup, effects and associated costs. Proc. 1977 Oil Spill Conference. Washington, D.C.: API. pp. 45-54.

Bobra, A. M., and D. Mackay. 1984. Acute toxicity of dispersants, physically and chemically dispersed crude oil (fresh and weathered) to *Daphnia magna* at 5° and 20°C. Unpublished paper. Esso Resources Ltd., Calgary, Alberta.

Bobra, A. M., and P. T. Chung. 1986. A catalog of oil properties, 2d ed. Environmental Protection Service Report EE-77. Ottawa: Environment Canada.

Bobra, A., D. Mackay, and W. Y. Shiu. 1979. Distribution of hydrocarbons among oil, water and vapor phases during oil dispersant toxicity tests. Bull. Environ. Contam. Toxicol. 23(4-5):558-565.

Bobra, A. M., S. Abernethy, P. G. Wells, and D. Mackay. 1984. Recent toxicity studies at the University of Toronto. Proc. 7th Annual Arctic Marine Oilspill Program Technical Seminar. Edmonton, Alberta: Environment Canada. pp. 82-90.

Bobra, A. M., W. Y. Shiu, D. Mackay, and R. H. Goodman. 1987. Acute toxicity of dispersed fresh and weathered crude oil and dispersants to *Daphnia magna*. Unpublished paper given at ASTM meeting, Williamsburg, Virgina, October.

Bocard, C. 1985. L'Operation Protecmar VI. Bull. du Cedre 22:6-10.

Bocard, C., and C. Gatellier. 1981. Breaking of fresh and weathered emulsions by chemicals. Proc. 1981 Oil Spill Conference. Washington, D.C.: API. pp. 601-607.

_____. 1982. Protecmar III. Report IFP 30482. Institut Francais du Petrole, Paris.

Bocard, C., and G. Castaing. 1986. Dispersant effectiveness evaluation in a dynamic flow-through system: The IFP dilution test. Oil and Chem. Pollut. 3(6):433-444.

Bocard, C., C. Gatellier, J. Croquette, and F. Merlin. 1981. Operation Protecmar. Spill Tech. Newsletter (March-April):54-85.

Bocard, C., J. Ducreaux, and C. Gatellier. 1983. Protecmar IV. Report IFT 31478. Institut Francais du Petrole, Paris.

Bocard, C., G. Castaing, and C. Gatellier. 1984. Chemical oil dispersion in trials at sea and in laboratory tests: The key role of dilution processes. *In* Oil Spill Chemical Dispersants: Research, Experience, and Recommendations, T. E. Allen, ed. STP 840. Philadelphia: ASTM. pp. 125-142.

Bocard, C., G. Castaing, J. Ducreux, C. Gatellier, J. Croquette, and F. Merlin. 1987. Protecmar: The French experience from a seven-year dispersant offshore trials program. Proc. 1987 Oil Spill Conference. Washington, D.C.: API. pp. 225-229.

Boehm, P. D. 1983. Long-term fate of crude oil in the Arctic nearshore environment— the BIOS experiments. Proc. 6th Arctic Marine Oilspill Program Technical Seminar. Edmonton, Alberta: Environmental Protection Service. pp. 280-291.

_____. 1984. The comparative fate of chemically dispersed and untreated oils in an Arctic nearshore environment. *In* Oil Spill Chemical Dispersants: Research, Experience, and Recommendations, T. E. Allen, ed. STP 840. Philadelphia: ASTM. pp. 338-360.

_____. 1985. Transport and transformation processes regarding hydrocarbon and metal pollutants in offshore sedimentary environments. *In* The Long-term Effects of Offshore Oil and Gas Development, D. F. Boesch and N. N. Rabalais, eds. Essex, U.K.: Elsevier Applied Science Publishers.

Boehm, P. D., and D. L. Fiest. 1982. Subsurface distributions of petroleum from an offshore well blowout. The Ixtoc I blowout, Bay of Campeche. Environ. Sci. Technol. 16:67-74.

Boehm, P. D., D. L. Fiest, D. Mackay, and S. Paterson. 1981. Physical-chemical weathering of petroleum hydrocarbons from the Ixtoc I blowout—chemical measurements and a weathering model. Proc. 1981 Oil Spill Conference. Washington, D.C.: API. pp. 453-460.

Boehm, P. D., W. Steinhauer, D. Cobb, S. Duffy, and J. Brown. 1984. Chemistry 2: analytical biogeochemistry—1983 study results. Baffin Island Oil Spill Working Report 83-2. Ottawa: Environmental Protection Service. 139 pp.

Boehm, P. D., W. Steinhauer, A. Requejo, D. Cobb, S. Duffy, and J. Brown. 1985. Comparative fate of chemically dispersed and untreated oil in the Arctic: Baffin Island Oil Spill studies 1980-83. Proc. 1985 Oil Spill Conference. Washington, D.C.: API. pp. 561-569.

Boehm, P. D., M. S. Steinhauer, D. R. Green, B. Fowler, B. Humphrey, D. L. Fiest, and W. J. Cretney. 1987. Comparative fate of chemically dispersed and beached crude oil in subtidal sediments of the arctic nearshore. Arctic 40(suppl. 1):133-148.

Boesch, D. F., and N. N. Rabalais. 1987. Long Term Environmental Effects of Offshore Oil and Gas Development. Barking, Essex, U.K.: Elsevier Applied Science Publishers.

Boney, A. D. 1968. Experiments with some detergents and certain intertidal algae. *In* The Biological Effects of Oil Pollution on Littoral Communities, J. D. Carthy and D. R. Arthur, eds. Supplement to Vol. 2 of Field Studies. London: Field Studies Council. pp. 55-72.

Borseth, J., T. Aunaas, M. Ekker, and K. E. Zachariassen. 1986. A comparison of in vivo and in situ exposure of marine fish eggs to Statfjord A + B crude oil topped at 15°C and this oil premixed with Finasol OSR-5. Abstract presented at International Seminar on Chemical and Natural Dispersion of Oil on Sea, Heimdal, Norway, November.

Bowden, K. F. 1965. Horizontal mixing in the sea due to a shearing current. J. Fluid Mech. 21:83-95.

Bouwmeester, R. J. B., and R. B. Wallace. 1986a. Oil entrainment by breaking waves. Proc. 9th Arctic Marine Oilspill Program Technical Seminar. Edmonton, Alberta: Environmental Protection Service. pp 39-49.

_____. 1986b. Dispersion of oil on a water surface due to wind and wave action. Department of Civil Engineering, Michigan State University quoted by R. B. Wallace, T. M. Rushlow, and R. J. B. Bouwmeester. 1986. In situ measurement of oil droplets using an image processing system. Proc. 9th Arctic Marine Oilspill Program Technical Seminar, pp. 421-429. Ottawa: Environmental Protection Service. This latter paper describes the equipment but gives no numerical results.

Braaten, B., A. Granmo, and R. Lange. 1972. Tissue swelling in *Mytilus edulis* L. induced by exposure to a nonionic surface active agent. Norw. J. Zool. 20:137-140.

Brannon, E. L., T. P. Quinn, R. P. Whitman, A. E. Nevissi, R. E. Nakatani, and C. D. McAuliffe. 1986. Homing of adult chinook salmon after brief exposure to whole and dispersed crude oil. Trans. Am. Fish. Soc. 115:823-827.

Bratbak, G., M. Heldal, G. Knutsen, T. Lien, and S. Norland. 1982. Correlation of dispersant effectiveness and toxicity of oil dispersants towards the alga *Chlamydomonas reinhardti.* Mar. Pollut. Bull. 13(10):351-353.

Bridie, A. L., T. H. Wander, W. Zegveld, and H. B. van der Heijde. 1980. Formation, prevention, and breaking of sea water in crude oil emulsions (chocolate mousses). Mar. Poll. Bull. 11:343-348.

Brochu, C., E. Pelletier, G. Caron, and J. E. Desnoyers. 1987. Dispersion of crude oil in seawater: The role of synthetic surfactants. Oil and Chem. Pollut. 3(1986/87):257-279.

Brown, H. M., R. H. Goodman, and G. P. Canevari. 1987. Where has all the oil gone? Dispersed oil detection in a wave basin and at sea. Proc. 1987 Oil Spill Conference. Washington, D.C.: API. pp. 307-312.

Buist, I. 1979. An experimental study on the dispersion of oil slicks into the water column. M.A.Sc. Thesis, University of Toronto.

Buist, I. A., and S. L. Ross. 1986. A study of chemicals to inhibit emulsification and promote dispersion of oil spills. International Seminar on Chemical and Natural Dispersion of Oil on the Sea. Heimdal, Norway, November.

———. 1987. Emulsion inhibitors: A new concept in oil spill treatment. Proc. 1987 Oil Spill Conference. Washington, D.C.: API. pp. 217-222.

Bunch, J. N. 1987. Effects of petroleum releases on bacterial numbers and micro-heterotrophic activity in the water and sediment of an arctic marine ecosystem. Arctic 40(Suppl. 1):172-183.

Bunch, J. N., C. Bedard, and T. Cartier. 1983. Abundance and Activity of Heterotrophic Marine Bacteria in Selected Bays at Cape Hatt, N.W.T.: Effects of Oil Spills, 1981. Canadian Manuscript Report of Fisheries and Aquatic Sciences, No. 1708. 96 pp.

Bunch, J. N., C. Bedard, and F. Dugre. 1985. Environmental chemistry data from an arctic marine environment at Cape Hatt, N.W.T.: 1981-1982. Canadian Data Report of Fisheries and Aquatic Sciences 578. Ste.-Anne-de-Bellevue, Quebec: Arctic Biological Station. 83 pp.

Bunch, J. N., and R. C. Harland. 1976. Biodegradation of Crude Petroleum by the Indigenous Microbial Flora of the Beaufort Sea. Beaufort Sea Technical Report No. 10, Beaufort Sea Project, Dept. Environ. Victoria, B.C. 52 pp.

Burns, W., and M. J. Herz. 1976. Development and field testing of a light aircraft oil surveillance system (LOASS). Report, NASA CR-2739 CG-D-1-76. Washington, D.C.: NASA.

Burwood, R., and G. C. Spears. 1974. Photo-oxidation as a factor in the environmental dispersal of crude oil. Estuar. Coast. Mar. Sci. 2:117-135.

Butler, R. G., and D. B. Peakall. 1982. Effect of oil dispersant mixtures on the basal metabolic rate of ducks. Bull. Environ. Contam. Toxicol. 29:520-524.

Butler, R. G., W. Trivelpiece, D. Miller, P. Bishop, C. D'Amico, M. D'Amico, G. Lamert, and D. Peakall. 1979. Further studies of the effects of petroleum hydrocarbons on marine birds. Bull. Mt. Desert Is. Biol. Lab. 19:33-35.

Butler, R. G., W. Trivelpiece, and D. S. Miller. 1982. The effects of oil, dispersant, and emulsions on the survival and behavior of an estuarine teleost and an intertidal amphipod. Environ. Res. 27(2):266-276.

Butler, R. G., A. Harfenist, F. A. Leighton, and D. B. Peakall. 1988. Impact of sublethal oil and emulsion exposure on the reproduction success of Leach's Storm Petrels: short and long-term effects. J. Appl. Ecol. 25:125-143.

Byford, D. C., and P. J. Green. 1984. A view of the Mackay and Labofina laboratory tests for assessing dispersant effectiveness with regard to performance at sea. *In* Oil Spill Chemical Dispersants: Research, Experience, and Recommendations, T. E. Allen, ed. STP 840. Philadelphia: ASTM. pp. 69-86.

Byford, D. C., P. J. Green, and A. Lewis. 1983. Factors influencing the performance and selection of low-temperature dispersants. Proceedings of the Arctic Marine Oilspill Program Technical Seminar. Ottawa: Environmental Protection Service. pp. 140-150.

Byford, D. C., P. R. Laskey, and A. Lewis. 1984. Effect of low temperature and varying energy input on the droplet size distribution of oils treated with dispersants. Proc. Seventh Annual Arctic Marine Oilspill Program Technical Seminar. Ottawa: Environment Canada. pp. 208-228.

Cairns, J. Jr. 1983. Are single species toxicity tests done alone adequate for estimating environmental hazard? Hydrobiologia 100:47-57.

Cairns, J. Jr., K. L. Dickson, and A. W. Maki, eds. 1978. Estimating the Hazard of Chemical Substances to Aquatic Life. Publication 657. Philadelphia: ASTM. 278 pp.

Calamari, D., and J. S. Alabaster. 1980. An approach to theoretical models in evaluating the effects of mixtures of toxicants in the aquatic environment. Chemosphere 9:533-538.

Calamari, D., and R. Marchetti. 1973. The toxicity of mixtures of metals and surfactants to rainbow trout (*Salmo gairdneri* Rich.). Water Res. 7:1453-1464.

California State Water Resources Control Board. 1971. Evaluating oil spill cleanup agents. Development of testing procedures and criteria. Publication No. 43. Sacramento, California. 150 pp.

Canadian Offshore Aerial Applications Task Force (COAATF). 1986. The Effectiveness of Three Aerially Applied Dispersants in an Offshore Field Trial. Calgary: Canadian Offshore Oil Spill Research Association. 145 pp.

Canevari, G. P. 1969. The role of chemical dispersants in oil cleanup. *In* Oil on the Sea, D. P. Hoult, ed. New York: Plenum Press. pp. 29-51.

_____. 1971. Oil spill dispersants—current status and future outlook. *In* Proc. of Joint Conference on Prevention and Control of Oil Spills. Washington, D.C.: API. pp. 263-270.

_____. 1982. The formulation of an effective demulsifier for oil spill emulsions. Mar. Pollut. Bull 13(2):49-54.

_____. 1985. The effect of crude oil composition on dispersant performance. Proc. 1985 Oil Spill Conference. Washington, D.C.: API. pp. 441-444.

_____. 1986. Oil spill dispersants: Mechanism, history, chemistry. Unpublished position paper presented to Committee on Effectiveness of Oil Spill Dispersants, National Research Council, Washington, D.C.

_____. 1987. Basic study reveals how different crude oils influence dispersant performance. Proc. 1987 Oil Spill Conference. Washington, D.C.: API. pp. 293-296.

Canevari, G. P., and G. P. Lindblom. 1976. Some dissenting remarks on "deleterious effects of Corexit 9527 on fertilization and development." Mar. Pollut. Bull. 7(7):127-128.

Capuzzo, J. M., and B. A. Lancaster. 1982. Physiological effects of petroleum hydrocarbons on larval lobsters (*Homarus americanus*): Hydrocarbon accumulation and interference with lipid metabolism. *In* Physiological Mechanisms of Marine Pollutant Toxicity. New York: Academic Press. pp. 477-501.

Carr, R. S., and O. Linden. 1984. Bioenergetic responses of *Gammarus salinus* and *Mytilus edulis* to oil and oil dispersants in a model ecosystem. Mar. Ecol.-Progr. Ser. 19(3):285-292.

Carr, R. S., J. N. Neff, and P. D. Boehm. 1985. Large-scale continuous flow exposure systems for studying the fate and effects of chemically and physically dispersed oil on benthic marine communities (Abstract). Proc. 1985 Oil Spill Conference. Washington, D.C.: API. 641 pp.

_____. 1986. A study of the fate and effects of chemically and physically dispersed oil on benthic marine communities using large-scale continuous-flow exposures systems. *In* Pollution and Physiology of Marine Organisms, F. J. Vernberg, W. Vernberg, A. Calabrese, and F. P. Thurberg, eds. Charleston: University of South Carolina Press.

Cashion, B. S. 1982. Draft No. 6, ASTM method for spill dispersants. Unpublished document, ASTM, Philadelphia.

Chan, K.-Y., and S. Y. Chiu. 1985. The effects of diesel oil and oil dispersants on growth, photosynthesis and respiration of *Chlorella salina*. Arch. Environ. Contam. Toxicol. 14(3):325-332.

Chapman, P. 1980. The South African Oil Pollution Research Program. Spill Technol. Newsletter 5(3):72-75.

_____. 1981. The sea fisheries oil toxicity programme. *In* Chemical Dispersion of Oil Spills: An International Research Symposium, D. Mackay, P. G. Wells, and S. Paterson, eds. Publication No. EE-17. Toronto: University of Toronto, Institute for Environmental Studies. pp. 181-184.

_____. 1985. Oil concentrations in seawater following dispersion with and without the use of chemical dispersants. A review of published data. Special report of the Sea Fisheries Institute 2:1-23. Capetown, South Africa.

Chau, E., and D. Mackay. 1985. Multihit dipsersion of oil spills. Report to Environment Canada, Environmental Protection Service, Ottawa.

Chau, E., A. Chau, W. Y. Shiu, and D. Mackay. 1986. Multi-hit dispersion of oil spills. Report EE-72. Ottawa: Environment Canada. 45 pp.

Clean Seas, Inc. 1981. Oil Spill Cleanup Manual. Santa Barbara, California: Woodward Clyde and Associates.

Cline, J. D., K. Kelly-Hansen, and C. N. Katz. 1982. The production and dispersion of dissolved methane in the southeastern Bering Sea. NOAA/MMS Alaska Office, Outer Continental Shelf Assessment Program, Final Reports of Principal Investigators 51:309-417. RU0153. Washington, D.C.: NOAA.

Coachman, L. K., and R. L. Charnell. 1979. On lateral water mass interaction—a case study, Bristol Bay, Alaska. J. Phys. Oceanogr. 9:278-297.

Comotto, R. M., R. A. Kimerle, and R. D. Swisher. 1979. Bioconcentration and metabolism of linear alkylbenzene sulfonate by daphnids and fathead minnows. STP 667. Philadelphia: ASTM. pp. 232-250.

CONCAWE (Oil Companies' European Organisation for Environmental and Health Protection). 1986. Oil Spill Dispersant Efficiency Testing: Review and Practical Experience. Report No. 86/52. Den Haag, The Netherlands: CONCAWE. 49 pp.

Cook, C. B., and A. H. Knap. 1983. Effects of crude oil and chemical dispersant on photosynthesis in the brain coral *Diploria strigosa*. Mar. Biol. 78(1):21-28.

Cormack, D. 1983a. Review of United Kingdom oil spill response techniques and equipment. Proc. Oil Spill Conference. Washington, D.C.: API. pp. 15-18.

_____. 1983b. The use of aircraft for dispersant treatment of oil slicks at sea. Marine Pollution Control Unit, Department of Transport, London.

_____. 1983c. Response to Oil and Chemical Pollution. Chapter 4. London: Applied Science Publishers. 531 pp.

Cormack, D., and H. Parker. 1979. The use of aircraft for the clearance of oil spills at sea. Proc. 1979 Oil Spill Conference. Washington, D.C.: API. pp. 469-474.

Cormack, D., and J. A. Nichols. 1977. The concentrations of oil in sea water resulting from natural and chemically induced dispersion of oil slicks. Proc. 1977 Oil Spill Conference. Washington, D.C.: API. pp. 381-385.

Cormack, D., J. A. Nichols, and B. Lynch. 1978. Investigation of factors affecting the fate of North Sea oils discharged at sea. Part 1: Ekofisk Crude Oil, July 1975-Feb. 1978. Report No. LR 273 (OP). Warren Spring Laboratory. Stevenage, England.

Cormack, D. B., W. J. Lynch, and B. D. Dowsett. 1986/87. Evaluation of dispersant effectiveness. Oil and Chem. Pollut. 3:87-103.

Cormack, D., N. Hurford, and D. Tookey. 1987. Remote sensing techniques for detecting oil slicks at sea—a review of work carried out in the United Kingdom. Proc. 1987 Oil Spill Conference. Washington, D.C.: API. pp. 95-100.

Cornillon, P., and M. Spaulding. 1978. An oil spill fates model. *In* Environmental Assessment of Treated vs. Untreated Oil Spills, Second Interim Progress Report. Prepared for Department of Energy. Kingston: University of Rhode Island.

Costa, D. P., and G. L. Kooyman. 1982. Oxygen consumption, thermoregulation, and the effect of fur oiling and washing in the sea otter, *Enhydra lutria*. Can. J. Zool. 60(11):2761-2767.

Crapp, G. B. 1971a. Laboratory experiments with emulsifiers. *In* The Ecological Effects of Oil Pollution on Littoral Communities, E. B. Cowell, ed. Essex, U.K.: Applied Science Publishers Ltd. pp. 129-149.

_____. 1971b. Field experiments with oils and emulsifiers. *In* The Ecological Effects of Oil Pollution on Littoral Communities, E. B. Cowell, ed. Essex, U.K.: Applied Science Publishers Ltd. pp. 114-128.

_____. 1971c. The biological consequences of emulsifier cleansing. *In* The Ecological Effects of Oil Pollution on Littoral Communities, E. B. Cowell, ed. Essex, U.K.: Applied Science Publishers Ltd. pp. 150-168.

Cretney, W. J., R. W. MacDonald, C. S. Wong, D R. Green, B. Whitehouse, and G. G. Geesey. 1981. Biodegradation of a chemically dispersed crude oil. Proc. 1981 Oil Spill Conference. Washington, D.C.: API. pp. 37-43.

Cross, W. E., and D. H. Thomson. 1982. Macrobenthos—1981 study results. Baffin Island Oil Spill (BIOS) working report. EPS 81-3. Ottawa: Environment Canada. 105 pp.

Cross, W. E., D. H. Thomson, and A. R. Maltby. 1983. Macrobenthos—1982 study results. Baffin Island Oil Spill (BIOS) working report. EPS 82-3. Ottawa: Environment Canada. 135 pp.

Cross, W. E., D. H. Thomson, C. M. Martin, and M. F. Fabijan. 1984. Macrobenthos—1983 study results. Baffin Island Oil Spill Working Report 83-3. Ottawa: Environment Canada. 176 pp.

Croswell, W. F., J. C. Fedors, F. E. Hoges, R. N. Swift, and J. C. Johnson. 1983. Ocean experiments and remotely sensed images of chemically dispersed oil spills. IEEE Trans., Geoscience and Remote Sensing GE-21(1):2-15.

Crothers, J. H. 1983. Field experiments on the effects of crude oil and dispersant on the common animals and plants of rocky sea shores. Mar. Environ. Res. 8(4):215-240.

Crowley, S. 1984a. Shipboard spraying equipment for undiluted dispersant concentrates. Report LR 492 (OP)M, Warren Spring Laboratory, Stevenage, England. 13 pp.

———. 1984b. An assessment of the Mackay apparatus for testing oilspill dispersants. Oil and Petrochém. Pollut. 2(1):47-56.

CTGREF. 1981. Evaluation de la toxicite algue des dispersants pour hydrocarbures vis-a-vis des poissons. Protocole experimental. Paris: CTGREF.

Cuddeback, J. E. 1981. Personal communication cited by API Task Force, 1986. The Role of Chemical Dispersants in Oil Spill Control API Pub. No. 4425. 26 pp.

Cullinane, J. P., P. McCarthy, and A. Fletcher. 1975. The effect of oil pollution in Bantry Bay. Mar. Pollut. Bull. 6(11):173-176.

Czyzewska, K. 1976. The effect of detergents on larval development of a crab. Mar. Pollut. Bull. 7(6):108-112.

Daling, P. 1986. Laboratory effectiveness testing of oil spill dispersants—correlation studies between two test methods. International Seminar on Chemical and Natural Dispersion of Oil on the Sea, Heimdal, Norway, November.

Daling, P. S. 1988. A study of the chemical dispersability of fresh and weathered crude oils. Proc. 11th AMOP Seminar. Ottawa: Environment Canada. pp. 481-499.

Daling, P. S., and P. J. Brandvik. 1988. A study of the formation and stability of water-in-oil emulsions. Proc. 11th AMOP Seminar. Ottawa: Environment Canada. pp. 153-170.

Dekker, R., and G. W. N. M. van Moorsel. 1987. Effects of different oil doses, dispersants and dispersed oil on macrofauna in model tidal flat ecosystems. (Abstract). Talk given to TNO Conference on Oil Pollution. Dordrecht, The Netherlands: Martinus Nijhoff.

Delaune, R. A., C. J. Smith, W. H. Patrick, J. W. Fleeger, and M. D. Tolley. 1984. Effect of oil on salt marsh biota: Methods for restoration. Environ. Pollut. Ser. A. 36(3):207-227.

Delvigne, G. A. L. 1983. Sea measurements on natural and chemical dispersion of oil. Report No. M1933-1. Delft Hydraulics Laboratory, The Netherlands.

———. 1985. Experiments on natural and chemical dispersion of oil in laboratory and field circumstances. Proc. 1985 Oil Spill Conference. Washington, D.C.: API. pp. 507-514.

———. 1987. Droplet size distribution of naturally dispersed oil. *In* Fate and Effects of Oil in Marine Ecosystems, J. Kuiper and W. J. Van den Brink, eds. Dordrecht, The Netherlands: Martinus Nijhoff. pp. 29-40.

Demarquest, J. P., J. Croquette, F. Merlin, C. Bocard, G. Castaing, and C. Gatellier. 1985. Recent advances in dispersant effectiveness evaluation: Experimental and field aspects. Proc. 1985 Oil Spill Conference. Washington, D.C.: API. pp. 445-452.

Depledge, M. H. 1984. Changes in cardiac activity, oxygen uptake and perfusion indices in *Carcinus maenas* exposed to crude oil and dispersant. Comp. Biochem. Physiol. C. Comp. Pharmacol. Toxicol. 78(2):461-466.

Dickinson, A., D. Mackay, and D. McWatt. 1985. Report on the Beaufort Sea small scale oil spill dispersant trial. Report EE-58. Ottawa: Environment Canada.

Division Qualité des Eaux, Peche et Pisculture. 1979. Evaluation de la toxicite algue des dispersants pour hydrocarbures, vis-a-vis des poissons. Protocols experimental. Paris: CTGREF, Peche et Pisculture.

Dodd, E. N. 1974. Oil and dispersants: Chemical considerations. *In* Ecological Aspects of Toxicity Testing of Oils and Dispersants, L. R. Beynon and E. B. Cowell, eds. Essex, England: Applied Science Publishers Ltd. pp. 3-9.

Dodge, R. E., S. C. Wyers, H. R. Frith, A. H. Knap, S. R. Smith, and T. D. Sleeter. 1984. Effects of oil and oil dispersants on the skeletal growth of the hermatypic coral, *Diploria strigosa*. Coral Reefs 3(4):191-198.

Dodge, R. E., A. H. Knap, S. C. Wyers, H. R. Frith, T. D. Sleeter, and S. R. Smith. 1985a. The effect of dispersed oil on the calcification rate of the reef-building coral *Diploria strigosa*. Proc. 5th International Coral Reef Congress, Tahiti, Vol. 6, pp. 453-457.

_____. 1985b. The effects of oil and oil dispersants on hermatypic and skeletal growth (extension rate). Coral Reefs 4.

Doe, K. G., and G. W. Harris, 1976. Selection of a standard marine species for toxicity testing of oil spill dispersants in Canada. I. Literature review and project proposal. Unpublished report. Halifax: Environment Canada, Environmental Protection Service, Toxicity Programs Division.

Doe, K. G., and P. G. Wells. 1978. Acute aquatic toxicity and dispersing effectiveness of oil spill dispersants: Results of a Canadian oil dispersant testing program (1973-1977). *In* Chemical Dispersants for the Control of Oil Spills, L. T. McCarthy, Jr., G. P. Lindblom, and H. F. Walter, eds. STP 659. Philadelphia: ASTM. pp. 50-65.

Doe, K. G., G. W. Harris, and P. G. Wells. 1978. A selected bibliography on oil spill dispersants. Economic and Technical Review Report EPS-3-EC-78-2. Ottawa: Fisheries and Environment Canada, Environmental Protection Service.

Dowsett, B. O., and D. Cormack. 1986. Position paper on application of dispersants at sea. Prepared by the U.K. Marine Pollution Control Unit for the Committee on Effectiveness of Oil Spill Dispersants, National Research Council, Washington, D.C.

Drewa, G., Z. Zbytniewski, and F. Pautsch. 1977. The effect of detergent "solo" and crude oil on the activities of Cathepsin D and acid phosphatase in hemolymph of *Crangon crangon* L. Polskie Archiwum Hydrobiologii 24(2):279-284.

Dutka, B. J., and K. K. Kwan. 1984. Study of long-term effects of oil and oil dispersant mixtures on freshwater microbial populations in manmade ponds. Sci. Total Environ. 35(2):135-148.

Dutka, B. J., J. Sherry, B. F. Scott, and K. K. Kwan. 1980. Effects of oil-dispersant mixtures on freshwater microbial populations. Can. Res. 13(5):58-62.

Duval, W. S., L. A. Harwood, and R. P. Fink. 1980. The sublethal effects of physically and chemically dispersed crude oil on the physiology and behaviour of the estuarine isopod *Gnorimosphaderoma oregonensis*. Ottawa: Environment Canada. 66 pp.

_____. 1982. The sublethal effects of dispersed oil on an estuarine isopod. Tech. Development Report EPS-4-EC-82-1. Ottawa: Environment Canada. 72 pp.

Dye, C. W., and R. B. Frydenborg. 1980. Oil dispersants and the environmental consequences of their usage: A literature review. Technical Series 5, No. 1. Tallahassee: Florida Department of Environmental Regulation. 51 pp.

Eastin, W. C., Jr., and B. A. Rattner. 1982. Effect of dispersant and crude oil ingestion on mallard ducklings (*Anas platyhynchos*). Bull. Environ. Contamin. Toxicol. 29:273-278.

Eganhouse, R. P., D. L. Blumfield, and I. R. Kaplan. 1983. Long-chain alkylbenzenes as molecular tracers of domestic wastes in the marine environment. Environ. Sci. Technol. 17:523-530.

Eicke, H., and G. D. Parfitt. 1987. Interfacial Phenomena in Apolar Media. New York and Basel: Marcel Dekker.

Eimhjellen, K., O. Nilssen, T. Sommer, and E. Sondstad. 1983. Microbiology, 2. Biodegradation of oil—1982 study results. Baffin Island Oil Spill (BIOS) Working Report 82-6. Ottawa: Environment Canada. 41 pp.

Eisler, R. 1975. Toxic, sublethal, and latent effects of petroleum on Red Sea macrofauna. Proc. 1975 Conference on Prevention and Control of Oil Pollution. Washington, D.C.: API. pp. 535-540.

Ekker, M., and B. M. Jenssen. 1986. Biological effects of chemical treatment of oil spills at sea: Seabird studies. Abstract. Internat. Seminar on Chemical and Natural Dispersion of Oil on Sea, Heimbal, Norway.

Elgershuizen, J. H. W. B., and H. A. M. de Kruijk. 1976. Toxicity of crude oils and a dispersant to the stoney coral *Madracis mirabilis*. Mar. Pollut. Bull. 7(2):22-25.

Elliott, A. J. 1986. The influence of near-surface current shears on the spreading of oil. Paper presented at the International Seminar on Chemical and Natural Dispersion of Oil on the Sea, Heimdal, Norway, November.

Elliott, A. J., N. Hurford, and C. J. Penn. 1986. Shear Diffusion and the Spreading of Oil Slicks. Mar. Poll. Bull. 17:308-313.

Engelhardt, F. R. 1981. Oil pollution in polar bears: Exposure and clinical effects. Proc. 4th Arctic Marine Oilspill Program Technical Seminar. Ottawa: Environment Canada. pp. 139-179.

_____. 1983. Petroleum effects on marine mammals. Aquat. Toxicol. 4:199-217.

_____. 1985. Effects of petroleum on marine mammals. *In* Petroleum Effects in the Arctic Environment, F. R. Engelhardt, ed. New York: Elsevier Applied Science Publishers. pp. 217-243.

Engelhardt, F. R., C. Mageau, E. S. Gilfillan, and P. D. Boehm. 1984. Effects of acute and long-term exposure to dispersed oil in benthic invertebrates. Proc. Arctic Marine Oil Spill Technical Seminar. Ottawa: Environment Canada. pp. 367-392.

Engelhardt, F. R., E. S. Gilfillan, P. D. Boehm, and C. Mageau. 1985. Metabolic effects and hydrocarbon fate in arctic bivalves exposed to dispersed petroleum. Mar. Environ. Res. 17(2-4):245-259.

Environment Canada. 1973. Guidelines on the use and acceptability of oil spill dispersants. EPS-1-EE-73-1. Ottawa: Environment Canada. 54 pp.

_____. 1976. Toxicity and effectiveness acceptability ratings for Synperonic O.S.D. 20. Unpublished report. Environment Canada, Environmental Protection Service, Halifax, Nova Scotia. 83 pp.

_____. 1984. Guidelines on the use and acceptability of oil spill dispersants, 2d ed. EPS-1-EP-84-1. Ottawa: Environment Canada. 31 pp.

Ernst, R., and J. Arditti. 1980. Biological effects of surfactants. IV. Effects of non-ionics and amphoterics on HeLa cells. Toxicol. 15(3):233-242.

Esso Chemical Supply Company. 1978. Product Information, Corexit 9527 Oil Dispersant Concentrate. OFC-C78-1805E. Houston, Texas: Exxon.

Evans, G. W., M. Lyes, and A. P. M. Lockwood. 1977. Some effects of oil dispersants on the feeding behaviour of the brown shrimp, *Crangon crangon*. Mar. Behav. Physiol. 4(3):171-181.

Exxon Chemical Company. 1980. Governmental policies on the use of oil spill chemicals. Oil Spill Cleanup Manual, Vol. III. Houston, Texas: Exxon.

_____. 1985. Oil Spill Chemicals Applications Guide. Houston: Exxon Chemical Co. 4 pp.

Fabregas, J., C. Herrero, and M. Veiga. 1984. Effect of oil and dispersant on growth and chlorophyll A content of the marine microalga *Tetraselmis suecica*. Appl. Environ. Microbiol. 47(2):445-447.

Falk-Petersen, I. B. 1979. Toxic effects of aqueous extracts of Ekofisk crude oil, crude oil fractions and commercial oil products on the development of sea urchin eggs. Sarsia 64:161-169.

Falk-Petersen, I. B., and S. Lönning. 1984. Effects of hydrocarbons on eggs and larvae of marine organisms. *In* Ecotoxicological Testing for the Marine Environment, Vol. 2., G. Persoone, G. Jaspers, and C. Claus, eds. International Symposium on Ecotoxicological Testing for the Marine Environment, State University of Ghent, Ghent, and Institute for Marine Sciences Research, Bredene, Belgium. pp. 197-217.

Falk-Petersen, I. B., S. Lönning, and R. Jakobsen. 1983. Effects of oil and dispersants on plankton organisms. Astarte 12(2):45-48.

Faller, A. J. 1978. Experiments with controlled Langmuir Circulations. Science 201:618-620.

Fannelop, T. K., and K. Sjoen. 1980. Hydrodynamics of underwater blowouts. AIAA 18th Aerospace Sciences Meeting, Pasadena, Calif. New York: American Institute of Aeronautics and Astronautics. 45 pp.

Farke, H., and C. P. Guenther. 1984. Effects of oil and a dispersant on intertidal macrofauna in field experiments with Bremerhaven Caissons and in the laboratory. *In* Ecotoxicological Testing for the Marine Environment, Vol. 2, G. Persoone, E. Jaspers, and C. Claus, eds. International Symposium on Ecotoxocological Testing for the Marine Environment, State University of Ghent, Ghent, and Institute for Marine Science Research, Bredene, Belgium. pp. 219-235.

Farke, H., D. Blome, N. Theobald, and K. Wonneberger. 1985a. Field experiments with dispersed oil and a dispersant in an intertidal ecosystem: Fate and biological effects. Proc. 1985 Oil Spill Conference. Washington, D.C.: API. pp. 515-520.

Farke, H., K. Wonneberger, W. Gunkel, and G. Dahlmann. 1985b. Effects of oil and a dispersant on intertidal organisms in field experiments with a mesocosm, The Bremerhaven Caisson. Mar. Environ. Res. 15(2):97-114.

Farrington, J. W., and P. D. Boehm. 1987. Natural and Pollutant Organic Compounds in Georges Bank, R. Backus ed. Cambridge, Mass.: MIT Press. pp. 195-207.

Fast, O. 1985. Monitoring an oil spill experiment with the Swedish Maritime Surveillance System. Proc. 1985 Oil Spill Conference. Washington, D.C.: API. pp. 597-602.

_____. 1987. Swedish Coast Guard starts using third generation maritime surveillance system. Proc. 1987 Oil Spill Conference. Washington, D.C.: API. pp. 137-141.

Fay, J. A. 1971. Physical processes in the spread of oil on a water surface. Proc. 1971 Oil Spill Conference. Washington, D.C.: API. pp. 463-468.

Fingas, M. 1985. Effectiveness of oil spill dispersants. Spill Technol. Newsletter (July-Dec.):47-64.

_____. 1986. The effectiveness of oil spill dispersants. Unpublished report, Environmental Protection Service, Ottawa, Canada.

_____. 1988. Dispersant effectiveness at sea: A hypothesis to explain current problems with effectiveness. Proc. 11th Arctic Marine Oil Spill Pollution Technical Seminar. Ottawa: Environment Canada. pp. 455-479.

Fingas, M. F., W. S. Duval, and G. B. Stevenson. 1979. The basics of oil spill cleanup. Ottawa: Environment Canada. 155 pp.

Fingas, M. F., M. A. Bobra, and R. K. Velicogna. 1987. Laboratory studies on the chemical and natural dispersibility of oil. Proc. 1987 Oil Spill Conference. Washington, D.C.: API. pp. 241-246.

Fingerman, S. W. 1980. Differences in the effects of fuel oil, an oil dispersant and three polychlorinated biphenyls on fin regeneration in the Gulf Coast killifish, *Fundulus grandis*. Bull. Environ. Contam. Toxicol. 25(2):234-240.

Fisher, F. B., E. J. List, R. C. Y. Koh, J. Imberger, and N. H. Brooks. 1979. Mixing in Inland and Coastal Water. London: Academic Press. 483 pp.

Flaherty, L. M., and J. E. Riley. 1987. New frontiers for oil dispersants. Proc. 1987 Oil Spill Conference. Washington, D.C.: API. pp. 317-320.

Flaherty, L. M., J. E. Riley, and A. G. Hansen. 1987. A computerized tool for oil spill response decision making. Proc. 1987 Oil Spill Conference. Washington, D.C.: API. pp. 269-274.

Foght, J. M., and D. W. S. Westlake. 1982. Effect of the dispersant Corexit 9527 on the microbial degradation of Prudhoe Bay oil. Can. J. Microbiol. 28:117-122.

Foght, J. M., P. M. Fedorak, and D. W. S. Westlake. 1983. Effect of the dispersant Corexit 9527 on the microbial degradation of sulfur hetero cycles in Prudhoe Bay, Alaska USA oil. Can. J. Microbiol. 29(5):623-627.

Fontana, M. 1976. An aspect of coastal pollution—the combined effect of detergent and oil at sea on sea spray composition. Water, Air and Soil Poll. 5:269-280.

Foret, J.-P. 1975. Study of the long-term effects of several detergents on the developmental sequence of the sedentary polychaete *Capitella capitata* (*Fabricius*). Tethys 6:751-778.

Foy, M. G. 1982. Acute lethal toxicity of Prudhoe Bay crude oil and Corexit 9527 to arctic marine fish and invertebrates. Technology Development Report EPS-4-EC-82-3. Ottawa: Environment Canada. 62 pp.

Frank Ayles & Associates Ltd. 1983. Meter Controlled Dispersant Spray Unit. Technical Specification SP-1-83. London: Frank Ayles & Associates Ltd.

_____. 1984. Tanker-Terminal Dispersant Spray Unit. Advertising Brochure. London: Frank Ayles & Associates Ltd.

Franklin, F. L., and R. Lloyd. 1982. The toxicity of twenty-five oils in relation to the MAFF dispersant test. Fish. Res. Tech. Rep. No. 70. Dir. Fish. Res. of Great Britain. Burnham-on Crouch, U.K.: MAFF, Fisheries Laboratory. 13 pp.

_____. 1986. The relationship between oil droplet size and the toxicity of dispersant/oil mixtures in the standard MAFF "sea" test. Oil and Chem. Pollut. 3(1986/87):37-52.

Fraser, J. P., and D. W. Reed. 1982. The SOCK Skimmer at Ixtoc-I. Paper presented at the International Symposium Ixtoc-I, Mexico City, June 1982. Preprint.

Fucik, K. W., T. J. Bright, and K. S. Goodman. 1984. Measurements of damage, recovery, and rehabilitation of coral reefs exposed to oil. *In* Restoration of Habitats Impacted by Oil Spills, J. Cairns, Jr. and A. L. Buikema, eds. Stoneham, Mass.: Butterworth Publishers. 182 pp.

Games, L. M. 1983. Practical applications and comparisons of environmental exposure assessment models. STP 802. Philadelphia: ASTM. pp. 282-299.

Ganning, B., and U. Billing. 1974. Effects on community metabolism of oil and chemically dispersed oil on Baltic Bladder Wrack, *Fucus vesiculosus*. *In* Ecological Aspects of Toxicity Testing of Oils and Dispersants, L. R. Beynon and E. B. Cowell, eds. Essex, U.K.: Applied Science Publishers. pp. 53-61.

Garrett, B. P., R. M. Gershey, and D. R. Green. 1986. Remote sensing chronic oil discharges. Report EE-80. Ottawa: Environment Canada.

Garrett, W. D. 1969. Confinement and control of oil pollution on water with monomolecular surface films. Proc. Joint Conference on Prevention and Control of Oil Spills. Washington, D.C.: API. pp. 257-261.

Garver, D. R., and G. N. Williams. 1978. Advancements in oil spill trajectory modeling. Proc. Oceans 1978 Conference. Washington, D.C.: Marine Technology Society.

Gatellier, C. R., J. L. Oudin, P. Fusey, J. C. Lacaze, and M. L. Priou. 1973. Experimental ecosystems to measure fate of oil spills dispersed by surface active products. Proc. 1973 Oil Spill Conf. Washington, D.C.: API. pp. 497-504.

Genders, S. 1986. In situ detection and tracking of tracers in the water column. Proc. International Seminar on Chemical and Natural Dispersion of Oil on Sea, November 1986. Trondheim, Norway: SINTEF Group. 20 pp.

George, J. D. 1971. The effects of pollution by oil and oil-dispersants on the common intertidal polychaetes *Cirriformia tentaculata* and *Cirratulus cirratus*. J. Appl. Ecol. 8:411-420.

Geraci, J. R., and D. J. St. Aubin. 1980. Offshore petroleum resource development and marine mammals: A review and research recommendations. Mar. Fish. Rev. 42(11):1-12.

Getman, J. H. 1977. Performance tests of three fast current oil recovery devices. Proc. 1977 Oil Spill Conference. Washington, D.C.: API. pp. 257-261.

Getter, C. D., and T. G. Ballou. 1985. Field experiments on the effects of oil and dispersant on mangroves. Proc. 1985 Oil Spill Conference. Washington, D.C.: API. pp. 577-582.

Getter, C. D., T. G. Ballou, and C. B. Koons. 1985. Effects of dispersed oil on mangroves: Synthesis of a seven-year study. Mar. Poll. Bull. 16(8):318-324.

Giere, O. 1980. The impact of crude oil and oil dispersants on the marine oligochaete *Marionina subterranea*. Cahiers de Biologie Marine XXI:51-60.

Gilfillan, E. S., D. S. Page, S. A. Hanson, J. C. Foster, J. R. Hotham, D. Vallas, R. P. Gerber, and S. D. Pratt. 1983. Effect of spills of dispersed and non-dispersed oil on intertidal infaunal community structure. Proc. 1983 Oil Spill Conference. Washington, D.C.: API. pp. 457-463.

Gilfillan, E. S., D. S. Page, S. A. Hanson, J. Foster, J. Hotham, D. Vallas, and R. Gerber. 1984. Effects of test spills of chemically dispersed and non-dispersed oil on the activity of aspartate aminotransferase and glucose-6-phosphate dehydrogenase activity in two intertidal bivalves, *Mya arenaria* and *Mytilus edulis*. *In* Oil Spill Dispersants: Research, Experience, and Recommendations, T. E. Allen, ed. Philadelphia: ASTM. pp. 299-313.

Gilfillan, E. S., D. S. Page, S. A. Hanson, J. C. Foster, J. Hotham, D. Vallas, E. Pendergast, S. Hebert, S. D. Pratt, and R. Gerber. 1985. Tidal area dispersant experiment, Searsport, Maine: An overview. Proc. 1985 Oil Spill Conference. Washington, D.C.: API. pp. 553-559.

Gilfillan, E. S., D. S. Page, and J. C. Foster. 1986. Tidal Area Dispersant Project: Fate and Effects of Chemically Dispersed Oil in the Nearshore Benthic Environment. Final Report. Pub. No. 4440. Washington, D.C.: API. 215 pp.

Gill, S. D. 1978. Developing a dispersant spraying capability. Spill Technol. Newsletter 3(1):33-38.

Gill, S. D., and C. W. Ross. 1980. Aerial application of oil spill dispersants. Proc. 3rd Arctic Marine Oilspill Conference. Washington, D.C.: API. pp. 328-334.

_____. 1982. 1981 dispersant application field trial—St. John's Newfoundland. Proc. 5th Arctic Marine Oilspill Program Technical Seminar. Ottawa: Environment Canada. pp. 255-263.

Gill, S. D., R. H. Goodman, and J. Swiss. 1985. Halifax 1983 Sea Trial of oil spill dispersant concentrates. Proc. 1985 Oil Spill Conference. Washington, D.C.: API. pp. 479-482.

Gillot, A., and A. Charlier. 1986. Correlation results between IFP and WSL laboratory tests of dispersants. International Seminar on Chemical and Natural Dispersion of Oil on the Sea, Heimdal, Norway, November.

Godon, A., and J. H. Milgram. 1986. In-tank dispersant mixing study. Proc. 9th Arctic and Marine Oilspill Program Technical Seminar. Ottawa: Environment Canada, Conservation and Protection. pp. 543-562.

Goldacre, R. J. 1968. Effects of detergents and oils on the cell membrane. *In* The Biological Effects of Oil Pollution on Littoral Communities, J. D. Carthy and D. R. Arthur, eds. London: Field Studies Council. pp. 131-137.

Goodman, R. H., and M. R. MacNeill. 1984. The use of remote sensing in the determination of dispersant effectiveness. *In* Oil Spill Chemical Dispersants: Research, Experience and Recommendations, T. E. Allen, ed. STP 840. Philadelphia: ASTM. pp. 143-160.

Gordon, D. C. Jr., P. D. Keizer, W. R. Hardstaff, and D. G. Aldous. 1976. Fate of crude oil spilled in seawater contained in outdoor tanks. Environ. Sci. Tech. 10(6):580-585.

Grandy, N. J. 1984. The effects of oil and dispersants on subtidal red algae. Ph.D. Thesis, Department of Marine Biology, University of Liverpool. 178 pp.

Granmo, A., and S. Kollberg. 1976. Uptake pathways and elimination of a nonionic surfactant in cod (*Gadus morhua* L.). Water Res. 10:189-164.

Green, D., B. Humphrey, and B. Fowler. 1983. The use of flow-through fluorometry for tracking dispersed oil. Proc. 1983 Oil Spill Conference. Washington, D.C.: API. pp. 473-475.

Green, D. R., J. Buckley, and B. Humphrey. 1982. Fate of chemically dispersed oil in the sea: A report on two field experiments. Report No. 4-EC-82-5. Ottawa: Environment Canada. 125 pp.

Griffin, W. C. 1954. Calculation of HLB values of nonionic surfactants. J. Soc. Cosmet. Chemists 5:249-256.

Griffiths, R. P., T. M. McNamara, B. A. Caldwell, and R. Y. Morita. 1981. A field study on the acute effects of the dispersant Corexit 9527 on glucose uptake by marine microorganisms. Mar. Environ. Res. 5(2):83-92.

Gundlach, E. R., and M. O. Hayes. 1977. The *Urquiola* oil spill, La Coruna, Spain: Case history and discussion of methods of control and cleanup. Mar. Poll. Bull. 8(6):132-136.

Gunkel, W. 1974. Toxicity testing at the Biologische Anstalt Helgoland, West Germany. *In* Ecological Aspects of Toxicity Testing of Oils and Dispersants, L. R. Beynon and E. B. Cowell, eds. Essex, U.K.: Applied Science Publishers Ltd. 149 pp.

Gunkel, W., and G. Gassman. 1980. Oil, oil dispersants and related substances in the marine environment. Helgol. Meeresunter. 33(1-4):164-181.

Gyllenberg, G., and G. Lundquist. 1976. Some effects of emulsifiers and oil on two copepod species. Acta Zool. Fenn. 148:1-24.

Hagström, B. E., and S. Lönning. 1977. The effects of Esso Corexit 9527 on the fertilizing capacity of spermatozoa. Mar. Pollut. Bull. 8:136-138.

Hakkila, K., and A. Niemi. 1973. Effects of oil and emulsifiers on eggs and larvae of northern pike *Esox lucius* in brackish water. Aqua. Fennica. pp. 44-59.

Hansen, H. P. 1977. Photodegradation of hydrocarbon surface films. Rapp. p-v. reun. Cons. Int. Explor. Mer 171:101-106.

_____. 1975. Photochemical degradation of petroleum hydrocarbon surface films on seawater. Mar. Chem. 3:183-195.

Hargrave, B. T., and C. P. Newcombe. 1973. Crawling and respiration as indices of sublethal effects of oil and a dispersant on an intertidal snail *Littorina littorea*. J. Fish. Res. Board Can. 30(12):1789-1792.

Harris, G. W., and K. G. Doe. 1977. Methods used by Environment Canada in the testing of oil spill dispersants. Technology Development Report EPS-4-EC-77-6. Ottawa: Fisheries and Environment Canada, Environmental Protection Service. 31 pp.

Harris, G. W., and P. G. Wells. 1979. A laboratory study on the adhesion of crude oil to beach sand in the presence of a dispersant. Spill Technol. Newsletter 4:293-298.

Harris, G. W., P. G. Wells, A. Bobra, and F. Abbott. 1986. The comparative toxicology of oil spill dispersants to fish—implications for Canada's dispersant toxicity guidelines and the protection of marine environmental quality. Proc. 9th Arctic Marine Oilspill Program Technical Seminar. Ottawa: Environment Canada. pp. 485-495.

Harrison, W., M. A. Winnick, P. T. Y. Kwang, and D. Mackay. 1975. Crude oil spills: Disappearance of aromatic and aliphatic components from small sea-surface slicks. Environ. Sci. Tech. 9:231-134.

Hartwick, E. B., R. S. S. Wu, and D. B. Parker. 1979. The effect of crude oil and oil dispersant on littleneck clam populations. Unpublished report. Environment Canada, Environmental Emergencies Branch, Ottawa.

_____. 1982. Effects of a crude oil and an oil dispersant Corexit 9527 on populations of the littleneck clam *Protothaca staminea*. Mar Environ. Res. 6(4):291-306.

Harty, B., and A. McLachlan. 1982. Effects of water soluble fractions of crude oil and dispersants on nitrate generation by sandy beach microfauna. Mar. Pollut. Bull. 13(8):287-291.

Hayward, P. A. 1984. Marine ecotoxicological testing in the framework of international conventions. *In* Ecotoxicological Testing for the Marine Environment, Vol. 1, G. Persoone, E. Jaspers, and C. Claus, eds. Presented at the International Symposium on Ecotoxicological Testing for the Marine Environment, Ghent, Belgium, September 1983.

Hazel, C. R., F. Kopperdahl, N. Morgan, and W. Thomsen. 1971. Evaluating oil spill cleanup agents, development of testing procedures and criteria. Publication No. 43. Sacramento: California State Water Resources Control Board.

Healy, R. N., R. L. Reed, and D. G. Stenmark. 1976. Multiphase microemulsion systems. J. Soc. Petrol. Engrs. 16(3):147-160.

Heldal, M., S. Norland, T. Lien, and G. Knutsen. 1978. Acute toxicity of several oil dispersants towards the green algae *Chlamydomonas* and *Dunaliella*. Chemosphere 3:247-255.

Henry, L. de W. 1971. Marine oil spills: An Australian view. Mar. Pollut. Bull. 2(9):134-138.

Herz, M. J. 1986. Trouble on oiled water. California Waterfront Age 2(3):10-20.

Herz, M. J., and D. Kopec. 1985. Analysis of the *Puerto Rican* tanker accident: Recommendations for future oil spill response capability. Tech. Rept. 3. Tiburon, Calif.: Romberg Tiburon Center for Environmental Studies. 119 pp.

Hoge, F. E., and R. N. Swift. 1980. Oil film thickness measurements using airborne laser-induced water Raman backscatter. Appl. Optics 19:3269-3281.

Hoi-Chaw, L. 1986. Effects of oil on mangrove organisms. Preprint, First Asian Fisheries Forum, Manila, Phillipines. May 25-31. 12 pp.

Hoi-Chaw, L., and F. Meow-Chan. 1985. Field and laboratory studies on the toxicities of oils to mangroves. Proc. 1985 Oil Spill Conference. Washington, D.C.: API. pp. 539-546.

Hollinger, J. P., and R. A. Menella. 1973. Oil spills: Measurement of their distributions and volumes by multifrequency microwave radiometers. Science 181:54-56.

Hooper, C. H. 1981. The IXTOC I oil spill: The Federal scientific response. Boulder, Colo.: NOAA Hazardous Materials Response Project. 186 pp.

Hornstein, B. 1973. The visibility of oil-water discharges. Proc. 1973 Oil Spill Conference. Washington, D.C.: American Petroleum Institute. pp. 91-99.

Howarth, R. W. 1987. The potential effects of petroleum on marine organisms on Georges Bank. *In* Georges Bank, R. H. Backus and D. W. Bourne, eds. Cambridge, Mass.: Massachusetts Institute of Technology. pp. 540-551.

Hsaio, S. I. C., D. W. Kittle, and M. G. Foy. 1978. Effects of crude oils and the oil dispersant Corexit on primary production of arctic marine phytoplankton and seaweed. Environ. Pollut. 15:209-221.

Huang, J. C., and F. C. Monastero. 1982. Review of the state-of-the-art of oil spill simulation models. Final Report of Raytheon Ocean Systems Company. Washington, D.C.: API.

Hubbs Marine Research Institute. 1986. Sea Otter Oil Spill Mitigation Study. OSC Study, MMS-86-0009. Washington, D.C.: U.S. Department of the Interior, Minerals Management Service. 219 pp.

Hundahl, H., and N. K. Jojerslev. 1986. Optical methods in oceanography with special reference to oil fluorometry. Proc. International Seminar on Chemical and Natural Dispersion of Oil on the Sea, Heimbal. Trondheim, Norway: SINTEF Group. 20 pp.

Hurst, R. J., and N. A. Oritsland. 1982. Polar bears thermoregulation: The effect of oil on the insulative properties of fur. J. Therm. Biol. 7:201:208.

Hutcheson, M., and G. W. Harris. 1982. Sublethal effects of a water soluble fraction and chemically dispersed form of crude oil on energy partitioning in two arctic bivalves. EE-27. Ottawa: Environment Canada. 30 pp.

Inoue, K., K. Kaneko, and M. Yoshida. 1978. Adsorption of dodecylbenzenesulfonates by soil colloids and influence of soil colloid on their degradation. Sil. Sci. Plant Nutr. 24:91-102.

Intera Environmental Consultants Ltd. 1982. Remote sensing analysis of the oil spill dispersant sea trial. Report No. EE-40. Ottawa: Environment Canada. 79 pp.

———. 1984. State-of-the-art survey of oil spill detection, tracking, and remote sensing in cold climates. Report EE-50. Ottawa: Environment Canada.

Intera Environmental Technologies Ltd. 1984. Radar surveillance in support of 1983 COAATTF oil spill trials, Calgary, Alberta. Report R83-039, EE-51. Ottawa: Environment Canada. 48 pp.

International Maritime Organization (IMO). 1982. IMO/UNEP guidelines on oil spill dispersants and environmental considerations. London: IMO. 43 pp.

———. 1986. Mission Report of the Regional Advisor on Marine Pollution (Asia and the Pacific). Hong Kong. March 11-13.

International Petroleum Industry Environmental Conservation Association (IPIECA). 1980. Application and Environmental Effects of Oil Spill Chemicals. Publication 1/80. London: IPIECA. 20 pp.

International Tanker Owners Pollution Federation (ITOPF). 1981. Use of booms in combatting oil pollution. Technical Information Paper No. 2. London: ITOPF. 8 pp.

_____. 1982. Aerial application of oil spill dispersants. Technical Information Paper No. 3. London: International Tanker Owners Pollution Federation. 8 pp.

_____. 1983. Use of skimmers in combatting oil pollution. London: ITOPF. 7 pp.

_____. 1986. Fate of marine oil spills. Technical Information Paper No. 10. London: ITOPF. 8 pp.

Jasper, W. L., T. J. Kim, and M. P. Wilson. 1978. Drop size distributions in a treated oil-water system. In Chemical Dispersants for the Control of Oil Spills, STP 659, L. T. McCarthy, Jr., G. P. Lindblom, and H. F. Walter, eds. Philadelphia: ASTM. pp. 203-216.

Jeffery, P. G. 1973. Large-scale experiments on the spreading of oil at sea and disappearance by natural factors. Proc. 1973 Oil Spill Conference. Washington, D.C.: API. pp. 469-474.

Jeffery, P. G., and J. A. Nichols. 1974. Dispersants for oil spill cleanup operations at sea, on coastal waters and beaches. Report No. LR 193(OP). Stevenage, U.K.: Warren Spring Laboratory.

Jernelov, A., and O. Linden. 1981. Ixtoc 1: A case study of the world's largest oil spill. Ambio 10:299-306.

Johansen, O. 1984. The Halten Bank Experiment. Proc. 7th Arctic Marine Oilspill Program, Technical Seminar. Edmonton, Alberta: Environmental Protection Service. pp. 17-36.

_____. 1985a. Oil drift simulations, particle in fluid concept. Trondheim, Norway: Oceanographic Center, SINTEF. 14 pp.

_____. 1985b. Particle in fluid model for simulation of oil drift and spread. Part I: Basic concepts. Trondheim, Norway: Oceanographic Center, SINTEF. 34 pp.

_____. 1987. DOOSIM—A new simulation model for oil spill management. Proc. 1987 Oil Spill Conference. Washington, D.C.: API. pp. 529-532.

Johnson, J. C., C. D. McAuliffe, and R. A. Brown. 1978. Physical and chemical behavior of small crude oil slicks on the ocean. In Chemical Dispersants for the Control of Oil Spills, STP 659, L. T. McCarthy, Jr., G. P. Lindblom, and H. F. Walter, eds. Philadelphia: ASTM. pp. 141-158.

Jones, W. T. 1972. Air barriers as oil-spill containment devices. Soc. Petrol. Engrs. J. 12:126-142.

Jordon, R. E., and J. R. Payne. 1980. Fate and Weathering of Petroleum Spills in the Marine Environment. A Literature Reveiw and Synopsis. Ann Arbor, Mich.: Ann Arbor Science Publications.

Karichoff, S. W., D. S. Brown, and T. A. Scott. 1979. Sorption of hydrophobic pollutants on natural sediments. Water Res. 13:241-248.

Kerminen, S., P. Tulkki, K. Hakkila, and K. Haapala. 1971. Memorandum on the potential for use of emulsifiers in Finnish conditions. Finland: National Board of Waters, and the Archipelago Sea Research Institute, University of Turku. 17 pp.

Kiceniuk, J. W., W. R. Penrose, and W. R. Squires. 1978. Oil spill dispersants cause bradycardia in a marine fish. Mar. Pollut. Bull. 9(2):42-45.

Kim, H. H., and G. O. Hickman. 1975. An airborne laser fluorosensor for the detection of oil on water. NASA publication SP-375. Washington, D.C.: NASA. pp. 197-202.

Kimerle, R. A., K. J. Macek, G. H. Sleight II, and M. E. Burrows. 1981. Bioconcentration of linear alkylbenzene sulfonate (LAS) in bluegill (*Lepomis macrochirus*). Water Res. 15:251-256.

Klein, A. E., and N. Pilpel. 1974. The effects of artificial sunlight upon floating oils. Water Res. 8:79-83.

Knap, A. H. 1987. Effects of chemically dispersed oil on the brain coral, *Diploria strigosa*. Mar. Pollut. Bull. 18(3):119-122.

Knap, A. H., T. D. Sleeter, R. E. Dodge, S. C. Wyers, H. R. Frith, and S. R. Smith. 1983. The effects of oil spills and dispersant use on corals—a review and multidisciplinary experimental approach. Oil and Petrochem. Pollut. 1:157-169.

Knap, A. H., S. C. Wyers, R. E. Dodge, T. D. Sleeter, H. R. Frith, S. R. Smith, and C. B. Cook. 1985. The effects of chemically and physically dispersed oil on the brain coral *Diploria strigosa* (Dana)—a summary review. Proc. 1985 Oil Spill Conference. Washington, D.C.: API. pp. 547-551.

Kobayashi, N. 1981. Comparative toxicity of various chemicals, oil extracts and oil dispersant extracts to Canadian and Japanese sea urchin eggs. Public. Seto Mar. Biolog. Lab XXVI(1-3):123-133.

Kolnes, J. 1986. Injection of oil spill chemicals into a blowing well. Proc. International Seminar on Chemical and Natural Dispersion of Oil on Sea, November 1986. Trondheim, Norway: SINTEF Group. 21 pp.

Kolpack, R. L., N. B. Plutchak, and R. W. Stearns. 1977. Fate of oil in a water environment, Phase II: A dynamic model of the mass balance for released oil. Washington, D.C.: API.

Koons, C. B., and H. R. Gould. 1984. Worldwide status of research on fate and effect of oil in the marine environment—1982. Special Report. Houston, Texas: Exxon Production Research Company. 117 pp.

Kooyman, G. L., R. W. Davis, and M. A. Castellini. 1977. Thermal conductance of immersed pinniped and sea otter pelts before and after oiling with Prudhoe Bay Crude. *In* Fate and Effects of Petroleum Hydrocarbons in Marine Organisms and Ecosystems, D. A. Wolfe, ed. New York: Pergamon Press. pp. 151-157.

Kornberg, H., E. D. Acheson, C. Blake, Sir H. Chilver, R. B. Clark, J. W. Edmonds, G. E. Fogg, F. G. Larminie, J. R. Maddox, Lord Nathan, T. R. E. Southwood, A. Spinks, R. E. Thornton, M. Warnock, Baroness White, and D. G. T. Williams. 1981. Oil Pollution of the Sea. Eighth Report of the Royal Commission on Environmental Pollution. London: H.M. Stationery Office. 305 pp.

Kozarac, Z., D. Hrsak, B. Cosovic, and J. Vrzina. 1983. Electroanalytical determination of the biodegradation of nonionic surfactants. Environ. Sci. Technol. 17(5):268-272.

Kruth, D. J., E. Overton, and J. Murphy. 1987. Protecting an Island's Drinking Water and Desalination Plant. Washington, D.C.: API. 49 pp.

Kuhnhold, W. W. 1972. The influence of crude oils on fish fry. *In* Marine Pollution and Sea Life, M. Ruivo, ed. London: Fishing News Ltd. pp. 315-318.

Laake, M., K. Tjessem, and K. Rein. 1984. Fate of a tritiated Ekofisk crude oil in a controlled ecosystem experiment with North Sea plankton. Environ. Sci. Technol. 18(9):641-647.

Lacaze, J. C. 1973. Influence of illumination on the biodegradation of a nonionic surface-active agent used for the dispersion of oil spills at sea. Acad. Sci. Paris C.R. Serie D. 227:409-412.

———. 1974. Influence of the illumination of a nonionic surfactant used for the dispersion of oil spills at sea. EPA-TR-59-75. Research Triangle Park, N.C.: U.S. Environmental Protection Agency. 8 pp.

Lacaze, J. C., and O. Villedon de Naide. 1976. Influence of illumination on phytotoxicity of crude oil. Mar. Pollut. Bull. 7(4):73-76.

Ladner, L., and A. Hagström. 1975. Oilspill protection in the Baltic Sea. J. Water Pollut. Control. Fed. 47(4):796-809.

Lai, H. C., and M. C. Feng. 1984. Fate and Effects of Oil in the Mangrove Environment. Universiti Sains Malaysia. 252 pp.

_____. 1985. Field and laboratory studies on the toxicities of oils to mangroves. Proc. 1985 Oil Spill Conference. Washington, D.C.: API. pp. 539-546.

Lambert, G., and D. B. Peakall. 1981. Thermoregulatory metabolism in mallard ducks exposed to crude oil and dispersant. Proc. 4th Arctic Marine Oil Spill Program. Ottawa: Environment Canada. pp. 181-194.

Landrum, P. F., S. R. Nihart, B. J. Eadie, and L. R. Herche. 1987. Reduction in bioavailability of organic contaminants to the amphipod *Pontoporeia hoyi* by dissolved organic matter of sediment interstitial waters. Environ. Toxicol. Chem. 6:11-20.

Lane, P. A., J. H. Vandermeulen, M. J. Crowell, and D. G. Patriquin. 1987. Impact of experimentally dispersed crude oil on vegetation in a northwestern Atlantic salt marsh—preliminary observations. Proc. 1987 Oil Spill Conference. Washington, D.C.: API. pp. 509-514.

LaRoche, G., R. Eisler, and C. M. Tarzwell. 1970. Bioassay procedures for oil and oil dispersant toxicity evaluation. J. Water Pollut. Control Fed. 42(11):1982-1989.

Larson, R. J., R. D. Vashon, and L. M. Games. 1983. Biodegradation of trace concentrations of detergent chemicals in freshwater and estuarine systems. Biodegradation 5:235.

Lasday, A. H. 1985. Economic evaluation of dispersants to combat oil spills. Presented at the Spill and Hazardous Materials Conference-Workshop, New Haven, Connecticut, October 8-10, 1985, New Haven Harbor Petroleum Coop.

Latiff, S. A. 1969. Preliminary results of the experiments on the toxicity of oil counteracting agent (Esso Corexit 7664), with and without Iraq crude oil, for selected members of marine plankton. Arch. Fischereiwiss. 20(2/3):182-185

Layman, P. L. 1984. Detergent report: Brisk detergent activity changes picture for chemical suppliers. Chem. Eng. News 62:17.

Lee, K., and E. Levy. 1986. Biodegradation of petroleum in the marine environment: Limiting factors and methods of enhancement. Canad. Tech. Rep. of Fish and Aquat. Sci. No. 1442. Ottawa: Fisheries and Oceans.

Lee, K., C. S. Wong, W. J. Cretney, F. A. Whitney, T. R. Parsons, C. M. Lalli, and J. Wu. 1985. Microbial responses to crude oil and Corexit 9527: Seaflux enclosure study. Microbiol. Ecol. 11:337-351.

Lee, M., F. Martinelli, G. Lynch, and P. R. Morris. 1981. The use of dispersants on viscous fuel oils and water in crude oil emulsions. Proc. 1981 Oil Spill Conference. Washington, D.C.: API. pp. 31-35.

Legore, R. S., D. S. Marszalek, J. E. Hofmann, and J. E. Cuddeback. 1983. A field experiment to assess impact of chemically dispersed oil on Arabian Gulf corals. *In* Proc. Middle East Oil Technical Conference, Society of Petroleum Engineers, Manama, Bahrain. pp. 51-60.

Lehr, W. J., R. J. Fraga, M. S. Belen, and H. M. Cekirge. 1984. A new technique to estimate initial spill size using a modified Fay-type spreading formula. Mar. Pollut. Bull. 15:326-329.

Lehtinen, C., H. Orenius, and C. Hoglund. 1985. Miljoaspekter pa anvandandet au dispergeringsmedel vid oljebekampning. Naturvardsverket Informationsenheten, Solna, Sweden.

Lehtinen, C. M., and A.-M. Vesala. 1984. Effectiveness of oil spill dispersants at low salinities and low water temperatures. *In* Oil Spill Chemical Dispersants: Research, Experience, and Recommendations, T. E. Allen, ed. STP 840. Philadelphia: ASTM. pp. 108-121.

Leighton, F. A., R. G. Butler, and D. B. Peakall. 1985. Oil and arctic marine birds: An assessment of risk. *In* Petroleum Effects in the Arctic Environment, F. R. Engelhardt, ed. Essex, U.K.: Applied Science Publishers. pp. 183-215.

Lewis, A., D. C. Byford, and P. R. Laskey. 1985. The significance of dispersed oil droplet size in determining dispersant effectiveness under various conditions. Proc. 1985 Oil Spill Conference. Washington, D.C.: API. pp. 433-440.

Lewis, M. A., and V. T. Wee. 1983. Aquatic safety assessment for cationic surfactants. Environ. Toxicol. Chem. 2:105-118.

Levell, D. 1972. Effects of oil pollution and cleaning on sand and mud fauna—interim report. Annu. Rep. Oil. Pollut. Res. Unit Orielton. 32-34 pp.

———. 1973. The effects of oil pollution and cleaning on the fauna of soft substrates. Annu. Rep. Oil. Pollut. Res. Unit Orielton. pp. 72-73.

———. 1976. The effect of Kuwait crude oil and the dispersant BP1100X on the lugworm. Arenicola marina L. Pp. 131-185 *in* Marine Ecology and Oil Pollution, J. M. Baker, ed. Essex, U.K.: Applied Science Publishers. 565 pp.

Lichtenthaler, R. G., and P. S. Daling. 1983. Dispersion of chemically treated crude oil in Norwegian offshore waters. Proc. 1983 Oil Spill Conference. Washington, D.C.: API. pp. 7-14.

———. 1985. Aerial application of dispersants—comparison of slick behavior of chemically treated versus non-treated slicks. Proc. 1985 Oil Spill Conference. Washington, D.C.: API. pp. 471-478.

Lindblom, G. P. 1979. Logistic planning for oil spill chemical use. Proc. 1979 Oil Spill Conference. Washington, D.C.: API. pp. 453-458.

———. 1987. Measurement and prediction of depositional accuracy in dispersant spraying from large airplanes. Proc. 1987 Oil Spill Conference. Washington, D.C.: API. pp. 325-328.

Lindblom, G. P., and B. S. Cashion. 1983. Operation considerations for optimum deposition efficiency in aerial application of dispersants. Proc. 1983 Oil Spill Conference. Washington, D.C.: API. pp. 53-60.

Lindblom, G. P., and C. D. Barker. 1978. Evaluation of equipment for aerial spraying of oil dispersant chemicals. *In* Chemical Dispersants for the Control of Oil Spills, L. T. McCarthy, G. P. Lindblom, and H. F. Walter, eds. STP 659. Philadelphia: ASTM. pp. 169-179.

Lindblom, G. P., B. D. Emery, and M. A. Garcia Lara. 1981. Aerial application of dispersants at the Ixtoc I spill. Proc. 1981 Oil Spill Conference. Washington, D.C.: API. pp. 259-262.

Linden, O. 1974. Effects of oil spill dispersants on the early development of Baltic herring. Ann. Zool. Fenn. 11:141-148.

———. 1975. Acute effects of oil and oil/dispersant mixture on larvae of Baltic herring. Ambio. 4(3):130-133.

———. 1976. The influence of crude oil and mixtures of crude oil/dispersants on the ontogenic development of the Baltic herring, *Clupea harengus membras*. Ambio. 5(3):136-140.

———. 1981. Biological impact and effects on fisheries of oil spill in Bahrain, August-September 1980. Report IVL-B-607. Swedish Water and Air Pollution Research Laboratory, Goeteborg.

Linden, O., A. Rosemarin, A. Lindskog, C. Hoglund, and S. Johansson. 1985. Ecological effects of oil versus oil plus dispersant on the littoral ecosystem of the Baltic Sea. Proc. 1985 Oil Spill Conference. Washington, D.C.: API. pp. 485-490.

———. 1987. Effects of oil and oil dispersant on an enclosed marine ecosystem. Environ. Sci. Technol. 21:374-382.

Lindstedt-Siva, J., P. H. Albers, K. W. Fucik, and N. G. Maynard. 1984. Ecological considerations for the use of dispersants in oil spill response. *In* Oil Spill Chemical Dispersants: Research, Experience, and Recommendations, T. E. Allen, ed. STP 840. Philadelphia: ASTM. pp. 363-377.

Linton, T. L., and C. B. Koons. 1983. Oil dispersant field evaluation: Ixtoc blowout, Bay of Campeche, Mexico. Oil Petrochem. Pollut. 1:183-188.

Little, D. I., and D. L. Scales. 1987a. Effectiveness of a type III dispersant on low-energy shorelines. Proc. 1987 Oil Spill Conference. Washington, D.C.: API. pp. 263-268.

_____. 1987b. The persistence of oil stranded on sediment shorelines. Proc. 1987 Oil Spill Conference. Washington, D.C.: API. pp. 433-438.

Little, D., J. M. Baker, T. P. Abbiss, S. J. Rowland, and P. J. C. Tibbetts. 1980. The fate and effects of dispersant-treated compared with untreated crude oil, with particular reference to sheltered intertidal sediments. Proc. Symp. Chem. Disper. Oil Spills, Toronto. Publication EE-17. Toronto: University of Toronto Institute for Environmental Studies. pp. 117-151.

Little, D., J. M. Baker, T. P. Abbiss, S. J. Rowland, and P. J. C. Tibbetts. 1981. The fate and effects of dispersant-treated compared with untreated crude oil, with particular reference to sheltered intertidal sediments. Proc. 1981 Oil Spill Conf. Washington, D.C.: API. pp. 283-293.

Little, D. I., T. P. Abiss, J. M. Baker, B. M. Dicks, and P. J. C. Tibbets. 1986. The effects of chemical dispersants on the flux of stranded crude oil into fine intertidal sand. Rapports et Proces-Verbaux des Reunions Conseil International pour l'Exploration de la Mer. 186:219-233.

Lloyd, R. 1980. The role of acute toxicity tests with aquatic animals in chemical registration and notification schemes. Unpublished.

Longhurst, A., ed. 1982. Consultation on the consequences of offshore oil production on offshore fish stocks and fishing operations. Can. Tech. Rept. Fish. Aquatic Sci. 1096.

Lönning, S., and B. E. Hagström. 1975. The effects of oil dispersants on the cell in fertilization and development. Norw. J. Zool. 23:131-134.

_____. 1976. Deleterious effects of Corexit 9527 on fertilization and development. Mar. Pollut. Bull. 7:124-127.

Lönning, S., and I. B. Falk-Petersen. 1978. The effects of oil dispersants on marine eggs and larvae. Astarte 11:135-138.

Loostrom, B. 1986/87. The Swedish airborne remote sensing system for maritime surveillance. Oil Chem. Pollut. 3(3):209-229.

Lyes, M. C. 1977. The effects of surface active agents on the behavior of selected crustaceans. Ph.D. Thesis. University of Southampton, United Kingdom.

Macek, K. J. and S. F. Krzeminski. 1975. Susceptibility of bluegill sunfish (*Lepomis maerochirus*) to nonionic surfactants. Bull. Environ. Contam. Toxicol. 13(3):377-384.

Mackay, D. 1984. Uses and abuses of oil spill models. Proc. 7th Arctic Marine Oil Spill Program Technical Seminar. Ottawa: Environment Canada. pp. 1-17.

_____. 1985. Chemical dispersion, a mechanism and a model. Proc. 8th Annual Arctic Marine Oil Spill Program Technical Seminar. Ottawa: Environment Canada. pp. 260-268.

_____. 1986. A mathematical model of chemical dispersion of oil slicks. University of Toronto, draft report prepared for Environment Canada, Ottawa.

Mackay, D., and F. Szeto. 1981. The laboratory determination of dispersant effectiveness method development and results. Proc. 1981 Oil Spill Conference. Washington, D.C.: API. pp. 11-17.

Mackay, D., and K. Hossain. 1982. An Exploratory Study of Naturally and Chemically Dispersed Oil. Ottawa: Environmental Protection Service.

Mackay, D., and P. G. Wells. 1980. A study of some physico-chemical factors influencing the acute toxicity of oil dispersant mixtures. Unpublished manuscript. Institute for Environmental Studies, University of Toronto. 102 pp.

_____. 1981. Factors influencing the aquatic toxicity of chemically dispersed oils. Proc. Arctic Marine Oil Spill Program Technical Seminar. Ottawa: Environment Canada. pp. 445-467.

_____. 1983. Effectiveness, behavior, and toxicity of (oil spill) dispersants. Proc. 1983 Oil Spill Conference. Washington, D.C.: API. pp. 65-71.

Mackay, D., and P. J. Leinonen. 1977. Mathematical model of the behavior of oil spills on water with natural and chemical dispersion. Report No. EPS 3-EC-77-19. Ottawa: Environmental Protection Service.

Mackay, D., and R. Mascarenhas. 1979. Testing dispersants for effectiveness and toxicity. Spill Technol. Newsletter 4(4):242-244.

Mackay, D., J. S. Nadeau, and C. Ng. 1978. A small scale laboratory dispersant effectiveness test. In Chemical Dispersants for Control of Oil Spills, L. T. McCarthy, ed. STP 659. Philadelphia: ASTM. pp. 35-49.

Mackay, D., I. Buist, R. Mascarenhas, and S. Paterson. 1980a. Oil Spill Process and Models. EE-8. Ottawa: Environment Canada.

Mackay, D., S. Paterson, and K. Trudel. 1980b. A mathematical model of oil spill behavior. University of Toronto report. Ottawa: Environmental Protection Service.

Mackay, D., P. G. Wells, and S. Paterson. 1981. Chemical Dispersion of Oil Spills: An International Research Symposium. Publication No. EE-17. Institute for Environmental Studies, University of Toronto.

Mackay, D., S. Chang, and P. G. Wells. 1982. Calculation of oil concentrations under chemically dispersed slicks. Mar. Pollut. Bull. 13(8):278-283.

Mackay, D., A. Chau, K. Hossain, and M. Bobra. 1984. Measurement and prediction of the effectiveness of oil spill chemical dispersants. In Oil Spill Chemical Dispersants: Research, Experience, and Recommendations, T. E. Allen, ed. STP 840. Philadelphia: ASTM. pp. 38-54.

Mackay, D., A. Chau, and Y. C. Poon. 1986. A study of the mechanism of chemical dispersion of oil spills. Publication EE-76. Ottawa: Environmental Protection Service. 150 pp.

Mageau, C., F. R. Englehardt, E. S. Gilfillan, and P. D. Boehm. 1987. Effects of short-term exposure to dispersed oil in Arctic invertebrates. Arctic 40(Suppl.):162-171.

Mantel, L. H., and L. L. Farmer. 1983. Osmotic and ionic regulation. In The Biology of Crustacea, Vol. 5, Internal Anatomy and Physiological Regulation, L. Mantel, ed. New York: Academic Press. pp. 53-61.

Marking, L. L. 1985. Toxicity of chemical mixtures. In Fundamentals of Aquatic Toxicology: Methods and Applications, G. M. Rand and S. R. Petrocelli, eds. Washington, D.C.: Hemisphere Publishing. pp. 164-176.

Martinelli, F. N. 1980. Studies on the use of helicopters for oil spill clearance. Warren Spring Laboratory Report LR369(OP). Stevenage, U.K. 19 pp.

_____. 1984. The status of Warren Spring Laboratory's rolling flask test. In Oil Spill Chemical Dispersants: Research, Experience, and Recommendations, T. E. Allen, ed. STP 840. Philadelphia: ASTM. pp. 55-68.

Martinelli, F. N., and D. Cormack. 1979. Investigation of the effects of oil viscosity and water-in-oil emulsion formation on dispersant efficiency. Report No. LR 313(OP). Stevenage, U.K.: Warren Spring Laboratory. 10 pp.

Marty, D., A. Bianchi, and C. Gatellier. 1979. Effects of three oil spill dispersants on marine bacterial populations. I. Preliminary study. Quantitative evolution of aerobes. Mar. Pollut. Bull. 10:285-287.

McAuliffe, C. D. 1971. GC determination of solutes by multiple phase equilibrium. Chem. Technol. 1:46-51.

_____. 1974. Determination of C_1-C_{10} hydrocarbons in water. Marine Pollution Monitoring (Petroleum), National Bureau of Standards Special Publication 409. Washington, D.C.: U.S. Government Printing Office. pp. 121-125.

_____. 1977. Evaporation and solution of C_2 to C_{10} hydrocarbons from crude oils on the sea surface. Proceedings of Symposium on Fate and Effects of Petroleum Hydrocarbons in Marine Ecosystems and Organisms, D. A. Wolfe et al., eds. New York: Pergamon Press. pp. 363-372.

_____. 1986. Organism exposure to volatile hydrocarbons from untreated and chemically dispersed crude oils in field and laboratory. Ninth Arctic Marine Oil Program Technical Seminar. Edmonton, Alberta: Environment Canada, Environmental Protection Service. pp. 497-526.

_____. 1987a. Organism exposure to volatile/soluble hydrocarbons from crude oil spills—a field and laboratory comparison. Proc. 1987 Oil Spill Conference. Washington, D.C.: API. pp. 275-288.

_____. 1987b. Presentation at Japan Maritime Disaster Prevention Center Meeting, Tokyo. October 19, 1987.

McAuliffe, D. C., A. E. Smalley, R. D. Groover, W. M. Welsh, W S. Pickle, and G. E. Jones. 1975. Chevron Main Pass Block 41 oil spill: Chemical and biological investigations. Proc. Conference on Prevention and Control of Oil Pollution. Washington, D.C.: API. pp. 555-566.

McAuliffe, C. D., J. C. Johnson, S. H. Greene, G. P. Canevari, and T. D. Searl. 1980. Dispersion and weathering of chemically treated crude oils on the ocean. Environ. Sci. Technol. 14(12):1509-1518.

McAuliffe, C. D., B. L. Steelman, W. R. Leek, D. E. Fitzgerald, J. P. Ray, and C. D. Barker. 1981. The 1979 Southern California dispersant treated research oil spills. Proc. 1981 Oil Spill Conference. Washington D.C.: API. pp. 269-282.

McCarthy, L. T., G. P. Lindblom, and H. F. Walter, eds. 1978. Chemical Dispersants for the Control of Oil Spills. STP 659. Philadelphia: ASTM. 307 pp.

McCarthy, L. T. Jr., I. Wilder, and J. S. Dorrler. 1973. Standard dispersant effectiveness and toxicity tests. EPA-R2-73-201. Washington, D.C.: U.S. Environmental Protection Agency. 57 pp.

McColl, W. D., M. F. Fingas, R. A. E. McKibbon, and S. M. Till. 1987. CCRS remove sensing of the Beaufort Sea dispersant trials 1986. Proc. 10th Arctic Marine Oil Spill Program Technical Conference. Ottawa: Environment Canada. pp. 291-306.

McEwan, E. H., N. Aitchison, and P. E. Whitehead. 1974. Energy metabolism of oiled muskrats. Can. J. Zool. 52:1057-1062.

McKeown, B. A. 1981. Long-term sublethal and short-term high dose effects of physically and chemically dispersed oil on accumulation and clearance from various tissues of juvenile coho salmon, *Oncorhyncus kisutch*. Mar. Environ. Res. 5:295-300.

McKeown, B. A., and G. L. March. 1978. The effect of bunker C oil and oil dispersant on: 1 serum glucose, serum sodium and gill morphology in both freshwater and seawater acclimated rainbow trout (*Salmo gairdneri*). Water Res. 12(3):157-163.

McManus, D. A., and D. W. Connell. 1972. Toxicity of the oil dispersant, Corexit 7664, to certain Australian marine animals. Search 3(6):222-224.

Meeks, D. G. 1981. A view on the laboratory testing and assessment of oil spill dispersant efficiency. Proc. 1981 Oil Spill Conference. Washington, D.C.: API. pp. 19-29.

Meyers, R. J., and L. A. Onstad. 1986. Dispersant application systems. Unpublished report to the Marine Board, National Research Council, Washington, D.C.

Miller, D. S., D. Brier-Russell, F. A. Leighton, D. Phalen, D. A. Jeffrey, G. Lambert, and D. B. Peakall. 1981. Effects of crude oil, dispersant and an oil-dispersant emulsion on herring gulls. Bull. Mount Desert Island Biol. Sta. 21:50-53.

Miller, D. S., R. G. Butler, W. Trivelpiece, S. Janes-Butler, S. Green, G. Lambert, and D. B. Peakall. 1982. Crude oil ingestion by seabirds: Possible metabolic and reproductive effects. Bull. Mount Desert Island Biol. Sta. 22:137-138.

Ministry of Transportation of Japan. 1974. The testing standards for characteristics of oil spill substances. Hakusa (marine testing). No. 563. Tokyo: Ministry of Transportation of Japan, The Ship Bureau.

Moldan, A. G. S., and P. Chapman. 1983. Toxicity testing of oil spill dispersants in South Africa. S. Afr. J. Mar. Sci./S. Afr. Tydskr. Seewet. 1:145-152.

Mommaerts-Billiet, F. 1973. Growth and toxicity tests on the marine nanoplanktonic alga *Platymonas tetrathele* G.S. West in the presence of crude oil and emulsifiers. Environ. Pollut. 4(4):261-282.

Moniteq Ltd. 1985. Oil spill remote sensing equipment test protocol. Report EE-65. Ottawa: Environment Canada.

Moore, S. B., R. A. Diehl, J. M. Barnhardt, and G. B. Avery. 1986. Acute and chronic aquatic toxicities of textile surfactants. Papers, International Conference and Exhibition, AATC.

Moore, T. W. 1968. Dispersal of oil slicks in ports and at sea. *In* International Conference on Oil Pollution of the Sea, P. Barclay-Smith, ed. Report of proceedings, Rome, Italy. pp. 234-239.

Mori, K., T. Kobayashi, and T. Fujishima. 1983. Effects of the toxicity of mineral oil and solvent emulsifier upon the eggs of marine fish. Bull. Fac. Fish. Mie Univ. 10:15-23.

Mori, K., S. Kimura, K. Aoki, and Y. Saito. 1984. Effects of the toxicity of mineral oil and solvent emulsifier upon the larvae and young of marine fish. Bull. Fac. Fish., Mie Univ. 11:27-35.

Morris, P. R. 1981. Research into the limitations and application techniques of oil spill dispersants. *In* Chemical Dispersion of Oil Spills: An International Research Symposium. Publication No. EE-17. Toronto: Institute for Environmental Studies, University of Toronto. pp. 19-29.

Morris, P. R., and F. Martinelli. 1983. A specification for oil spill dispersants. Report LR 448 (OP), Warren Spring Laboratory, Stevenage England. 21 pp.

Morrow, J. E., R. L. Gritz, and M. P. Kirton. 1975. Effects of some components of crude oil on young coho salmon. Copeia 2:326-331.

Mount, D. E. 1985. Scientific problems in using multispecies toxicity tests for regulatory purposes. *In* Multispecies Toxicity Testing, J. Cairns, Jr., ed. New York: Pergamon Press. 261 pp.

Mulkins-Phillips, G. J., and J. E. Stewart. 1974. Effect of four dispersants on biodegradation and growth of bacteria on crude oil. Appl. Microbiol. 28:547-552.

Murray, S. P. 1975. Wind and current effects on large-scale oil slicks. Offshore Tech. Conf. 3:523-533.

Nadeau, S., and D. Mackay. 1978. Evaporation rates of complex hydrocarbon mixtures under environmental conditions. Spill Technol. Newsletter 3(March-April).

Nagell, B., M. Notini, and O. Grahn. 1974. Toxicity of four oil dispersants to some animals from the Baltic Sea. Mar. Biol. 28:237-243.

Nagy, E., B. F. Scott, and J. Hart. 1984. The fate of oil and oil-dispersant mixtures in freshwater ponds. Sci. Total Environ. 35:115-134.

Nakatani, R. E., E. L. Brannon, A. E. Nevissi, M. L. Landolt, D. G. Elliott, R. P. Whitman, S. P. Kaluzny, and T. P. Quinn. 1983. Effects of crude oil on chemoreception and homing in Pacific salmon. FRI-UW-83. Seattle: Fisheries Research Institute, University of Washington. 132 pp.

Nakatani, R. E., E. O. Salo, A. E. Nevissi, R. P. Whitman, B. P. Snyder, and S. P. Kaluzny. 1985. Effect of Prudhoe Bay crude oil on the homing of coho salmon in marine waters. Publication No. 4411. Washington, D.C.: API. 55 pp.

National Oceanic and Atmospheric Administration (NOAA). 1983. Assessing the Social Costs of Oil Spills: The Amoco Cadiz Case Study. Washington, D.C.: U.S. Department of Commerce.

National Research Council (NRC). 1985. Oil in the Sea: Inputs, Fates and Effects. Washington, D.C.: National Academy Press. 601 pp.

Neff, J. M. 1979. Polycyclic Aromatic Hydrocarbons in the Aquatic Environment: Sources, Fates, and Biological Effects. London: Applied Science Publishers. 262 pp.

Neff, J. M., and J. W. Anderson. 1981. Response of marine animals to petroleum and specific petroleum hydrocarbons. London: Applied Science Publishers. 177 pp.

Neff, J. M., R. E. Hillman, R. S. Carr, R. L. Buhl, and J. I. Lahey. 1987. Histopathologic and biochemical responses in arctic marine bivalve molluscs exposed to experimentally spilled oil. Arctic 40(Suppl. 1):220-229.

Nelson-Smith, A. 1972. Oil Pollution and Marine Ecology. London: Paul Elek Scientific Books Ltd. 260 pp.

———. 1978. The effects of oil pollution and emulsifier cleansing on shore life in south-west Britain. J. Appl. Ecol. 5(1):97-107.

———. 1980. Oil-spill chemicals. A bibliography on the nature, application, effects and testing of chemicals used against oil spilled in the marine environment. London: International Petroleum Industry Environmental Conservation Assoc. (IPIECA) 85 pp.

———. 1985. Supplement to 1980 bibliography IPIECA 73 pp.

Nes, H. 1983. Effectiveness and toxicity experiments with oil dispersants. PFO Project Report No. 1405. Trondheim: Norwegian Oil Pollution Research and Development Program (PFO), The Continental Shelf Institute.

———. 1984. Effectiveness of oil dispersants. Laboratory experiments. PFO Project Report No. 1410. Trondheim: Norwegian Oil Pollution Research and Development Program (PFO), The Continental Shelf Institute.

Nes, H., and S. Norland. 1983. Effectiveness and toxicity of oil dispersants. Proc. 6th Arctic Marine Oil Spill Program. Ottawa: Environmental Protection Service. pp. 132-139.

Nevissi, A. E., R. E. Nakatani, J. C. Fetzer, and C. D. McAuliffe. 1987. Uptake and loss of hydrocarbons by adult chinook and coho salmon exposed to untreated and chemically dispersed Prudhoe Bay crude oil. Salmon Tainting Project, Phase 1 draft report. Submitted to Cook Inlet Response Organization and Alaska Clean Seas.

Nichols, J. A., and H. D. Parker. 1985. Dispersants: Comparison of laboratory tests and field trials with practical experience at spills. Proc. 1985 Oil Spill Conference. Washington, D.C.: American Petroleum Institute. pp. 421-427.

Nichols, J. A., and I. C. White. 1979. Aerial application of dispersants in Bantry Bay following the *Betelgeuse* incident. Mar. Pollut. Bull. 10:193-197.

Norland, S., M. Heldal, T. Lien, and G. Knutsen. 1978. Toxicity testing with synchronized cultures of the green alga *Chlamydomonas*. Chemosphere 7(3):231-245.

Norton, M. G., and F. L. Franklin. 1980. Research into toxicity evaluation and control criteria of oil dispersants. Fisheries Research Technical Report No. 57. Ministry of Agriculture, Fisheries and Food, Directorate of Fisheries Research, Lowestoft, U.K. 20 pp.

Norton, M. G., F. L. Franklin, and R. A. A. Blackman. 1978. Toxicity testing in the United Kingdom for the evaluation of oil slick dispersants. *In* Chemical Dispersants for the Control of Oil Spills, L. T. McCarthy, G. P. Lindblom, and H. F. Walter, eds. STP 659. Philadelphia: ASTM. pp. 18-34.

Norwegian Ministry of Environment. 1980. Regulations concerning the composition and use of dispersants to combat oil spills. Oslo: Ministry of Environment. 8 pp.

Nuwayhid, M. A., S. P. Davies, and H. Y. Elder. 1980. Changes in the ultrastructure of the gill epithelium of *Patella vulgata* after exposure to North Sea crude oil and dispersants. J. Mar. Biol. Assoc. U.K. 60(2):439-448.

Oil Spill Intelligence Report. 1986a. Storage tank rupture causes massive spill in Panama. Oil Spill Intelligence Report IX(20):1-2.

_____. 1986b. Dispersant spray system delivered to BP spill response base in UK. Oil Spill Intelligence Report IX(50):2.

_____. 1986c. Dispersant tests conducted in Beaufort Sea. Oil Spill Intelligence Report IX(37):1.

_____. 1987a. Biegert completes ADDS-Pack training in Singapore. Oil Spill Intelligence Report X(31):1.

_____. 1987b. Freighter sinks and begins leaking fuel after collision off California. Oil Spill Intelligence Report X(38):1.

_____. 1987c. Oil seeping from sunken freighter remains at sea off California coast. Oil Spill Intelligence Report X(39):1.

Okubo, A. 1967. The effect of shear in an oscillatory current on horizontal diffusion from an instantaneous source. Limnol. Oceanog. 1:104-204.

Okubo, A. 1971. Oceanic diffusion diagrams. Deep-sea Res. 18:789.

O'Neill, R. A., R. A. Nevill, and V. Thompson. 1983. The Arctic Marine Oil Spill Program (AMOP) Remote Sensing Study. Report No. EPS 4-EC-3. Ottawa: Environment Protection Service.

Onstad, L. A., and Lindblom, G. P. 1987. Design and evaluation of a large boat-mounted dispersant spraying system and its integration with other application equipment. Presented at Symposium on Dispersants: New Ecological Approach Through the 90's, Williamsburg, Virginia, October 12-14, 1987, American Society for Testing and Materials.

Ordzie, C. J., and G. C. Garofalo. 1981. Lethal and sublethal effects of short-term acute doses of Kuwait crude oil and a dispersant Corexit 9527 on bay scallops *Argopecten irradians* and two predators at different temperatures. Mar. Environ. Res. 5(3):195-210.

Overton, E. B., J. R. Patel, and J. L. Laseter. 1979. Chemical characterization of mousse and selected environmental samples from the AMOCO CADIZ oil spill. Proc. 1979 Oil Spill Conference. Washington, D.C.: API. pp. 168-174.

Overton, E. B., J. L. Laseter, W. Mascarella, C. Raschke, I. Nairy, and J. W. Farrington. 1980. Photochemical oxidation of IXTOC-I oil. Proc. Symp. Researcher-Pierce Ixtoc-I Cruise, D. K. Atwood, ed. Washington, D.C.: NOAA. pp. 341-346.

Ozelsel, S. 1983. The combined effects of some dispersants and petroleum hydrocarbon derivatives on *Mytilus galloprovincialis*. Rev. Int. Oceanogr. Med. 72:37-44.

_____. 1981. The acute toxicity of several dispersants on *Palaemonetes pugi* O. (Crustacea, Decapoda). Rev. Int. Oceanogr. Med. 63-64:103-117.

Page, D. S., E. S. Gilfillan, J. C. Foster, J. R. Hotham, R. P. Gerber, D. Vallas, S. A. Hanson, E. Pendergast, S. Hebert, and L. Gonzalez. 1983. Long-term fate of dispersed and undispersed crude oil in two nearshore test spills. Proc. 1983 Oil Spill Conference. Washington, D.C.: API. pp. 465-471.

Page, D. S., J. C. Foster, J. R. Hotham, D. Vallas, E. S. Gilfillan, S. A. Hanson, and R. P. Gerber. 1984. Tidal area dispersant project: Fate of dispersed and undispersed oil in two nearshore test spills. *In* Oil Spill Chemical Dispersants: Research, Experience, and Recommendations, T. E. Allen, ed. STP 840. Philadelphia: ASTM. pp. 280-298.

Page, D. S., E. S. Gilfillan, J. C. Foster, E. Pendergast, L. Gonzalez, and D. Vallas. 1985. Compositional changes in dispersed crude oil in the water column during a nearshore test spill. Proc. 1985 Oil Spill Conference. Washington, D.C.: API. pp. 521-530.

Papineau, C. 1983. The sublethal effect of dispersants and oil emulsion on the gill atpase of *Palaemon serratus*. Oceanis 9(3):217.

Papineau, C., and G. Cheze. 1984. Histopathological modifications of gills of the shrimp *Palaemon serratus* subjected to the effect of petroleum dispersants and emulsions. Cah. Biol. Mar. 25(1):75-81.

Papineau, C., and Y. LeGal. 1983. Sublethal effect of petroleum dispersants and emulsions on the atpase of branches of *Palaemon serratus*. Revue Internat. Oceanograph. Med. 70-71(0):39-48.

Parsons, T. R. et al. 1984. An experimental marine ecosystem response to crude oil and Corexit 9527. 2. Biological effects. Mar. Environ. Res. 13:265-275.

Pastorak, R. A., P. Booth, J. H. Stern, L. L. Hornsby, and P. M. Chapman. 1985. Fate and effects of oil dispersants and chemically dispersed oil in the marine environment. OCS Study, MMS-85-0048. Washington, D.C.: Department of the Interior. 114 pp.

Pavia, R., and L. A. Onstad. 1985. Plans for integrating dispersant use in California. Proc. 1985 Oil Spill Conference. Washington, D.C.: API. pp. 85-88.

Pavia, R., and R. W. Smith. 1984. Development and implementation of guidelines for dispersant use: Regional rsponse teams. *In* Oil Spill Chemical Dispersants: Research, Experience, and Recommendations, T. E. Allen, ed. STP 840. Philadelphia: ASTM. pp. 378-389.

Payne, J. F. 1982. Metabolism of complex mixtures of oil spill surfactant compounds by a representative teleost (*Salmo gairdneri*), crustacean (*Cancer irroratus*), and mollusc (*Chlamys islandicus*). Bull. Environ. Contam. Toxicol. 28:277-280.

Payne, J. R., and C. R. Phillips. 1985. Petroleum spills in the marine environment. *In* The Chemistry and Formation of Water-in-Oil Emulsions and Tar Balls. Chelsea, Mich.: Lewis Publishers. 148 pp.

Payne, J. R., and G. D. McNabb, Jr. 1984. Weathering of petroleum in the marine environment. MTS J. 18(3):24-42.

Payne, J. R., B. E. Kirstein, G. D. McNabb, Jr., J. L. Lambach, C. de Oliveira, R. E. Jordan, and W. Hom. 1983. Multivariate analysis of petroleum hydrocarbon weathering in the subarctic marine environment. Proc. 1983 Oil Spill Conference. Washington, D.C.: API. pp. 423-434.

Payne, J. R., C. R. Phillips, M. Floyd, G. Longmire, J. Fernandez, and L. M. Flaherty. 1985. Estimating dispersant effectiveness under low temperature-low salinity conditions. Proc. 1985 Oil Spill Conf. Washington, D.C.: API. pp. 638.

Payne, J. R., C. R. Phillips, and W. Hom. 1987. Transport and transformations: Water column processes. In The Long-Term Effects of Offshore Oil and Gas Development: An Assessment and a Research Strategy, D. F. Boesch and N. N. Rabelais, eds. London: Elsevier Applied Science Publishers.

Peakall, D. B., and D. S. Miller. 1981. The use of combined laboratory and field studies to assess the impact of oil and dispersants on seabirds. In Chemical Dispersion of Oil Spills, D. Mackay, P. G. Wells, and S. Paterson, eds. Toronto: University of Toronto. pp. 67-69.

Peakall, D. B., D. J. Hallet, J. R. Bend, G. L. Foureman, and D. S. Miller. 1982. Toxicity of Prudhoe Bay crude oil and its aromatic fractions to nesting herring gulls. Environ. Res. 27:206-215.

Peakall, D. B., D. A. Jeffrey, and D. S. Miller. 1985. Weight loss of herring gulls exposed to oil and oil emulsion. Ambio. 14:108-110.

Peakall, D. B., P. G. Wells, and D. Mackay. 1987. A hazard assessment of chemically dispersed oil spills and seabirds. Mar. Environ. Res. 22:91-106.

Pearson, W. H. 1985. Oil effects on spawning behavior and reproduction in Pacific herring (Clupea harengus pallasi). Unpublished report, API, Washington, D.C.

Pelto, M. J., C. A. Manen, and B. E. Kirstein. 1983. A two-dimensional numerical model for predicting the concentrations of oil under a slick. Proc. 6th Arctic Marine Oil Spill Program Technical Seminar. Ottawa: Environment Canada. pp. 20-23.

Penrose, W. R., L. L. Dawe, and M. J. Sandeman. 1976. Analysis of oil, oil dispersants and metabolites during biodegradation in seawater. Manuscript Report, Department of the Environment, Fisheries and Marine Service, St. John's Newfoundland, Canada. 13 pp.

Percy, J. A. 1977. Effects of dispersed crude oil upon the respiratory metabolism of an arctic marine amphipod, Onismus (Boekisimus) affinis. In Fate and Effects of Petroleum Hydrocarbons in Marine Organisms and Ecosystems, D. A. Wolfe, ed. New York: Pergamon Press.

Percy, J. A., and P. G. Wells. 1984. Effects of petroleum in polar marine environments. MTS J. 18(3):51-61.

Perkins, E. J. 1972. Some methods of assessment of toxic effects upon marine invertebrates. Proc. Soc. Anal. Chem. 9(5):105-114.

Perkins, E. J., E. Gribbon, and J. W. M. Logan. 1973. Oil dispersant toxicity. Mar. Pollut. Bull. 4(6):90-93.

Petroleanos Mexicanos. 1980. Informe de los trabajos realizados para el control del poso Ixtoc I, el combate del derrame de petroleo y determinacion de sus efectos sobre el ambiente marino. Instituto Mexicano del Petroleano, Mexico, D.F. pp. 73-74.

Poliakoff, M. Z. 1969. Oil dispersing chemicals. A study of the composition, properties and use of chemicals for dispersing oil spills. Program No. 15080FHS 05/69, Contract No. DI-14-12-549. Federal Water Pollution Control Administration. Department of the Interior, Edison Water Quality Laboratory, Edison, N.J. 27 pp.

Port of Singapore Authority. 1976. Testing methods of oil dispersants. Unpublished manuscript. Chemistry Department, Port of Singapore Authority.

Portmann, J. E. 1969. A summary of the results of toxicity tests with 36 oil-dispersing mixtures. International Council for the Exploration of the Sea, Fisheries Improvement Committee. Paper C. M. 1969/E:9.

———. 1970. The toxicity of 120 substances to marine organisms. U.K. MAFF Shellfish Information Leaflet No. 19. Burnham-on-Crouch, Essex, England.

———. 1972. Results of acute toxicity tests with marine organisms, using a standard method. *In* Marine Pollution and Sea Life, M. Ruivo, ed. London: FAO and Fishing News Books Ltd. pp. 212-217.

Portmann, J. E., and P. M. Connor. 1968. The toxicity of several oil-spill removers to some species of fish and shellfish. Mar. Biol. 1(4):322-329.

Power, F. M. 1983. Long-term effects of oil dispersants on intertidal benthic invertebrates. Part 3. Toxicity to barnacles and bivalves of untreated and dispersant-treated fresh and weathered condensate. Oil Petrochem. Pollut. 1:171-181.

Raj, P. K., and R. Griffith. 1979. The survival of oil slicks on the ocean as a function of sea state limit. Proc. 1979 Oil Spill Conference. Washington, D.C.: API. pp. 719-724.

Rand, G. M., and S. R. Petrocelli, eds. 1985. Fundamentals of Aquatic Toxicology: Methods and Applications. Washington, D.C.: Hemisphere Publishing. 666 pp.

Regional Response Team Working Group. 1986. Dispersant guidelines for Alaska, Federal Region X, January 1986. Draft. U.S. Coast Guard, Juneau. 24 pp.

Renzoni, A. 1973. Influence of crude oil, derivatives and dispersants on larvae. Mar. Pollut. Bull. 4(1):9-13.

Rewick, R. T., K. A. Sabo, J. Gates, J. H. Smith, and L. T. McCarthy, Jr. 1981. An evaluation of oil spill dispersant testing requirements. Proc. 1981 Oil Spill Conference. Washington, D.C.: API. pp. 5-10.

Rewick, R. R., K. A. Sabo, and J. H. Smith. 1984. The drop-weight interfacial tension method for predicting dispersant performance. *In* Oil Spill Chemical Dispersants: Research, Experience, and Recommendations, T. E. Allen, ed. STP 840. Philadelphia: ASTM. pp. 94-107.

Rice, S. D., A. Moles, T. L. Taylor, and J. F. Karinen. 1979. Sensitivity of 39 Alaskan marine species to Cook Inlet crude oil and No. 2 fuel oil. Proc. 1979 Oil Spill Conference. Washington, D.C.: API. pp. 549-554.

Rice, S. D., S. Korn, and C. C. Brodersen. 1981. Toxicity of ballast-water treatment effluent to marine organisms at Port Valdez, Alaska. Proc. 1981 Oil Spill Conference. Washington, D.C.: API. pp. 55-61.

Richardson, M. G. 1979. "Esso Bernicia" incident, Shetland. Mar. Pollut. Bull. 10(4):97.

Rogerson, A., and J. Berger. 1981. The toxicity of the dispersant Corexit 9527 and oil-dispersant mixtures to ciliate protozoa. Chemosphere 10:33-39.

Rontani, J. S., J. S. Bertrand, F. Blanc, and G. Giusti. 1986. Accumulation of some monoaromatic compounds during the degradation of crude oil by marine bacteria. Mar. Chem. 18:1-7.

Rosen, M. J. 1978. Surfactants and Interfacial Phenomena. New York: John Wiley & Sons. 304 pp.

Ross, C. 1979. Dispersant research and development program. Proc. Arctic Marine Oil Spill Program Technical Seminar. Ottawa: Environment Canada.

Ross, C. W., P. B. Hildebrand, and A. A. Allen. 1978. Logistic requirements for aerial application of oil spill dispersants. *In* Chemical Dispersants for the Control of Oil Spills, L. T. McCarthy, G. P. Lindblom, and H. F. Walter, eds. STP 659. Philadelphia: ASTM. pp. 66-80.

Ross, S. L. Environmental Research Limited. 1982. Calibration of UV-IR line scanner for oil thickness using measured field data. Ottawa: Environment Protection Service.

Ross, S. L., I. A. Buist, E. Young, and L. Rinaldo. 1985. The use of emulsion inhibitors to control offshore oil spills: Part I. Proc. 8th Annual Arctic Marine Oil Spill Program Technical Seminar. Ottawa: Environment Canada, Environmental Protection Service. pp. 192-211.

Rowland, S. J., P. J. C. Tibbetts, D. Little, J. M. Baker, and T. P. Abbiss. 1981. The fate and effects of dispersant-treated compared with untreated crude oil, with particular reference to sheltered intertidal sediments. Proc. 1981 Oil Spill Conference. Washington, D.C.: API. pp. 283-293.

Sawada, N., and H. Ohtsu. 1975. Inhibitory effects of oil dispersants on the fertilization of sea urchin eggs. Mem. Ehime Univ., Series B., 7, 97-100.

Schalin, L. O. 1987. Rules and guidelines for the approval procedure for the use of oil spill dispersants. *In* Baltic Sea Environment Proceedings No. 22. Baltic Marine Environment Protection Commission, Helsinki Commission. pp. 218-230.

Schmidt, E. J., and R. A. Kimerle. 1981. New design and use of a fish metabolism chamber. *In* Aquatic Toxicology and Hazard Assessment: Fourth Conference, D. R. Branson and K. L. Dickson, eds. STP 737. Philadelphia: ASTM. pp. 436-438.

Scholten, M., and J. Kuiper. 1987. The effects of oil and chemically dispersed oil on natural phytoplankton communities. 1987 Oil Spill Conference. Washington, D.C.: API. pp. 255-257.

Schriel, R. C. 1987. Airborne surveillance: The role of remote sensing and visual observation. Oil and Chemical Pollution 3(3):181-190.

Scott, B. F., and B. Glooschenko. 1984. Impact of oil and oil dispersant mixtures on flora and water chemistry parameters in freshwater ponds. Sci. Total Environ. 35:169-190.

Scott, B. F., P. J. Wade, and W. D. Taylor. 1984. Impact of oil and oil dispersant mixtures on the fauna of fresh water ponds. Sci. Total Environ. 35(2):191-206.

Sergy, G. A., ed. 1985. The Baffin Island Oil Spill (BIOS) Project: A summary. Proc. 1985 Oil Spill Conference. Washington, D.C.: API. pp. 571-575.

_____. 1987. The Baffin Island Oil Spill (BIOS) Project. Arctic 40, Supplement 1. 279 pp.

Sekerah, A., and M. Foy. 1978. Acute lethal toxicity of Corexit 9527/Prudhoe Bay crude oil mixtures to selected arctic invertebrates. Spill Technol. Newsletter 3(2):37-41.

Shaw, D. G., and S. K. Reidy. 1979. Chemical and size fractionation of aqueous petroleum dispersions. Environ. Sci. Tech. 13:1259-1263.

Shelton, R. G. J. 1969. Dispersant toxicity test procedures. Proc. Joint Conference on Prevention and Control of Oil Spills. Washington, D.C.: API. pp. 187-191.

Shum, J. S., and J. H. Nash. 1987. Evaluation and calibration of a dispersant application system. Proc. 1987 Oil Spill Conference. Washington, D.C.: API. pp. 259-262.

Slade, G. J. 1982. Effect of Ixtoc I crude oil and Corexit 9527 dispersant on spot (*Leiostomus xanthurus*) egg mortality. Bull. Environ. Contam. Toxicol. 29:525-530.

Sleeter, T. D., and J. N. Butler. 1982. Petroleum hydrocarbons in zooplankton faecal pellets from the Sargasso Sea. Mar. Pollut. Bull. 13(2):54-56.

Smedley, J. B. 1981. Assessment of aerial application of oil spill dispersants. Proc. 1981 Oil Spill Conference. Washington, D.C.: API. pp. 253-257.

Smiley, B. D. 1982. The effects of oil on marine mammals. In Oil and Dispersants in Canadian Seas—Research Appraisal and Recommendations, J. B. Sprague, J. H. Vandermeulen, and P. G. Wells, eds. Economic and Technical Review Report EPS-3-EC-82-2. Ottawa: Environment Canada. pp. 113-123.

Smith, C. J., R. D. DeLaune, and W. H. Patrick, Jr. 1981. A method for determining stress in wetland plant communities following an oil spill. Environ. Pollut. Ser. A 26:297-304.

Smith, C. J., R. D. Delaune, W. H. Patrick, Jr., and J. W. Fleeger. 1984. Impact of dispersed and undispersed oil entering a Gulf Coast salt marsh. Environ. Toxicol. Chem. 3(4):609-616.

Smith, D. D., and G. H. Holliday. 1979. API/SC-PCO Southern California 1978 oil spill test program. Proc. 1979 Oil Spill Conference. Washington, D.C.: API. pp. 403-406.

Smith, J. E. 1968. Torrey Canyon Pollution and Marine Life. New York: Cambridge University Press. 196 pp.

Smith, R. W., and R. Pavia. 1983. Dispersant use guidelines for federal regions IX and X. Proc. 1983 Oil Spill Conference. Washington, D.C.: API. pp. 3-6.

Sorstrom, S. E. 1986. The 1985 full scale experimental oil spill at Haltenbanken, Norway. International Seminar on Chemical and Natural Dispersion of Oil on the Sea. Oceanographic Center, SINTEF Group, Trondheim, Norway. Session 1, Paper 2.

Southward, A. J., and E. C. Southward. 1978. Recolonization of rocky shores in Cornwall after use of toxic dispersants to clean up the Torrey Canyon spill. J. Fish. Res. Board Can. 35:682-706.

Spaulding, M. L. 1986. A state of the art review of oil spill trajectory and fate modeling. International Seminar on Chemical and Natural Dispersion of Oil on the Sea. Oceanographic Center, SINTEF Group, Trondheim, Norway. Session 4, Paper 1.

Spooner, M. F. 1969. Some ecological effects of marine oil pollution. Proc. Joint Conference on Prevention and Control of Oil Spills. Washington, D.C.: API. pp. 313-316.

Spooner, M. F., and C. J. Corkett. 1974. A method for testing the toxicity of suspended oil droplets on planktonic copepods used at Plymouth. In Ecological aspects of toxicity testing of oils and dispersants, L. R. Beynon and E. B. Cowell, eds. Essex, England: Applied Science Publishers Ltd. pp. 69-74.

————. 1979. Effects of Kuwait oils on feeding rates of copepods. Mar. Poll. Bull. 10(7):197-202.

Sprague, J. B. 1970. Measurement of pollutant toxicity to fish. II. Utilizing and applying bioassay results. Water Res. 4:3-32.

Sprague, J. B., J. H. Vandermeulen, and P. G. Wells. 1982. Oil and Dispersants in Canadian Seas—Research Appraisal and Recommendations. EPS 3-EC-82-2. Ottawa: Environment Canada, Environmental Protection Service.

Spraying Systems Co. 1984. Catalog No. 38. Spraying Systems, Wheaton, Ill.

————. 1985. TeeJet Buyers Guide Bulletin 225. Spraying Systems, Wheaton, Ill.

Stacey, M. L. 1983. Review of U.K. contingency planning and resource capability. Proc. 1983 Oil Spill Conference. Washington, D.C.: API. pp. 195-197.

Struhsaker, J. W., M. B. Eldridge, and T. Escheverria. 1974. Effects of benzene (a water-soluble component of crude oil) on eggs and larvae of Pacific herring and northern anchovy. *In* Pollution and Physiology of Marine Organisms, J. F. Vernberg and W. B. Vernberg, eds. New York: Academic Press. pp. 253-284.

Swedmark, M. 1974. Toxicity testing at Kristineberg Zoological Station. *In* Ecological Aspects of Toxicity Testing of Oils and Diseprsants, L. R. Beynon and E. B. Cowell, eds. Essex, England: Applied Science Publishers. pp. 41-51.

Swedmark, M., B. Braaten, E. Emanuelsson, and A. Granmo. 1971. Biological effects of surface active agents on marine animals. Mar. Biol. 9(3):183-201.

Swedmark, M., A. Granmo, and S. Kollberg. 1973. Effects of oil dispersants and oil emulsions on marine animals. Water Res. 7:1649-1672.

Swiss, J. J., and S. D. Gill. 1984. Planning, development, and execution of the 1983 East Coast dispersant trials. Proc. 7th Annual Arctic Marine Oil Spill Program Technical Seminar. Ottawa: Environment Canada. pp. 443-453.

Swiss, J. J., N. Vanderkooy, S. D. Gill, and R. H. Goodman. 1987a. Poster abstract, Beaufort Sea dispersant trial. Proc. 1987 Oil Spill Conference. Washington, D.C.: API. 634 pp.

Swiss, J. J., N. Vanderkooy, S. D. Gill, R. H. Goodman, and H. M. Brown. 1987b. Beaufort Sea oil spill dispersant trial. Proc. 10th Arctic Marine Oilspill Program Technical Seminar. Ottawa: Environment Canada. pp. 307-328.

Takada, T., and R. Ishiwatari. 1987. Linear alkylbenzenes in urban riverine environments in Tokyo: Distribution, source, and behavior. Environ. Sci. Technol. 21:875-883.

Tanaka, Y. 1979. Effects of surfactants on cleavage and further development of sea urchin embryos. 2. Disturbance in the arrangement of cortical vesicles and change in cortical appearance. Development, Growth and Differentiation 21(4):331-342.

Tarzwell, C. M. 1969. Standard methods for determination of relative toxicity of oil dispersants and mixtures of dispersants and various oils to aquatic life. Proc. Joint Conference on Prevention and Control of Oil Spills. Washington, D.C.: API. pp. 179-186.

————. 1970. Comments on standard methods for the determination of the relative toxicity of oil dispersants and various oils to aquatic organisms. Proc. Industry-Government Seminar on Oil Spill Treating Agents. Publication No. 4055. Washington, D.C.: API. pp. 80-85.

————. 1971. Toxicity of oil and oil dispersant mixtures to aquatic life. *In* Water Pollution by Oil, P. Hepple, ed. Essex, U.K.: Applied Science Publishers. pp. 263-272.

Teal, J. M., and R. W. Howarth. 1984. Oil Spill Studies: A Review of Ecological Effects. Environ. Mgmt., Vol. 8, No. 1. pp. 27-44.

Teas, H. J. 1979. Silviculture with saline water. *In* The Biosaline Concept, A. Hollaender, ed. New York: Plenum. pp. 117-161.

Teas, H. J., E. O. Duerr, and J. R. Wilcox. 1987. Effects of South Louisiana crude oil and dispersants on *Rhizophora* mangroves. Mar. Pollut. Bull. 18:122-124.

Thebeau, L. C., and T. W. Kana. 1981. Onshore impacts and cleanup during the Burmah Agate oil spill—November 1979. Proc. 1981 Oil Spill Conference. Washington, D.C.: API. pp. 139-145.

Thelin, I. 1981. Effects in culture of two crude oils and one oil dispersant on zygotes and germlings of *Fucus serratus, Linnaeus, Fucales, Phaeophyceae*. Botanica Marina 24(10):515-519.

Thompson, G. B. 1980. Manual on the reference method for cooperative assessment of oil and oil-dispersant toxicity in Southeast Asia (draft). Manila: South China Sea Fisheries Development and Coordinating Programme. 18 pp.

———. 1985. The dispersant option: Environmental considerations. Proc. Australian National Oil Spill Conference, Sydney. Australian Institute of Petroleum Ltd. and Australia Federal Department of Transportation.

Thompson, G. B., and J. M. McEnally. 1985. Coastal resource atlas for oil spills in Port Jackson. State Pollution Control Commission, New South Wales, Australia. 10 pp.

Thompson, G. B., and R. S. S. Wu. 1981. Toxicity testing of oil slick dispersants in Hong Kong. Mar. Pollut. Bull. 12(7):233-237.

Thorhaug, A., and J. H. Marcus. 1985. Effects of dispersant and oil on subtropical and tropical seagrasses. Proc. 1985 Oil Spill Conference. Washington, D.C.: API. pp. 497-501.

———. 1987a. Preliminary mortality effects of seven dispersants on subtropical/tropical seagrasses. Proc. 1987 Oil Spill Conference. Washington, D.C.: API. pp. 223-224.

———. 1987b. Oil spill clean-up: The effect of three dispersants on three subtropical/tropical seagrasses. Mar. Pollut. Bull. 18(3):124-126.

Thorhaug, A., J. Marcus, and F. Booker. 1986. Oil and dispersed oil on subtropical and tropical sea grasses in laboratory studies. Mar. Pollut. Bull. 17(8):357-361.

Throndsen, J. 1982. Oil pollution and plankton dynamics. 3. Effects on flagellate communities in controlled ecosystem experiments in Lindaspollene, Norway, June 1980 and 1981. Sarsia 67:163-169.

To, N. M., H. M. Brown, and R. H. Goodman. 1987. Data analysis and modeling of dispersant effectiveness in cold water. Proc. 1987 Oil Spill Conference. Washington, D.C.: API. pp. 303-306.

Tokuda, H. 1977a. Fundamental studies on the influence of oil pollution upon marine organisms—II. Lethal concentrations of oil spill emulsifier components for marine phytoplankton. Bull. Jap. Soc. Sci. Fish. 43(1):103-106.

———. 1977b. Fundamental studies on the influence of oil pollution upon marine organisms—III. Effects of oil-spill emulsifiers and surfactants on the growth of Porphyra-laver. Bull. Jap. Soc. Scient. Fish. 43:587-593.

———. 1979. Fundamental studies on the influence of oil pollution upon marine organisms. IV. The toxicity of mixtures of oil products and oil-spill emulsifiers to phytoplankton. Nippon Suisan Gakkaishi 45(10):1289-1291.

Tokuda, H., and S. Arasaki. 1977. Fundamental studies on the influence of oil pollution upon marine organisms. I. Lethal concentrations of oil-spill emulsifiers for some marine phytoplankton. Bull. Jap. Soc. Sci. Fish. 43(1):97-102.

Tracy, H. B., R. A. Lee, C. E. Woelke, and G. Sanborne. 1969. Relative toxicities and dispersing evaluations of eleven oil-dispersing products. J. Water Pollut. Control Fed. 41(12):2062-2069.

Tramier, B., G. H. R. Aston, M. Durrieu, A. Lepain, J. A. C. M. van Oudenhoven, R. Robinson, K. W. Sedladek, and P. Sibra. 1981. A Field Guide to Coastal Oil Spill Control and Clean-up Techniques. CONCAWE Report No. 9/81. Den Haag, The Netherlands: CONCAWE.

Traxler, R. W., and L. S. Bhattacharya. 1978. Effect of a chemical dispersant on microbial utilization of petroleum hydrocarbons. In Chemical Dispersants for the Control of Oil Spills, L. T. McCarthy, Jr., G. P. Lindblom, and H. F. Walter, eds. STP 659. Philadelphia: ASTM. pp. 181-187.

Traxler, R. W., L. S. Bhattacharya, P. Griffin, P. Pohlot, G. Garofalo, K. Kulkarni, and M. P. Wilson, Jr. 1983. Microbial response to dispersant-treated oil in ecosystems. *In* Biodeterioration 5, T. A. Oxley and S. Barry, eds. Proc. 5th International Biodeterioration Symposium, Aberdeen, U.K. pp. 282-294.

Trudel, B. K. 1978. The effect of crude oil and crude oil/Corexit 9527 suspensions on carbon fixation by a natural marine phytoplankton community. Spill Technol. Newsletter 3(2):56-64.

—. 1984. A mathematical model for predicting the ecological impact of treated and untreated oil spills. *In* Oil Spill Chemical Dispersants: Research, Experience, and Recommendations, T. E. Allen, ed. STP 840. Philadelphia: ASTM. pp. 390-413.

Trudel, B. K., and S. L. Ross. 1987. Method for making dispersant-use decisions based on environmental impact considerations. Proc. 1987 Oil Spill Conference. Washington, D.C.: API. pp. 211-216.

Trudel, B. K., S. L. Ross, and L. C. Oddy. 1983. Workbook on Dispersant Use Decision-Making: The Environmental Impact Aspects. Prepared for the Environmental Protection Service, Environment Canada, Ottawa. 59 pp.

Una, G. V., and M. J. N. Garcia. 1983. Biodegradation of non-ionic dispersants in sea-water. Eur. J. Appl. Microbiol. Biotechnol. 18(5):315-319.

U.S. Coast Guard (USCG). 1987. Polluting Incidents In and Around U.S. Waters. Calendar Year 1983 and 1984. COMDTINST M16450.2G. Washington, D.C.: U.S. Department of Transportation.

U.S. Department of the Navy. 1973. Military specification. Emulsifier, Oil-Slick MIL-E-22864B (Navy). 1973-713-145/520. Washington, D.C.: Government Printing Office. 9 pp.

U.S. Environmental Protection Agency. 1981. Handbook for Oil Spill Protection Cleanup Priorities. EPA-600/8-81-002. Cincinnati: Municipal Environmental Research Laboratory.

—. 1984. National Oil and Hazardous Substances Pollution Contingency Plan: Final Rule. 40 CFR Part 300, Federal Register 49(129):29192-29207.

U.S. Fish and Wildlife Service (USFWS) and University of California at Santa Cruz. 1985. Draft Environmental Impact Statement for Proposed Translocation of Southern Sea Otters. DES 86-33. Sacramento: USFWS.

Vacca-Torelli, M., A. L. Geraci, and A. Risitano. 1987. Dispersant application by hydrofoil: High speed control and cleanup of large oil spills. Proc. 1987 Oil Spill Conference. Washington, D.C.: API. pp. 75-79.

van Oudenhoven, J. A. C. M. 1983. The Hasbah 6 (Saudi Arabia) blowout: The effects of an international oil spill as experienced in Qatar. 1983 Oil Spill Conference. Washington, D.C.: API. pp. 381-388.

van Oudenhoven, J. A. C. M., V. Draper, G. P. Ebbon, P. D. Holmes, and J. L. Nooyen. 1983. Characteristics of petroleum and its behavior at sea. Report No. 8/83, CONCAWE. Den Haag, Netherlands. 47 pp.

Vashchenko, M. A. 1978. Influence of dispersants on the embryonic development of the sea-urchin *Strongylocentrotus nudus*. Soviet J. Mar. Biol. 4(5):848-853.

Venezia, L. D., and V. U. Fossato. 1977. Characteristics of suspensions of Kuwait oil and Corexit 7664 and their short- and long-term effects on *Tisbe bulbisetosa* (copepoda: *Harpacticoida*). Mar. Biol. 42(3):233-237.

Verriopoulos, G., and M. Moraitou-Apostolopoulou. 1982. Comparative toxicity of oil (Tunisian crude oil, Zarzaitine type), oil dispersant (Finasol OSR-2) and oil/dispersant mixture on *Artemia salina*. Workshop on Pollution of the Mediterranean, Cannes. Monaco: CIESM. pp. 743-774.

_____. 1983. Comparative toxicity of oil (Tunisian crude oil Zarzaitine type), oil dispersant Finasol OSR-2 and oil-dispersant mixture on *Artemia salina*. J. Etud. Pollut. Mar. 0(6):743-748.

Verriopoulos, G., M. Moraitou-Apostolopoulou, and A. Xatzispirou. 1986. Evaluation of metabolic responses of *Artemia salina* to oil and oil dispersant as a potential indicator of toxicant stress. Bull. Environ. Contam. Toxicol. 36(3):444-451.

Villedon de Naide, O. 1979. Studies on the toxicity of petroleum products and dispersants in the marine environment by measuring the photosynthetic activity of a test-organism. Paris: Conservatoire National Des Arts et Metiers. 70 pp.

Wardley-Smith, J., ed. 1976. The Control of Oil Pollution on the Sea and Inland Waters. London: Graham and Trotman.

Weber, D. D., D. J. Maynard, W. D. Gronlund, and V. Konchin. 1981. Avoidance reactions of migrating adult salmon to petroleum hydrocarbons. Can. J. Fish. Aquat. Sci. 38:779-781.

Wells, P. G. 1978. Biological hazards of controlling marine oil spills with chemicals. A case for caution. Discussion paper for the Gordon Research Conference, Plymouth, N.H., July 17-21.

_____. 1981. Chemical dispersion of oil spills—an international research symposium. Summary—toxicology session. *In* Chemical Dispersion of Oil Spills: An International Research Symposium, D. Mackay, P. G. Wells, and S. Paterson, eds. Publication No. EE-17. Toronto: Institute for Environmental Studies, University of Toronto.

_____. 1982a. Dispersant evaluations—toxicity. A discussion paper. Unpublished report, Environment Canada, Environmental Emergencies Branch, Ottawa. 35 pp.

_____. 1982b. Zooplankton. *In* Oil and Dispersants in Canadian Seas: Research Appraisal and Recommendations, J. B. Sprague, J. H. Vandermeulen, and P. G. Wells, eds. EPS 3-EC-82-2. Ottawa: Environmental Protection Service. pp. 65-80.

_____. 1984. The toxicity of oil spill dispersants to marine organisms: A current perspective. *In* Oil Spill Chemical Dispersants: Research, Experience, and Recommendations, T. E. Allen, ed. STP 840. Philadelphia: ASTM. pp. 177-202.

_____. 1985. Lethal and sublethal effects of dispersants and oil-dispersant mixutres on marine organisms—a synopsis. Spill Technol. Newsletter 10(1-3):11-25.

Wells, P. G., and G. W. Harris. 1979. Dispersing effectiveness of some oil dispersants: Tests with the "Mackay Apparatus" and Venezuelan Lago Medio crude oil. Spill Technol. Newsletter 4(4):232-241.

_____. 1980. The acute toxicity of dispersants and chemically dispersed oil. Proc. Arctic Marine Oil Spill Program (AMOP) Technical Seminar. Edmonton, Alberta: Environment Canada. pp. 144-157.

Wells, P. G., and J. A. Percy. 1985. Effects of oil on arctic invertebrates. *In* Petroleum Effects in the Arctic Environment, F. R. Engelhardt, ed. London: Elsevier. pp. 101-156.

Wells, P. G., and J. B. Sprague. 1976. Effects of crude oil on American lobster (*Homarus americanus*) larvae in the laboratory. J. Fish. Res. Board Can. 33(7):1604-1614.

Wells, P. G., and K. G. Doe. 1976. Results of the E.P.S. oil dispersant testing program: Concentrates, effectiveness testing and toxicity to marine organisms. Spill Technol. Newsletter 1:9-16.

Wells, P. G., and P. D. Keizer. 1975. Effectiveness and toxicity of an oil dispersant in large outdoor salt water tanks. Mar. Pollut. Bull. 6(10):153-157.

Wells, P. G., S. Abernethy, and D. Mackay. 1982. Study of oil water partitioning of a chemical dispersant using an acute bioassay with marine crustaceans. Chemosphere 11(11):1071-1086.

————. 1984a. The effectiveness of dispersants and the acute toxicity of chemically dispersed crude oils—studies with the Mackay-Steelman-Nadeau apparatus and marine copepods. In Proc. of the 7th Annual Arctic Marine Oilspill Program Technical Seminar. Ottawa: Environment Canada. pp. 47-59.

————. 1985. Acute toxicity of solvents and surfactants of dispersants to two planktonic crustaceans. Proc. 8th Annual Arctic Marine Oil Spill Program Technical Seminar. Ottawa: Environment Canada. pp. 228-240.

Wells, P. G., J. W. Anderson, and D. Mackay. 1984b. Uniform methods for exposure regimes in aquatic toxicity experiments with chemically dispersed oils. In Oil Spill Chemical Dispersants: Research, Experience, and Recommendations, T. E. Allen, ed. STP 840. Philadelphia: ASTM. pp. 23-37.

Westergaard, R. H. 1983. Dispersion: The stepchild in Norwegian oil preparedness. Norsk Oljerevy/Norw. Oil Rev. 9(5):35-41.

Westlake, D. W. S. 1982. Microorganisms and the degradation of oil under northern marine conditions. In Oil and Dispersants in Canadian Seas—Research Appraisal and Recommendations, J. B. Sprague, J. H. Vandermeulen, and P. G. Wells, eds. Report EPS-3-EC-82-2. Ottawa: Environment Canada. pp. 47-53.

Wheeler, R. B. 1978. The Fate of Petroleum in the Marine Environment. Special Report, Exxon Production Research Company, Houston, Texas. 32 pp.

White, I. C. 1976. Toxicity testing of oils and oil dispersants. In Lectures presented at the 4th FAO/SIDA Training Course on Aquatic Pollution in Relation to Protection of Living Resources. Bioassays and Toxicity Testing. Rome: FAO. pp. 168-180.

White, I. C., and J. A. Nichols. 1983. The cost of oil spills. Proc. 1983 Oil Spill Conference. Washington, D.C. API. pp. 541-544.

White, J. R., R. E. Schmidt, and W. E. Plage. 1979. The Aireye remote sensing system for oil spill surveillance. Proc. 1979 Oil Spill Conference. Washington, D.C.: API. pp. 129-136.

Whitney, F. A. 1984. The effects and fate of chemically dispersed crude oil in a marine ecosystem enclosure data report and methods. Can. Data Rep. Hydrogr. Ocean Sci. 0(29),I-VIII: 1-78.

Wildish, D. J. 1974. Arrestant effect of polyoxyethylene esters on swimming in the winter flounder. Water Res. 8:579-583.

————. 1972. Acute toxicity of polyoxyethylene esters and polyoxyethylene ethers to S. Salar and G. oceanicus. Water Res. 6:759-762.

Williams, T. D. 1978. Chemical immobilization, baseline parameters and oil contamination in the sea otter. Report No. MMC-77-06. Washington, D.C.: U.S. Marine Mammal Commission. 18 p.

Wilson, D. E., J. C. H. Mungall, S. Pace, P. Carpenter, H. Teas, T. Goddard, R. Whitaker, and P. Kinney. 1981. Numerical trajectory modeling and associated field measurements in the Beaufort Sea and Chukchi Sea nearshore areas. NOAA/OCSEAP Final Report.

Wilson, K. W. 1974. Toxicity testing for ranking oils and oil dispersants. In Ecological Aspects of Toxicity Testing of Oils and Dispersants, L. R. Beynon and E. B. Cowell, eds. Barking, Essex, U.K.: Applied Science Publishers. pp. 11-22.

————. 1976. Effects of oil dispersants on the developing embryos of marine fish. Mar. Biol. 36:259-268.

_____. 1977. Acute toxicity of oil dispersants to marine fish larvae. Mar. Biol. 40:65-74.

_____. 1981. Licensing of oil dispersants—what kind of toxicity data do we need? *In* Chemical Dispersion of Oil Spills—An International Research Symposium, D. Mackay, P. G. Wells, and S. Paterson, eds. Publication No. EE-17. Toronto: University of Toronto, Institute for Environmental Studies. pp. 173-177.

_____. 1984. Policies on the use of dispersants—the role of toxicity testing programmes for oil dispersant chemicals. *In* Combatting Oil Pollution in the Kuwait Action Plan Region. Geneva: United Nations Environment Programme. pp. 371-387.

Wilson, M. P. Jr. 1980. Assessment of treated vs. untreated oil spills. Final Report, Contract E(11-1)4047. U.S. Department of Energy, University Energy Center, University of Rhode Island, Kingston, RI.

Wilson, K. W., E. B. Cowell, and L. R. Beynon. 1973. The toxicity testing of oils and dispersants: A European view. Proc. Joint Conference on Prevention and Control of Oil Spills. Washington, D.C.: API. pp. 255-261.

_____. 1974. The toxicity testing of oils and dispersants: A European view. *In* Ecological Aspects of Toxicity Testing of Oils and Dispersants, L. R. Beynon and E. B. Cowell, eds. Essex, U.K.: Applied Science Publishers. pp. 129-141.

Western Oil and Gas Association (WOGA). 1986. Sea Otter Oil Spill Contingency Plan. Los Angeles: WOGA.

Wu, R. S. S. 1981. Differences in the toxicities of an oil dispersant and a surface active agent to some marine animals and their implications in the choice of species in toxicity testing. Mar. Environ. Res. 5(2):157-163.

Wyers, S. C. 1985. Sexual reproduction of the coral *Diploria strigosa* (Scleratinia, Faviidae) in Bermuda: Research in progress. Proc. 5th International Coral Reef Congress, Tahiti, Vol. 4, pp. 301-502.

Wyers, S. C., H. R. Frith, R. E. Dodge, S. R. Smith, A. H. Knap, and T. D. Sleeter. 1986. Behavioral effects of chemically dispersed oil and subsequent recovery in *Diploria strigosa* (*Dana*). Mar. Ecol. 7(1):23-42.

Zagorski, W. and D. Mackay. 1981. Formation of water-in-oil emulsions. Proc. 4th Arctic Marine Oilspill Program Technical Seminar. Edmonton, Alberta: Environment Protection Service. pp. 75-86.

Zawadzki, D., J. D. Stieb, and S. McGee, Jr. 1987. Considerations for dispersant use: tank vessel *Puerto Rican* incident. Proc. 1987 Oil Spill Conference. Washington, D.C.: API. pp. 341-345.

Zeeck, E., P. Franke, S. Gross, W. Gross, M. Niedrig, and S. Schroer. 1984. Experimental investigations about effects of crude oil and dispersed crude oil in tidal flat environments. 4. Bacterial oil degradation as determined by fluorescence spectroscopy. Senckenb. Marit. 16(1-6):57-68.

Zillioux, E. J., H. R. Foulk, J. C. Prager, and J. A. Cardin. 1973. Using *Artemia* to assay oil dispersant toxicities. J. Water Pollut. Control Fed. 45(11):2389-2396.

SOURCES OF UNPUBLISHED DATA AND PRIVATE COMMUNICATIONS

Araujo, Rosalina P. A., CETESB (State Environmental Protection Agency of São Paulo), São Paulo, Brazil

Canevari, G. P., Canevari Associates, Cranford, New Jersey

Cashion, Bryan, Exxon Research and Engineering, Florham Park, New Jersey

Chapman, Piers, Sea Fisheries Institute, Cape Town, Republic of South Africa

Cormack, D., Dr., deputy director (Environment), Warren Spring Laboratory, Stevenage, Herfordshire, England

Fingas, Mervin, head, Chemistry and Physics Section Technical Services Branch, Environmental Protection Service, Ottawa, Ontario, Canada

Flaherty, Michael L., EPA retired, Potomac, Maryland

Fraser, John P., Shell Oil Company, Houston, Texas

Goldstein, Bernard J., director, Environmental Health Center, NMDNJ Rutgers Medical School, Cataway, New Jersey

Heldal, M., ADR, Institute for General Microbiology, University of Bergen, Norway

Humphrey, Alan, Environmental Response Team, U.S. Environmental Protection Agency, Edison, New Jersey

International Tanker Owners Pollution Federation (ITOPF) Ltd., London, England

Jones, Colin M., Consulting Engineers, Arlington, Virginia

Knutsen, G., Department of Microbiology and Plant Physiology, University of Bergen, Norway

Kolde, Harold E., Environmental Monitoring Support Laboratory, U.S. Environmental Protection Agency, Cincinnati, Ohio

Kvita, P., Rensselaer Polytechnic Institute, Troy, New York

Lavache, Cmdr. Mark L., chief, Pollution Response Branch, U.S. Coast Guard, Washington, D.C.

Leendertse, J. J., RAND Corporation, Santa Monica, California

Lien, T. Department of Microbiology and Plant Physiology, University of Bergen, Norway

Lindblom, Gordon P., Exxon, Inc. (retired), Houston, Texas

Lloyd, Dr. Richard, Ministry of Agriculture, Fisheries, and Food, Fisheries Laboratory, Essex, England

Mackay, Prof. Donald, Department of Chemical Engineering and Applied Chemistry Institute for Environmental Studies, University of Toronto, Ontario, Canada

McAuliffe, Clayton D., Clayton McAuliffe and Associates, Inc., Environmental Consulting, Fullerton, California

McGibbon, S., Sea Fisheries Research Institute, Cape Town, Republic of South Africa

Meyers, Robert J., Robert J. Meyers and Associates, Houston, Texas

Nichols, Joseph A., The International Tanker Owners Pollution Federation Ltd., London, England

Norland, S., Department of Microbiology and Plant Physiology, University of Bergen, Norway

Norton, Michael G., Ministry of Agriculture, Fisheries, and Food, Fisheries Laboratory, Essex, England

Onstad, L. A., Clean Seas, Carpenteria, California

Pirani, Kenneth C., Materials Testing Laboratories, Department of Defense, Alexandria, Australia

Ross, Sydney, Rensselaer Polytechnic Institute, Troy, New York

Ross, S. L., S. L. Ross Environmental Research Ltd, Ottawa, Ontario, Canada

Tennyson, E., U.S. Minerals Management Service, Washington, D.C.

Thompson, G. B., State Pollution Control Commission, New South Wales, Sydney, Australia

Trudel, B. K., S. L. Ross Environment Research Ltd., Ottawa, Ontario, Canada

Wells, Peter G., Conservation and Protection, Environment Canada, Dartmouth, Nova Scotia

White, Rudolph C., Marketing Department, American Petroleum Institute, Washington, D.C.

Wilson, Jack E., Ecology and Environment, Inc., Arlington, Virginia

Zobell, Claude E., Scripps Institution of Oceanography, University of California-San Diego, La Jolla, California

APPENDIXES

A
Dispersant Products Information

TABLE A-1 Product Schedule of the Environmental Protection Agency's National Contingency Plan, July 1987

Bulletin Number	Dispersant Product Name	Dispersant Manufacturer	Previous Acceptance Date
1	BP1100X (hydrocarbon solvent based)	BP Detergents Ltd. Pumpherston Works Livingston, West Lothian Scotland EH53 0LQ Tel: 0506 31111 Telex: 72278 (John R. Nicol)	10/20/77
2	Cold Clean 500 (water based)	Essex Fire and Safety Co. P.O. Box 87709 Houston, TX 77287 (713) 641-3616 (Virginia A. Watters)	10/7/77
3	Conco Dispersant K (concentrate)	Continential Chemical Co. 270 Clifton Blvd. Clinton, NJ 07011-3686 (202) 472-5000 (P. D. Turits)	4/25/78
4	Corexit 7664 (water based)	Exxon Chemical Co. 8230 Stedman Street Houston, TX 77029 (713) 670-1702 (Don Jacques)	11/1/78

TABLE A-1 (Continued)

Bulletin Number	Previous Dispersant Product Name	Dispersant Manufacturer	Acceptance Date
5	Corexit 8667 (hydrocarbon solvent based)	Exxon Chemical Co. 8230 Stedman Street Houston, TX 77029 (713) 670-1702 (Don Jacques)	11/1/78
6	Corexit 9527 (hydrocarbon solvent based)	Exxon Chemical Co. 8230 Stedman Street Houston, TX 77029 (713) 670-1702 (Don Jacques)	3/10/78
7	EC. O ATALN 'TOL AT7 (water based)	ASPRA, Inc. 4401 - 23rd Avenue West Seattle, WA 98199 (206) 284-9838 (A. I. Janofsky)	11/13/70
8	Finasol OSR-7 (water-based concentrate)	American Petrofina, Inc. P.O. Box 2159 Dallas, TX 75221 (214) 750-2640 (Jerry W. Johnson)	5/21/80
9	Gold Crew Dispersant (water-based concentrate)	Ara Chem, Inc. P.O. Box 5031 San Diego, CA 92105-0001 (619) 286-4131 (Rita Jimenez McNeely)	8/31/77
10	Magnotox (water-based concentrate)	Magnus Maritec Int'l., Inc. 150 Roosevelt Place P.O. Box 150 Palisades Park, NJ 07650 (201) 592-0700 (Andreas C. Ladjias)	7/1/81
11	OFC D-609 (concentrate)	Chem Link Petroleum, Inc. P.O. Box 370 Sand Springs, OK 74063 (918) 245-2224 (Glenn D. Fielder)	8/20/79
12	Oil Spill Eliminator N/T No. 4 (hydrocarbon solvent based)	Petrocon Marine and Chemical Corp. 243-44th Street Brooklyn, NY 11232 (212) 499-3111 (Frank B. Sidoti)	5/21/80

TABLE A-1 (Continued)

Bulletin Number	Previous Dispersant Product Name	Dispersant Manufacturer	Acceptance Date
13	OSD/LT Oil Spill Dispersant (concentrate)	Drew Chemical Corp. 1 Drew Chemical Plaza P.O. Box 157 Boonton, NJ 07005 (202) 263-7817 (Rochelle Galiber Asbell)	5/11/79
14	Petro-Green ADP-7	Petro-Green, Inc. 3952 Candlenut Lane P.O. Box 814665 Dallas, TX 75381 (214) 484-7336 (Arnold Paddock)	9/30/84
15	Petromend, MP-900-W (water-based concentrate)	Petromend, Inc. P.O. Box 47532 8300 Sovereign Row Dallas, TX 75247 (214) 630-1330 (Alan Cohn)	9/30/84
16	Proform-Pollution Control Agent (water-based concentrate)	Proform Products Corp. 220 California Avenue Suite 100 Palo Alto, CA 94306 (415) 321-5207 (Rudolf Kruska)	5/9/79
17[a]			
18	Slik-A-Way (water-based)	MI-DEE Products, Inc. 5253 Springdale Avenue Pleasanton, CA 94566 (415) 846-8166 (Paul Spellman)	10/5/78
19	Dispersant 11 (concentrate)	Dubois Chemicals 1100 Dubois Tower Cincinnati, OH 45202 (513) 762-6894 (W. N. Grawe)	10/16/84
20	Topsall No. 30 (oil and petroleum cleaning agent)	Stutton North Corp. P.O. Box 724 Mandeville, LA 70448 (504) 626-3900 (Sid Studin)	1/7/85

TABLE A-1 (Continued)

Bulletin Number	Previous Dispersant Product Name	Dispersant Manufacturer	Acceptance Date
21	Corexit 9550 (hydrocarbon solvent based)	Exxon Chemical Co. 8230 Stedman Street Houston, TX 77029 (713) 670-1702 (Don Jacques)	5/22/85
22	Jansolv-60 Dispersant (principally water-based with some solvent)	Sunshine Technology Corp. 2475 Albany Avenue West Hartford, CT 06117 (203 232-9227 (Stephan Kaufmann)	7/9/85
23	Ruffnek (oil and petroleum cleaning agent)	Malter International Corp. 80 First Street Gretna, LA 70053 (504) 362-3232 (Laboratory)	7/16/85
24	NEOS AB 3000 (hydrocarbon solvent based)	NEOS Company Limited 8th Floor, Kanden Building 2-1, Kano-cho 6-chome Chuo-ku, Kobe 650, Japan (078) 331-9381 (S. Miyoshi)	4/22/85
25	Crudex (organic surfactant based)	Environmental Security, Inc. 352 Abbeyville Road Lancaster, PA 17603 (717) 392-1251 (Jay Greene)	6/4/86
26	Bio Solve (water based)	Metra Chem Corp. 792 Hartford Pike Shrewsbury, MA 01545 (617) 845-1193 (Mario J. Genduso)	12/22/86
27	NK-3 (water based)	GFC Chemical Co. P.O. Box 80537 Lafayette, LA 70598-0537 (318) 234-8262 (Joe Winkler)	2/19/87
28	Enersperse 700 (solvent based)	BP Detergents Ltd. Drumshoreland Road Pumpherston Works Livingston, West Lothian Scotland EH53 OLQ (011-44-5016) 31111 (David Kerr)	7/27/87

TABLE A-1 (Continued)

Bulletin Number	Previous Dispersant Product Name	Dispersant Manufacturer	Acceptance Date
29	Slickgone NS (solvent based)	DASIC International Ltd. Winchester Hill - Romsey Hampshire S041 7YD England, U.K. (0794) 512419 (John L. Belk)	2/22/88
30	Mare Clean 505 (solvent based)	Mitsubishi International Corp. International Specialty Chemicals Department 520 Madison Avenue New York, NY 10022 (212) 605-2533 (K. Komatsu [FER])	2/23/88

[a]Sea Master NS-555, bulletin number 17, is no longer manufactured and has been removed from the U.S. EPA Product Schedule.

TABLE A-2 List of Dispersants Approved by the Environmental Protection Service of Canada, February 1987

Product Name	Manufacturer
Corexit 9527	Exxon/Esso Chemical
Corexit 9550	Exxon/Esso Chemical
Corexit CRX-8	Exxon/Esso Chemical
Drew Oil Dispersant LT	Drew Chemical
Enersperse 1100X	British Petroleum/ PetroCan chemicals
Enersperse 700	British Petroleum/ PetroCan Chemicals
Gamelin 2000	Gamelin Chemical
Oilsperse 43	Diachem Industries
SlickGone LTS	Dasic International

SOURCE: Eco/Log Week, February 6, 1987.

TABLE A-3 Composition of Some Currently Produced Oil Spill Dispersants[a]

Base	Product	Composition
Hydrocarbon[b] based	BP1100X (British Petroleum)	Fatty acid ester (15 wt%) Hydrocarbon solvent (84%) (Aromatics in solvent < 3%) Additives (1%)
	Corexit 8667 (Exxon)	Oxyalkylated surfactant in nonaromatic hydrocarbon solvent Amber liquid, mild odor
	Corexit 9550	Mixture of surfactant ester in mixed oxygenated and nonaromatic hydrocarbon solvents; contains glycol ethers Amber liquid
Water based[c]	Corexit 7664	Nonionic polyhydric alcohol ester of fatty acid + 20% isopropyl alcohol Ester surfactants in mixed oxygenated solvents Light-amber liquid, slight alcohol odor
	Corexit 9527	Glycols Yellow liquid, unique odor
Concentrate[d]	BP1100D	Surfactants have infrared spectra identical to those in BP1100X Amber-colored clear liquid with odor

[a] Chemical composition terminology is that of the manufacturer.
[b] Water-immiscible, can be used on high-viscosity crude and residual oils,
weathered crudes, waxy crudes, and water-in-oil emulsions.
[c] Can be used on light-distillate fuels and low-viscosity crude and product.
[d] Can be used on all types of oil.

SOURCE: Wells, 1984.

TABLE A-4 Some Patented Oil Spill Dispersant Formulations[a]

Assignee	Patent Number	Active Ingredients	Solvent
ICI	Brit. 1,338,385 Nov. 73	Esters of tall oil fatty acid and polyethylene glycol monomethyl ether	
ICI	Brit. 1,338,391 1973	Esters of tall oil fatty acid and polyethylene glycol monobenzyl or monnonylphenyl ether	
BP	Brit. 1,361,179 July 74	Ethoxylated fatty alcohol 1.6-6.0 mol EO/mol alcohol	Kerosene, isopropanol
Exxon	U.S. appl. 457,098 (April 74?)	Sorbitan monooleate Ethoxylated sorbitan monooleate Na dioctyl sulfosuccinate	
Exxon	U.S. 3,793,218 (Feb. 74?)	As above	
IFP	Fr. appl. 74/03.364 Feb.74	Ethoxylated esters of saccharine and oleic acid	
IFP	Fr. appl. 74/43.200 1974	Esters of ethoxylated trimethylolpropane monooleate and glutamic acid	
Dasic Intern	Brit. 1,404,648 May 75	Polyethylene glycol monooleate (9%) sorbitan monooleate (3%), monopropylene glycol (2%) [% or ratio?]	Kerosene
BP	Brit. 1,399,80? July 75	Polyethylene glycol esters of oleic acid	Kerosene
IFP	Belg. 886,656 (1975?)	Polyethylene glycol monooleate Ethoxylated sorbitan oleate Ethoxylated octyl phosphate Ethoxylated tridecylphosphate	
BP	Brit. 1,419,803	Polyethylene glycol esters	Propoxylated
Lankro	Fr. 2,330,653 June 77	Monoester of polyethylene glycol and a fatty acid (R < 10 carbons, EO = 2)	
PCUK	Fr. appl. 78/27.894	Ethoxylated/propoxylated esters of an alcohol ethoxylate and phosphoric acid	
Institute for Oceanology, USSR	Ger. Off. 2,626,552 Dec. 78	Diester of an ethoxylated alcohol and phosphoric acid	
Institute for Oceanology, USSR	U.S. 4,197,197 April 80	Diester of an ethoxylated alcohol and phosphoric acid (R = 7 to 12 carbons, EO = 4 to 6)	
PCUK	Fr. 2,394,602 Jan. 79	Mono and diesters of an alcohol ethoxylate and phosphoric acid	
ICI	Brit. appl. 79/17, 106 May 79	Hexamethylene diamine Ethylene/propylene oxide condensate, nonionic surfactant	Hydrocarbon
Shell	Brit. 1,557,182 Dec. 79	Mixtures of polyethylene glycol monooleate, e.g., diethylene glycol monooleate, polyethylene glycol monooleate	
PCUK	Fr. 2,439,817 (1980?)	Ethoxylated/propoxylateld esters of an ethoxylated/propoxylated alcohol and phosphoric acid	
Labofina	Fr. 2,479,251 (1980?)	Tall oil esters (35%), ethyldioxitol (47%), sorbitan monolaurate (7%) water (10%), calcium sulfonate (1%)	

TABLE A-4 (Continued)

Assignee	Patent Number	Active Ingredients	Solvent
Cosden Tech. (Finasol)	U.S. 469,603 Sept. 84	Sorbitan monooleate Ethoxylated sorbitan monooleate Sodium diethylhexyl sulfosuccinate	
Exxon	U.S. 4,502,962 March 85 (1976?)	Sorbitan monooleate Ethoxylated sorbitan monooleate Ethoxylated sorbitan trioleate Sodium ditridecyl sulfosuccinate of fatty acids and aryl, alky., alkyl-ether, or alkyl-aryl sulfate/sulfonate; e.g., Na lauryl sulfate, alkyl benzene sulfonates	Hydrocarbon Butanol (m.w. 90-250)

KEY: ICI--Imperial Chemical Industries, United Kingdom; BP--British Petroleum, United Kingdom; IFP--Institute Francaise du Petrole, France; and PCUK--Produits Chemique Ugine Kuhlman, France.

[a]Chemical composition terminology is that of the manufacturer.

SOURCES: Wells, 1984; Canevari, private communication.

B
Case Studies

Although experience gained in treating large marine oil spills from tanker accidents should give additional insight into the factors governing dispersant effectiveness, the few documented cases are disappointing. Field studies of accidental major oil spills where dispersants were used generally provide incomplete or equivocal information because less than half of the spilled oil was sprayed and the studies usually did not distinguish between exposure to dispersed oil and undispersed oil. Nevertheless, at least two studies following major spills focused on waters where oil was dispersed: the Main Pass Block 41 *Platform C* and *Ixtoc I* spills in the Gulf of Mexico. Other spills that were studied, including the *Amoco Cadiz* (Spooner, 1978), the *Urquiola* (Gundlach et al., 1978), and the *Florida* (Sanders et al., 1981), offer little useful data with which to evaluate the ecological effects of dispersant use. Only a brief review of relevant facts are included here.

TORREY CANYON, 1967

The effects of the massive use of dispersants to clean up the large oil spill resulting from the *Torrey Canyon* accident were studied by Smith (1968) and Southward and Southward (1978), among others.

Ten thousand tons* (about 75,000 bbl) of first-generation toxic dispersants were applied to 14,000 tons (about 105,000 bbl) crude oil stranded on rocky coastline in Cornwall, England. The oil alone was not very toxic, although some mortality of limpets and barnacles occurred. Use of dispersants, however, caused extensive mortalities of animals and algae, proportional to dispersant dose. The widely reported "rapid" recovery of shorelines following the spill was largely not true: recolonization of affected areas followed a natural succession pattern, but at a slower rate. Recovery was slowest where exposure to oil and dispersant was highest and 10 years later was still not complete in some areas. Herbivores were affected more than plants, and some community structures were altered.

The major problem may not have been the dispersants themselves, but the way in which they were used. Clearly, their major purpose was aesthetic, not ecological. The objective was to remove visible oil from shorelines, without regard for ecological consequences.

The results from the *Torrey Canyon* spill are a major reason for concern about the toxicity of dispersants. The surfactants were in aromatic hydrocarbon solvents, which are fairly effective in dispersing oil, but are highly toxic to marine and shore organisms when sprayed in high concentrations directly on beaches. Biological surveys showed that heavy excessive mortalities resulted where these early formulations were used in large quantities. In fact, oil contamination alone resulted in fewer adverse biological effects on shore areas than where the dispersants were used.

MAIN PASS BLOCK 41, *PLATFORM C*, 1972

Gulf of Mexico, Main Pass Block 41, *Platform C,* located about 11 mi east of the Mississippi River Delta in 39 ft of water, discharged about 65,000 bbl of crude oil over 3 weeks in March 1970. During that time, about 2,000 bbl of two chemical dispersants were sprayed in diluted form by fire monitors from a barge. The dilute dispersant was applied to the platform and immediate surrounding waters to keep the steel structure water-wet, both for safety of personnel working on the platform and to keep the structure from melting if the oil and gas reignited. No attempt was purposely made to treat all of the slick with spray boats or aircraft.

---

*Maritime spills are reported in long tons (2,240 lb = 7.5 bbl of oil).

It was estimated that 25 to 30 percent of the oil evaporated in the first 24 hr, 10 to 20 percent was recovered from the water surface, less than 1 percent dissolved, and less than 1 percent was identified in sediments within a 5-mi radius of the platform. Some oil was emulsified, observed as a "widening creamy-yellow near-surface plume as it moved with the water from the platform," which became too diffused to observe from about 1 mi away (McAuliffe et al., 1975).

Over 3 days when the oil was estimated to be spilling at a rate of 1,500 bbl/day, water samples were collected near the platform and from the dispersed oil plume until it could no longer be seen (McAuliffe et al., 1975). The highest oil concentration in a sample of the plume closest to the platform was 60 mg/liter. One mile away, oil concentration decreased to 1 mg/liter. Estimates of oil dispersed ranged between 4 and 66 percent.

Neither the untreated oil nor the dispersed oil appear to have had adverse effects on marine life (spilled oil did not strand). Over 550 species of benthic organisms were identified in 233 samples from that region. The numbers of species and individuals organisms showed lower values in some samples near the platform. However, seasonal variations, bottom sediment type, and other environmental parameters, including Mississippi River discharges, made it impossible to determine whether these locations had been affected by the spilled oil.

There was no correlation between number of species, number of individuals, or other biological parameters and the hydrocarbon content of the sediments for samples within a 10-mi radius of the platform. This lack of correlation suggests lack of significant effect of oil on benthic organisms, which would have been most likely to be affected. Extensive trawl samples showed no alteration in the annual life cycle of commercially important shrimp, blue crabs were observed throughout the area, and the number of fish species was comparable to a prior survey.

ELENI V, 1978

The *Eleni V* carried a heavy fuel oil with viscosity 5,000 cSt at 20° C. Colliding with the *Roseline* off the southeast coast of England in 1978, she released 7,500 tons (56,000 bbl) of oil. A total of 900 tons (6,800 bbl) of dispersant concentrate BP1100D and 10 percent Dasic LTD were applied by 22 vessels over 3 weeks, but because of high viscosity they were ineffective, and virtually all the oil washed ashore (Nichols and Parker, 1985).

HASBAH 6, 1978

The *Hasbah 6* well blowout off northern Saudi Arabia released 14,000 tons (105,000 bbl) of heavy crude oil, which was treated with dispersant by boat and helicopter. Calm seas, the high viscosity of weathered crude making up the slick, and inadequate preparation for such a large spill diminished the effectiveness of the effort (Kornberg et al., 1981; Oil Spill Intelligence Report, 1980).

IXTOC I, 1979-1980

The *Ixtoc I* well blowout occurred in the southern Gulf of Mexico in June 1979, spilling an estimated 30,000 bbl/day between June and December. Over these 6 months, spray booms 25-ft long on each aircraft wing, with 228 nozzle bodies, pumped 400 gal/min per pump. Some aircraft were equipped with two such systems, thus doubling the maximum rate available (Lindblom et al., 1981).

Observers saw results within several hours, and by the following day no oil was visible in the treated area. This included oil that had drifted 500 mi and had been on the water for 4 to 6 months before spraying (Lindblom et al., 1981). Dispersal of oil at 4 to 6 months was visually confirmed by the air crews on return trips (Lindblom et al., 1981).

Slicks on very calm water to which dispersants were applied were usually dispersed by overnight storms. A small amount of oil was stranded north of Tampico, Mexico while aerial operations were confined to the southern Gulf. After redeployment of aircraft along the shoreline to the U.S. border, no oil came ashore in Mexico until hurricane Henri arrived in late September and grounded the airplanes. At that time oil stranded on the western Yucatan shoreline, but lagoons were protected from oiling by the rapid outflow of fresh water from heavy rains (Lindblom et al., 1981).

Field evaluation by the Gulf Universities Research Consortium (Linton and Koons, 1983) indicated that dispersant "effectively dispersed the *Ixtoc I* crude from the surface into the upper 3 m of the water column." Limited biological studies during and following the spill did not reveal adverse effects. For example, shrimp landings in subsequent years were unchanged or greater than previous yearly catch statistics.

BETELGUESE, 1979

The 1979 spill of 7,000 bbl of mixed Arabian crude from the tanker *Betelgeuse* in Bantry Bay, southwest Ireland was probably the first significant oil spill in Europe for which dispersants were applied from aircraft in preference to vessels (Nichols and White, 1979). Thirty-five metric tons (260 bbl) of BP1100WD was believed to have protected shorelines successfully with an application rate of 2 to 3 gal/acre (19 to 28 liter/ha) (Nichols and Parker, 1985; Nichols and White, 1979). On the other hand, winds up to Beaufort force 4 were present and much of the effective dispersal may have been due to wind turbulence.

PUERTO RICAN, 1984

In 1984, after an explosion and fire, the tanker *Puerto Rican* broke in two, spilling 25,000 to 35,000 bbl of refined petroleum products, mostly lubricating oils, in the Point Reyes-Farallon Island Marine Sanctuary, 25 mi west of San Francisco Bay. Within 10 hr the Regional Response Team, acting under the State of California Contingency Plan, granted final approval for dispersant use based on an apparent threat to marine mammals (Herz, 1986; Herz and Kopec, 1985). Dispersant application was delayed 4 to 5 hr, however, because wind and wave conditions prevented a sampling vessel from reaching the scene (the Regional Contingency Plan required samples).

An airplane sprayed about 2,500 gal (60 bbl) of Corexit 9527 on about half of the spill. Dispersion estimates by aerial observers ranged from none to 20 to 30 percent. During 2 to 3 weeks, oil skimmers collected 1,500 bbl of oil-water emulsion (50 percent), accounting for only 2 to 3 percent of the amount spilled.

C

Conversion Factors

Acres	×	0.4047	=	Hectares
Acres	×	4.047×10^{-3}	=	Square Kilometers
Barrels	×	42.0	=	Gallons (U.S.)
Barrels	×	35.0	=	Gallons (Imperial)
Cubic Feet	×	2.832×10^{-2}	=	Cubic Meters
Feet	×	0.3048	=	Meters
Gallons (U.S.)	×	0.833	=	Gallons (Imperial)
Gallons (U.S.)	×	3.785	=	Liters
Gallons (U.S.)/Acre	×	9.353	=	Liters/Hectare
Hectares	×	2.471	=	Acres
Kilometers	×	0.6214	=	Miles (Statute)
Kilometers	×	0.5396	=	Miles (Nautical)
Kilometers/Hour	×	0.5396	=	Knots
Knots	×	1.852	=	Kilometers/Hour
Knots	×	1.151	=	Statute Miles/Hour
Knots	×	0.515	=	Meters/Second
Liters	×	0.264	=	Gallons (U.S.)
Liters/Hectare	×	0.1	=	Cubic Meters/ Square Kilometer
Meters	×	3.281	=	Feet
Miles (Nautical)	×	6.076×10^3	=	Feet
Miles (Nautical)	×	1.852	=	Kilometers
Miles (Nautical)	×	1.1516	=	Miles (Statute)

Miles (Statute)	×	5.280×10^3	=	Feet
Miles (Statute)	×	1.609	=	Kilometers
Miles (Statute)/Hour	×	0.8684	=	Knots
Tons (U.S.) (2,000 Lbs)	×	.907	=	Tons (Metric)
Tons (Metric)	×	1.102	=	Tons (U.S.)
Tons (Metric)	×	294	=	Gallons U.S. (Avg. Oil)
Tons (Metric)	×	7.5	=	Barrels

Glossary and Acronyms

ABS. Alkyl benzene sulfonate.

Acute. Having a sudden onset, lasting a short time, of a stimulus severe enough to induce a response rapidly. Can be used to define either the exposure or the response to an exposure (effect). The duration of an acute aquatic toxicity test is generally 4 days or less and mortality is the response measured.

AMOP. Arctic Marine Oilspill Program.

BIOS. Baffin Island Oil Spill Project.

Chronic. Involving a stimulus that is lingering; often signifies periods from several weeks to years, depending on the reproductive life cycle of the aquatic species. Can be used to define either the exposure or the response to an exposure (effect). Chronic exposure typically induces a biological reponse of relatively slow progress and long continuance.

CMC. Critical micelle concentration.

CONCAWE. Conservation of Clean Air and Water in Europe.

DOOSIM. Dispersion of Oil on Sea Simulation (model).

EC_{50}. Median effective concentration; the concentration of material in water to which test organisms are exposed that is estimated to be effective in producing some sublethal response in 50 percent of the test organisms (Rand and Petrocelli, 1985). See also, LC_{50}.

FAO. Food and Agriculture Organisation of the United Nations.

HLB. Hydrophile-lipophile balance.

IFP. Institut Francaise du Petrol (French Institute of Petroleum).

ITOPF. International Tanker Owners Pollution Federation.

IKU. Continental Shelf Institute, Trondheim, Norway.

LAB. Linear alkyl benzenes.

LAS. Linear alkylbenzene sulfonates.

LC_{50}. Median lethal concentration; the concentration of material in water to which test organisms are exposed that is estimated to be lethal to 50 percent of the test organisms (Rand and Petrocelli, 1985). See also EC_{50}.

Lethal. Causing death by direct action. Death of aquatic organisms is the cessation of all visible signs of biological activity.

MNS. Mackay-Nadeau-Steelman test.

Mousse. Water in oil emulsion.

NOEL. No-observable-effect level.

OHMSETT. Oil and Hazardous Materials Simulated Environmental Test Tank (U.S. EPA).

OSC. On-scene coordinator.

SPM. Suspended particulate matter.

Sublethal. Below the concentration that directly causes death. Exposure to sublethal concentrations of a material may produce less obvious effects on behavior, biochemical or physiological functions, and histology of organisms.

Toxicity. The inherent potential of the capacity of a material to cause adverse effects in a living organism (Rand and Petrocelli, 1985). It is usually expressed as an effect concentration (EC_{50}, LC_{50}) at a specific time, or as an effect time (ET_{50}, LT_{50}) at a specific concentration. Effect concentrations (e.g., 4-day LC_{50}s) in this review are expressed as parts per million (ppm) or parts per billion (ppb), the units are used interchangeably with mg/liter and μg/liter, respectively, minor differences in exact concentrations notwithstanding.

Toxicity Threshold Concentration. A concentration above which some effect or response will be produced and below which it will not.

UNEP. United Nations Environment Programme.

VMD. Volume mean diameter.

WSF. Water-soluble fraction.

WSL. Warren Spring Laboratory.

Index

327